Handbook of Experimental Pharmacology

Volume 227

Editor-in-Chief
W. Rosenthal, Jena

Editorial Board
J.E. Barrett, Philadelphia
V. Flockerzi, Homburg
M.A. Frohman, Stony Brook, NY
P. Geppetti, Florence
F.B. Hofmann, München
M.C. Michel, Ingelheim
P. Moore, Singapore
C.P. Page, London
A.M. Thorburn, Aurora, CO
K. Wang, Beijing

More information about this series at
http://www.springer.com/series/164

Hans-Georg Schaible
Editor

Pain Control

 Springer

Editor
Hans-Georg Schaible
Jena University Hospital
Institute of Physiology/Neurophysiology
Jena
Thüringen
Germany

ISSN 0171-2004 ISSN 1865-0325 (electronic)
Handbook of Experimental Pharmacology
ISBN 978-3-662-46449-6 ISBN 978-3-662-46450-2 (eBook)
DOI 10.1007/978-3-662-46450-2

Library of Congress Control Number: 2015936907

Springer Heidelberg New York Dordrecht London
© Springer-Verlag Berlin Heidelberg 2015
This work is subject to copyright. All rights are reserved by the Publisher, whether the whole or part of the material is concerned, specifically the rights of translation, reprinting, reuse of illustrations, recitation, broadcasting, reproduction on microfilms or in any other physical way, and transmission or information storage and retrieval, electronic adaptation, computer software, or by similar or dissimilar methodology now known or hereafter developed.
The use of general descriptive names, registered names, trademarks, service marks, etc. in this publication does not imply, even in the absence of a specific statement, that such names are exempt from the relevant protective laws and regulations and therefore free for general use.
The publisher, the authors and the editors are safe to assume that the advice and information in this book are believed to be true and accurate at the date of publication. Neither the publisher nor the authors or the editors give a warranty, express or implied, with respect to the material contained herein or for any errors or omissions that may have been made.

Printed on acid-free paper

Springer-Verlag GmbH Berlin Heidelberg is part of Springer Science+Business Media (www.springer.com)

Preface

Current pain treatment is successful in many patients, but nevertheless numerous problems have to be solved because still about 20 % of the people in the population suffer from chronic pain. A major aim of pain research is, therefore, to clarify the neuronal mechanisms which are involved in the generation and maintenance of different pain states, and to identify the mechanisms which can be targeted for pain treatment. This volume on pain control addresses neuronal pain mechanisms at the peripheral, spinal, and supraspinal level which are thought to significantly contribute to pain and which may be the basis for the development of new treatment principles. Chapters on nociceptive mechanisms in the peripheral nociceptive system address the concept of hyperalgesic priming, the role of voltage-gated sodium channels in different inflammatory and neuropathic pain states, the hyperalgesic effects of NGF in different tissues and in inflammatory and neuropathic pain states, and the contribution of proteinase-activated receptors (PAR) to the development of pain in several chronic pain conditions. Chapters on nociceptive mechanisms in the spinal cord address the particular role of NO and of glial cell activation in the generation and maintenance of inflammatory and neuropathic pain, and they discuss the potential role of local inhibitory interneurons, of the endogenous endocannabinoid system, and the importance of non-neuronal immune mechanisms in opioid signaling in the control of pain. Furthermore, it is presented how spinal mechanisms contribute to the expression of peripheral inflammation. At the supraspinal level, the role of the amygdala and their connections to the medial prefrontal cortex in pain states are addressed. A particular chapter discusses the experimental methods to test central sensitization of the nociceptive system in humans. Finally, differences and similarities of the neuronal systems of pain and itch are reported. Altogether, the chapters demonstrate that both the concentration on single key molecules of nociception and the interference with disease-related mediators may provide novel approaches of pain treatment.

Jena, Germany   Hans-Georg Schaible

Contents

Emerging Concepts of Pain Therapy Based on Neuronal Mechanisms. . . . 1
Hans-Georg Schaible

The Pharmacology of Nociceptor Priming. 15
Ram Kandasamy and Theodore J. Price

Sodium Channels and Pain. 39
Abdella M. Habib, John N. Wood, and James J. Cox

Role of Nerve Growth Factor in Pain. 57
Kazue Mizumura and Shiori Murase

Central Sensitization in Humans: Assessment and Pharmacology. 79
Lars Arendt-Nielsen

Nitric Oxide-Mediated Pain Processing in the Spinal Cord. 103
Achim Schmidtko

The Role of the Endocannabinoid System in Pain. 119
Stephen G. Woodhams, Devi Rani Sagar, James J. Burston,
and Victoria Chapman

The Role of Glia in the Spinal Cord in Neuropathic
and Inflammatory Pain. 145
Elizabeth Amy Old, Anna K. Clark, and Marzia Malcangio

Plasticity of Inhibition in the Spinal Cord. 171
Andrew J. Todd

Modulation of Peripheral Inflammation by the Spinal Cord. 191
Linda S. Sorkin

The Relationship Between Opioids and Immune Signalling
in the Spinal Cord. 207
Jacob Thomas, Sanam Mustafa, Jacinta Johnson, Lauren Nicotra,
and Mark Hutchinson

The Role of Proteases in Pain. 239
Jason J. McDougall and Milind M. Muley

Amygdala Pain Mechanisms 261
Volker Neugebauer

Itch and Pain Differences and Commonalities 285
Martin Schmelz

Index 303

Emerging Concepts of Pain Therapy Based on Neuronal Mechanisms

Hans-Georg Schaible

Contents

1 Pathophysiological Background 3
   1.1 Types of Pain 3
   1.2 The Nociceptive System 4
   1.3 Neuronal Mechanisms of Pathophysiologic Nociceptive and Neuropathic Pain 5
   1.4 Molecular Mechanisms of Pain 8
2 Conclusion 11
References 11

Abstract

Current pain treatment is successful in many patients, but nevertheless numerous problems have to be solved because still about 20 % of the people in the population suffer from chronic pain. A major aim of pain research is, therefore, to clarify the neuronal mechanisms which are involved in the generation and maintenance of different pain states and to identify the mechanisms which can be targeted for pain treatment. This volume on pain control addresses neuronal pain mechanisms at the peripheral, spinal, and supraspinal level which are thought to significantly contribute to pain and which may be the basis for the development of new treatment principles. This introductory chapter addresses the types of pain which are currently defined based on the etiopathologic considerations, namely physiologic nociceptive pain, pathophysiologic nociceptive pain, and neuropathic pain. It briefly describes the structures and neurons of the nociceptive system, and it addresses molecular mechanisms of nociception which may become targets for pharmaceutical intervention. It will provide a frame for the chapters which address a number of important topics. Such topics

---

H.-G. Schaible (✉)
Institute of Physiology 1/Neurophysiology, Jena University Hospital, Friedrich Schiller University of Jena, Teichgraben 8, Jena 07740, Germany
e-mail: Hans-Georg.Schaible@med.uni-jena.de

© Springer-Verlag Berlin Heidelberg 2015
H.-G. Schaible (ed.), *Pain Control*, Handbook of Experimental Pharmacology 227,
DOI 10.1007/978-3-662-46450-2_1

are the concept of hyperalgesic priming, the role of voltage-gated sodium channels and nerve growth factor (NGF) in different inflammatory and neuropathic pain states, the hyperalgesic effects of NGF in different tissues, the contribution of proteinase-activated receptors (PARs) to the development of pain in several chronic pain conditions, the role of spinal NO and of glial cell activation in the generation and maintenance of inflammatory and neuropathic pain, the potential role of spinal inhibitory interneurons, the endogenous endocannabinoid system, and the importance of nonneuronal immune mechanisms in opioid signaling in the control of pain, the influence of spinal mechanisms on the expression of peripheral inflammation, the role of the amygdala and their connections to the medial prefrontal cortex in pain states, the experimental methods to test central sensitization of the nociceptive system in humans, and differences and similarities of the neuronal systems of pain and itch. Finally it will be discussed that both the concentration on single key molecules of nociception and the interference with disease-related mediators may provide novel approaches of pain treatment.

Keywords

Nociceptive pain • Neuropathic pain • Nociceptive system • Peripheral sensitization • Central sensitization • Nociceptor • Pain mechanisms

Pain therapy is an important need in most fields of medicine because numerous diseases are associated with significant pain. Although pain treatment is successful in many patients, numerous problems still have to be solved. An impressive fact is that about 20 % of the people in the population suffer from chronic pain. According to epidemiological studies, chronic pain is most frequent in the musculoskeletal system, and osteoarthritis pain and low back pain are the leading causes (Breivik et al. 2006).

There are numerous reasons for the existence of chronic pain and the failure of pain therapy. Concerning drug therapy, we have only a limited spectrum of drugs which are available for pain treatment. It is largely based on the use of nonsteroidal anti-inflammatory drugs (NSAIDs) which inhibit prostaglandin synthesis and on the use of opioids. In addition, for the treatment of neuropathic pain, drugs are used which reduce the neuronal excitability. The available drugs may not be sufficient to achieve long-lasting pain relief. Furthermore, they have side effects which limit their use in the long term. Intense pain research is, therefore, necessary to improve the situation.

Pain research has several aims. A major first aim is to clarify the neuronal mechanisms which are involved in the generation and maintenance of pain. Based on the numerous studies in different disciplines, it is quite clear that pain is the result of complex mechanisms which interact in many ways. Pain is determined by neurophysiological mechanisms in the nociceptive system as well as by other components such as psychological and social factors. The key to better pain therapy is an advanced understanding of the processes which are integrated in order to produce the clinical symptom pain. The second major aim is to identify the mechanisms which can be targeted for pain treatment. However, due to the complexity of factors contributing to pain, pain treatment is not limited to the use of drugs. For the treatment of chronic pain, rather interdisciplinary approaches are suitable which include drug therapy, physiotherapy, psychotherapy, and others.

This volume on pain control addresses neuronal pain mechanisms at the peripheral, spinal, and supraspinal level which are thought to significantly contribute to pain and which may be the basis for the development of new treatment principles. Naturally it will not be possible to cover all relevant areas in this volume.

Related to the sensation of pain is the sensation of itch (see Schmelz 2015). Pain and itch are generally regarded antagonistic as painful stimuli such as scratching suppress itch. Several findings are in agreement with the specificity theory for itch, but there are also considerable overlaps of mechanisms of pain and itch, and therefore, research concepts should address the common mechanisms.

1 Pathophysiological Background

1.1 Types of Pain

From the etiopathological point of view, currently three types of pain are distinguished (Schaible and Richter 2004). If noxious stimuli threaten normal tissue, **physiologic nociceptive pain** is elicited. Usually intense mechanical (noxious pressure, noxious movements, etc.) or thermal stimuli (noxious heat, noxious cold) are necessary to activate the nociceptive system. This type of pain protects the body from being damaged.

In the course of inflammation or tissue injury, **pathophysiologic nociceptive pain** is evoked. It is characterized by mechanical and/or thermal allodynia and hyperalgesia. The threshold for elicitation of pain is lowered into the normally innocuous range, with the consequence that normally non-painful stimuli elicit pain. Pathophysiologic nociceptive pain is the most frequent cause for seeking medical treatment. The nociceptive system undergoes significant changes, but overall, its functions are intact. This pain is often dependent on stimulation, i.e., evoked by load. Pathophysiologic nociceptive pain has the purpose to prevent the tissue from further damage and to support healing processes. Under suitable conditions, it disappears after successful healing.

The third type of pain results from damage or disease of neurons of the nociceptive system. In this case nerve fibers themselves are afflicted, and therefore, this type of pain is called **neuropathic pain**. This form of pain is abnormal, often aberrant, because it does not signal tissue inflammation or tissue injury, and it may be combined with loss of the normal nerve fiber function. Neuropathic pain is useless because it does not serve as a warning signal for body protection. Damage or diseases of the peripheral as well as of the central nociceptive system can elicit neuropathic pain.

1.2 The Nociceptive System

Pain is produced by the activation of the nociceptive system, the part of the nervous system which is specialized for the detection and processing of noxious stimuli. In the brain the nociceptive system cooperates with other systems allowing bidirectional interactions between the nociceptive and other systems. The peripheral nociceptive system provides the sensors for noxious stimuli; the central nervous system processes the nociceptive input and produces the conscious sensation of pain.

The *peripheral nociceptive system* is composed of the nociceptive nerve fibers which innervate the tissue. Peripheral nociceptors are either C-fibers or A∂-fibers, and their sensory endings in the tissue are free nerve endings. Most nociceptive sensory fibers are polymodal and respond to noxious mechanical and thermal stimuli as well as to a variety of chemical stimuli. The excitation threshold of these nerve fibers is near or at the noxious (tissue damaging) range, and the fibers encode noxious stimuli of different intensities by their discharge frequencies. In order to sense noxious stimuli, nociceptive sensory endings are equipped with ion channels which open upon the application of noxious stimuli. Some of these transduction molecules were identified, but there are still numerous gaps in knowledge (see Sect. 1.4). By opening such ion channels, noxious stimuli depolarize the sensory neurons. If the so-evoked depolarizing sensor potentials reach a sufficiently high amplitude, they trigger the opening of sodium channels and elicit action potentials which propagate along the nerve fiber and cause synaptic activation of nociceptive neurons in the spinal cord (or of the brain stem for nociceptive input from the head) (Schaible and Richter 2004).

The *central nociceptive system* consists of the nociceptive neurons in the spinal cord and in different supraspinal structures which are activated by noxious stimuli. *Nociceptive neurons in the spinal cord* form either ascending tracts which transmit the nociceptive information to the thalamus and the brain stem, or they are local interneurons which activate neurons within the same or adjacent segments. The spinothalamic tract ascends to the ventrobasal complex of the thalamus which is a relay nucleus on the way to the sensory cortex. Branches of the spinothalamic tract or other ascending tracts project to the brain stem, e.g., to the parabrachial nucleus which forms a pathway to the amygdala (Bushnell et al. 2013). They form also connections to brain stem nuclei which are the origin of the descending inhibitory and excitatory systems (Ossipov et al. 2010).

Nociceptive neurons in the thalamocortical system generate the conscious pain experience. Currently a distinction is made between the lateral system and the medial system. The lateral system consists of neurons in the ventrobasal nucleus of the thalamus and in the cortical areas S1 and S2, i.e., the somatosensory cortex. The activation of these neurons is thought to generate the sensory discriminative component of pain, i.e., the sensory analysis of the noxious stimulus. The medial system consists of neurons in the medial part of the thalamus and projections to the insula, the anterior cingulate cortex, and the forebrain. These pathways generate the affective emotional component of pain, the unpleasantness and the suffering, and they are involved in the generation of behavioral responses to pain (Treede et al. 1999; Vogt 2005). Nociceptive stimuli also activate the amygdala which is a major site for the generation of fear (Duvarci and Pare 2014). The thalamocortical nociceptive system interacts with numerous other systems which are involved in brain functions, e.g., neuronal circuits which are involved in the generation of emotions and others (Bushnell et al. 2013). A well-known consequence of such interactions is the occurrence of depression during pain states.

The brain stem forms a *descending system* which generates *descending inhibition* and *descending excitation*. The nucleus of origin of descending inhibition is the periaqueductal gray which projects to the rostroventral medulla. From there tracts descend to the spinal cord where they influence the spinal nociceptive processing. The descending inhibitory system serves as an endogenous pain control system which keeps the nociceptive system under control. It can be activated from the brain and is, e.g., active during placebo responses (Ossipov et al. 2010).

In the chapter on **itch**, Schmelz addresses the differences and similarities of the neuronal systems of pain and itch. Separate specific pathways for itch and pain processing have been uncovered, and several molecular markers at the primary afferent and spinal level have been established in mice that identify neurons involved in the processing of histaminergic and non-histaminergic itch. However, in addition to broadly overlapping mediators of itch and pain, there is also an evidence for overlapping functions in primary afferents. Nociceptive primary afferents can provoke itch when activated very locally in the epidermis, and sensitization of both nociceptors and pruriceptors has been found following local nerve growth factor (NGF) application in volunteers. Thus, the mechanisms that underlie the development of chronic itch and pain including spontaneous activity and sensitization of primary afferents as well as spinal cord sensitization may well overlap to a great extent (Schmelz 2015).

1.3 Neuronal Mechanisms of Pathophysiologic Nociceptive and Neuropathic Pain

In clinically relevant pain states, the nociceptive system undergoes significant changes at the peripheral as well as the central level. Pathophysiologic nociceptive pain and neuropathic pain involve different as well as common mechanisms. Figure 1 displays major changes which are observed in chronic pain states.

At the peripheral level distinct processes were observed which characterize pathophysiologic nociceptive and neuropathic pain. The hallmark of pathophysiologic nociceptive pain, e.g., pain during inflammation or after tissue injury, is **peripheral sensitization**. Nociceptive nerve fibers exhibit a lowering of their excitation threshold for the response to mechanical and/or thermal stimuli and increased firing frequencies during the application of stimuli of noxious intensities. Such processes were characterized in the skin, muscle, joint, and visceral organs (Schaible and Richter 2004). Molecular mechanisms of peripheral sensitization are addressed in Sect. 1.4 (see below). More recently, the concept of **hyperalgesic**

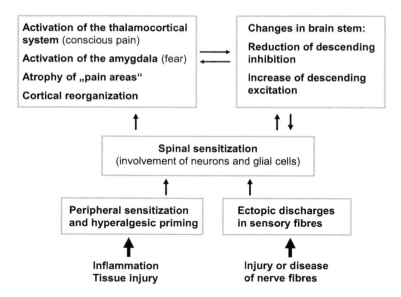

Fig. 1 Changes in the nociceptive system during pathophysiologic nociceptive pain and neuropathic pain. Spinal sensitization and increased hyperexcitability at the supraspinal level form the process of central sensitization

priming was introduced (see Kandasamy and Price 2015). Priming arises from an initial injury and results in the development of a remarkable susceptibility to normally subthreshold noxious inputs causing a prolonged pain state in primed animals. Priming increases the sensitization process which is evoked by sensitizing mediators. As an example, application of prostaglandin E2 to normal tissue causes a short-lasting sensitization of nociceptors if applied before injury or priming. However, if the neurons were primed, e.g., by interleukin-6, NGF, and other priming stimuli, prostaglandin E2 will cause a long-lasting sensitization (see Kandasamy and Price 2015).

A frequent process of neuropathic pain at the peripheral level is the generation of **ectopic discharges**. These action potentials can be elicited at the lesion site of the nerve fibers, but they can also be generated in the soma of the lesioned neurons (Devor 2009). Underlying mechanisms are changes in the expression of ion channels, actions of inflammatory mediators on lesioned fibers, and effects of the sympathetic nervous system on lesioned nerve fibers. In the latter case the neuropathic pain may be sympathetically maintained (Schaible and Richter 2004).

Peripheral nociceptive processes often trigger changes in the spinal cord which are called **central sensitization**. The changes in the spinal cord provide a gain of the nociceptive processing at the spinal site (Cervero 2009; Woolf and Salter 2000). Nociceptive spinal cord neurons which receive increased input from inflamed regions show the following phenomena: a lowering of threshold, increased responses to innocuous and noxious stimuli, and an expansion of the receptive fields (Schaible et al. 2009). In the sensitized state, more spinal cord neurons show responses to a stimulus applied to a specific peripheral site. These changes reflect an increase of the synaptic processing including the suprathreshold activation of synapses which may be too weak in the normal state to depolarize the neuron sufficiently. In many aspects these processes are similar to the **long-term potentiation** which was characterized as a major process of memory formation in the hippocampus (Sandkühler 2000). Central sensitization is also thought to occur in neuropathic pain states.

Several cell types may contribute to the spinal sensitization. First, the sensitization of peripheral nociceptors increases the sensory input into the spinal cord, thus providing a stronger presynaptic component of synaptic activation. Second, postsynaptic spinal cord neurons are rendered hyperexcitable by the activation of NMDA and other receptors (Sandkühler 2000; Woolf and Salter 2000). Third, glial cells may be activated and produce cytokines and other mediators which facilitate the spinal processing. Glial cells are strongly activated in neuropathic pain states, but they may also contribute to inflammatory pain (McMahon and Malcangio 2009). The involvement of glial cells in pain states is addressed by Old et al. (2015). Fourth, the activity of inhibitory interneurons may be reduced. The inhibitory neurons in the spinal cord and the mechanisms by which the inhibitory control is decreased or lost are addressed by Todd (2015). The spinal sensitization and the resulting thalamocortical processing are thought to underlie the observation that in many pain states the pain becomes widespread (Phillips and Clauw 2013). The significance of central sensitization in humans under clinically relevant conditions, and the experimental methods to test central sensitization in humans, will be addressed by Arendt-Nielsen (2015).

Ascending nociceptive information activates the thalamocortical system. During chronic pain states significant changes of this system were observed in patients using fMRI. Remarkably, many chronic pain states such as chronic osteoarthritic pain are associated with a so-called **atrophy** of the regions in which pain is processed. The underlying cellular mechanisms have not been identified. Interestingly, this atrophy seems to be reversible because after successful treatment of pain, the brain structures show a normalization (Bushnell et al. 2013; Gwilym et al. 2010; Rodriguez-Raecke et al. 2009). Under neuropathic conditions the cortex may show a **reorganization** with significant changes in the cortical maps. Such changes were, e.g., observed during phantom limb pain.

As already mentioned, ascending tracts not only activate the thalamocortical system. They also activate the amygdala via the parabrachial nucleus. Further input to the amygdala is provided by the nerve fibers from the thalamus and from the cortex (Duvarci and Pare 2014). The amygdala is key nuclei in the generation of **fear**, and they can be activated in pain conditions (Kulkarni et al. 2007). In this volume, the role of the amygdala and their connections to the medial prefrontal cortex (mPFC) in pain states will be addressed by Neugebauer (2015). Pain-related mPFC deactivation results in cognitive deficits and the failure of inhibitory control of amygdala processing. Impaired cortical control allows the uncontrolled persistence of amygdala pain mechanisms.

Neural pathways descending from the brain stem mediate inhibition and facilitation of nociceptive spinal cord neurons (Ossipov et al. 2010; Vanegas and Schaible 2004). During severe chronic pain, a **reduction of descending inhibition**, in particular the diffuse inhibitory noxious control (DNIC), was reported (Kosek and Ordeberg 2000; Lewis et al. 2012). In addition, **descending facilitation** may contribute to pain, in particular during neuropathic pain (Vanegas and Schaible 2004). Thus, descending inhibitory systems from the brain stem may be less effective and/or descending excitatory systems from the brain stem may be overactive during chronic pain. These changes may be (partly) reversible after successful pain treatment (Kosek and Ordeberg 2000).

Effects of the nervous system on inflammation. It must be noted that the importance of the nociceptive nervous system extends beyond the generation of pain. The nervous system is able to influence inflammatory processes in the organs. Such influences are mediated by the efferent effects of nociceptive sensory afferents which produce neurogenic inflammation, by fibers of the sympathetic and parasympathetic nervous system, and by neuroendocrine influences (Schaible and Straub 2014). Spinal hyperexcitability is not only important for pain generation (see above). It plays also a role in the regulation of joint inflammation (Waldburger and Firestein 2010). In this volume this topic will be addressed by Sorkin (2015). Both pro- and anti-inflammatory feedback loops can involve just the peripheral nerves and the spinal cord or can also include more complex, supraspinal structures such as the vagal nuclei and the hypothalamic–pituitary axis.

1.4 Molecular Mechanisms of Pain

Molecular mechanisms of nociception are of considerable interest for pharmacologic approaches, and therefore, they are particularly addressed in this volume. The peripheral nociceptor as well as the spinal cord and the amygdala are in the focus.

Nociception in the periphery consists of two elementary processes, the transduction of stimuli (the generation of a sensor potential by the impact of a noxious stimulus) and the transformation of the sensor potential into a series of action potentials. Noxious stimuli are mechanical or thermal (heat and cold), and also some chemical mediators (e.g., bradykinin or H^+) cause pain. The chemosensitivity of nociceptors is particularly important for the process of sensitization (and priming).

For the transduction of thermal stimuli into sensor potentials, ion channels of the transient receptor potential (TRP) family are responsible. While the involvement of TRPV1, TRPV2, and TRPM8 in the sensation of noxious heat (TRPV1 and TRPV2) and innocuous cold (TRPM8) has been established, the significance of other TRP channels in thermo(noci)ception is not that clear. For two TRP channels (TRPA1 and TRPV4), a role in mechanical hyperalgesia is being discussed (Kwan et al. 2009; Levine and Alessandri-Haber 2007; Malsch et al. 2014; Segond von Banchet et al. 2013) although these channels may not be the transduction molecules involved in the "normal mechanonociception." The current knowledge on the involvement of TRP ion channels in the sensation of noxious heat and noxious cold and of the involvement of these ion channels in the generation of thermal hyperalgesia has been summarized (Basbaum et al. 2009; Julius 2013; Stein et al. 2009) and is not the topic of this volume.

Some chemicals can also open ion channels. For example, H^+ triggers the opening of acid-sensing ion channels (ASICs), and capsaicin opens TRPV1. Most mediators, however, activate membrane receptors and are thereby involved in the sensitization of nociceptive neurons (see below).

The sensor potential triggers the generation of action potentials. For action potentials voltage-gated sodium channels are essential. In nociceptive neurons, mainly the sodium channels $Na_v1.7$, $Na_v1.8$, and $Na_v1.9$, and under neuropathic conditions $Na_v1.3$, are expressed (Waxman and Zamponi 2014). $Na_v1.7$ is activated by slow, subtle depolarization close to the resting potential, and it thus sets the gain on nociceptors. $Na_v1.8$, which shows depolarized voltage dependence, produces most of the current responsible for the action potential upstroke, and it supports repetitive firing. $Na_v1.9$ does not contribute to the action potential upstroke but depolarizes the cells and prolongs and enhances small depolarization thus enhancing excitability (Waxman and Zamponi 2014). In this volume Habib et al. (2015) will address the role of these ion channels in different inflammatory and neuropathic pain states. They show that particular Na_v ion channels are involved in different pathophysiologic states. Because Na^+ channel blockers are thought to be promising targets for new analgesics (Gold 2008), such knowledge is important for the understanding of which blocker might be suitable under the particular conditions.

When neurons are sensitized both the channels of transduction and the voltage-gated ion channels, in particular the Na^+ channels, show changes such that the excitability is enhanced (Linley et al. 2010; Schaible et al. 2011). Some mediators such as prostaglandin E2 change the opening properties of TRPV1 and of sodium channels such that weaker stimuli are sufficient to open the ion channels. The effect of prostaglandin E2 is mediated by G protein-coupled receptors which activate second messengers in the nociceptors (Hucho and Levine 2007), and these second messenger systems change the opening properties of the ion channels.

While prostaglandins are known for a long time as sensitizing molecules, more recent research revealed a number of other receptor types in nociceptive sensory neurons which are of great importance for the sensitization. It was shown that proinflammatory cytokines such as TNF-α, interleukin-6, and interleukin-17 induce a persistent state of sensitization in C-fibers (Brenn et al. 2007; Richter et al. 2010, 2012). Cytokines are thought to play a significant role in the generation of inflammatory and neuropathic role (Schaible et al. 2010; Sommer and Kress 2004; Üceyler et al. 2009). Interleukin-6 is thought to be an important molecule of hyperalgesic priming (see Kandasamy and Price 2015).

NGF and its receptor trkA were discovered as suitable targets for pain treatment. A single application of an antibody to NGF was shown to provide significant pain relief in osteoarthritis for several weeks (Lane et al. 2010). NGF has a variety of actions on nonneuronal cells and sensory neurons which regulate the excitability in the long term (Bennett 2007). In this volume Mizumura and Murase (2015) address the hyperalgesic effects of NGF in different tissues and in inflammatory and neuropathic pain states, and they address the mechanisms involved.

Proteinase-activated receptors (PARs) are a family of G protein-coupled receptor that is activated by extracellular cleavage of the receptor in the N-terminal domain. This slicing of the receptor exposes a tethered ligand which binds to a specific docking point on the receptor surface to initiate intracellular signaling. McDougall and Muley summarize how serine proteinases activate PARs leading to the development of pain in several chronic pain conditions. The potential of PARs as a drug target for pain relief is discussed (McDougall and Muley 2015).

Excitatory synaptic transmission in the spinal cord under basal conditions is mediated by the transmitter glutamate, the transmitter of nociceptive sensory neurons. Central sensitization is also dependent on glutamate, in particular acting on NMDA receptors. However, numerous other transmitters and mediators are involved in the complex signaling in the spinal cord (e.g., NK1 receptors for substance and CGRP receptors) (Woolf and Salter 2000). Other mediators such as spinal prostaglandins contribute to spinal sensitization (Bär et al. 2004). The particular role of NO to nociceptive spinal cord signaling will be addressed by Schmidtko (2015). The role of mediators involved in glial cell activation and functions will be addressed by Old et al. (2015).

Under normal conditions, excitatory and inhibitory synaptic mechanisms are presumably in a balanced activity state. Such inhibition is provided by specific local inhibitory interneurons (see Todd 2015), but it may also be provided by mediators which act in a feedback manner from activated neurons. Such inhibitory control is, e.g., provided by endocannabinoids which are addressed in this volume by Woodhams et al. (2015). Cannabinoid 1 (CB_1) receptors are found at presynaptic sites throughout the peripheral and central nervous systems, while the CB_2 receptor is found principally (but not exclusively) on immune cells. The endocannabinoid (EC) system is now known to be one of the key endogenous systems regulating pain sensation, with modulatory actions at all stages of pain processing pathways. As already discussed, pain states may involve a reduction of inhibitory mechanisms.

A particular interesting aspect is that some mediators may exert excitatory as well as inhibitory actions, depending on the functional context. An example is the change of GABAergic inhibitory mechanisms in neuropathic pain states (see Todd 2015). However, even mediators such as prostaglandin E2 which are usually considered excitatory may provide antinociception when pain pathways are activated, by the activation of receptor subtypes which are coupled to inhibitory signaling pathways (Natura et al. 2013). In this volume Schmidtko (2015) reports about both the pro- and antinociceptive effects of NO signaling resulting from a different downstream signaling.

Spinal cord mechanisms may even alter the antinociceptive effect of potent analgesic drugs. Opioids are considered the gold standard for the treatment of moderate to severe pain. However, heterogeneity in analgesic efficacy, poor potency, and side effects are associated with opioid use. Traditionally opioids are thought to exhibit their analgesic actions via the activation of the neuronal G protein-coupled opioid receptors. However, neuronal activity of opioids cannot fully explain the initiation and maintenance of opioid tolerance, hyperalgesia, and allodynia. In this volume Thomas et al. (2015) report the importance of nonneuronal mechanisms in opioid signaling, paying particular attention to the relationship of opioids and immune signaling.

Abnormally enhanced output from the CeLC of the **amygdala** is also the consequence of an imbalance between excitatory and inhibitory mechanisms (see Neugebauer 2015). Impaired inhibitory control mediated by a cluster of GABAergic interneurons in the intercalated cell masses (ITC) allows the development of glutamate- and neuropeptide-driven synaptic plasticity of excitatory inputs from the brain stem (parabrachial area) and from the lateral–basolateral amygdala network (LA-BLA, site of integration of polymodal sensory information).

2 Conclusion

It is increasingly evident how many different neuronal and molecular mechanisms contribute to the expression of pain, in particular in clinically relevant pain states. We begin to understand some mechanisms of pain vulnerability (Denk et al. 2014). The complexity of pain processing and related neuronal events puts a considerable challenge to the development of new therapeutic strategies. Is the focus on single key molecules such as a particular sodium channel an appropriate therapeutical approach or should one aim to interfere with disease-related mediators such as NGF or cytokines? The answer to this crucial question is not straightforward. Both types of drugs have been proven useful in medical therapy. Local anesthetics targeting specifically sodium channels can interrupt pain (usually for a short time only), but on the other hand, the use of antibodies to particular cytokines which have numerous actions is extremely potent in the therapy of rheumatic diseases such as rheumatoid arthritis. Thus, future pain therapy should provide effective treatments using either specific drugs with the aim of interfering with specific nociceptive processes or using drugs which have the potency of long-term modification of pain mechanisms.

References

Arendt-Nielsen L (2015) Central sensitization in humans: assessment and pharmacology. In: Schaible H-G (ed) Pain control. Springer, Berlin, pp 79–102

Bär K-J, Natura G, Telleria-Diaz A, Teschner P, Vogel R, Vasquez E, Schaible H-G, Ebersberger A (2004) Changes in the effect of spinal prostaglandin E_2 during inflammation—prostaglandin E (EP1-EP4) receptors in spinal nociceptive processing of input from the normal or inflamed knee joint. J Neurosci 24:642–651

Basbaum AI, Bautista DM, Scherrer G, Julius D (2009) Cellular and molecular mechanisms of pain. Cell 139:267–284

Bennett D (2007) NGF, sensitization of nociceptors. In: Schmidt RF, Willis WD (eds) Encyclopedia of pain, vol 2. Springer, Berlin, pp 1338–1342

Breivik H, Beverly C, Ventafridda V, Cohen R, Gallacher D (2006) Survey of chronic pain in Europe: prevalence, impact on daily life, and treatment. Eur J Pain 10:287–333

Brenn D, Richter F, Schaible H-G (2007) Sensitization of unmyelinated sensory fibres of the joint nerve to mechanical stimuli by interleukin-6 in the rat. An inflammatory mechanism of joint pain. Arthritis Rheum 56:351–359

Bushnell MC, Ceko M, Low LA (2013) Cognitive and emotional control of pain and its disruption in chronic pain. Nat Rev Neurosci 14:502–511

Cervero F (2009) Spinal cord hyperexcitability and its role in pain and hyperalgesia. Exp Brain Res 196:129–137

Denk F, McMahon SB, Tracey I (2014) Pain vulnerability: a neurobiological perspective. Nat Neurosci 17:192–200

Devor M (2009) Ectopic discharge in Aβ afferents as a source of neuropathic pain. Exp Brain Res 196:115–128

Duvarci S, Pare D (2014) Amygdala microcircuits controlling learned fear. Neuron 82:966–980

Gold MS (2008) Na^+ channel blockers for the treatment of pain: context is everything, almost. Exp Neurol 210:1–6

Gwilym SE, Filippini N, Douaud G, Carr AJ, Tracey I (2010) Thalamic atrophy associated with painful osteoarthritis of the hip is reversible after arthroplasty. Arthritis Rheum 62:2930–2940

Habib AM, Wood JN, Cox JJ (2015) Sodium channels and pain. In: Schaible H-G (ed) Pain control. Springer, Berlin, pp 39–56

Hucho T, Levine JD (2007) Signaling pathways in sensitization: toward a nociceptor cell biology. Neuron 55:365–376

Julius D (2013) TRP channels and pain. Annu Rev Cell Dev Biol 29:355–384

Kandasamy R, Price TJ (2015) The pharmacology of nociceptor priming. In: Schaible H-G (ed) Pain control. Springer, Berlin, pp 15–37

Kosek E, Ordeberg G (2000) Lack of pressure pain modulation by heterotopic noxious conditioning stimulation in patients with painful osteoarthritis before, but not following surgical pain relief. Pain 88:69–78

Kulkarni B, Bentley DE, Elliott R, Julyan PJ, Boger E, Watson A, Boyle Y, El-Deredy W, Jones AKP (2007) Arthritic pain is processed in brain areas concerned with emotions and fear. Arthritis Rheum 56:1345–1354

Kwan KY, Glazer JM, Corey DP, Rice FL, Stucky CL (2009) TRPA1 modulates mechanotransduction in cutaneous sensory neurons. J Neurosci 29:4808–4819

Lane NE, Schnitzer TJ, Birbara CA, Mokhtarani M, Shelton DL, Smith MD, Brown MT (2010) Tanezumab for the treatment of pain from osteoarthritis of the knee. N Engl J Med 363:1521–1531

Levine JD, Alessandri-Haber N (2007) TRP channels: targets for the relief of pain. Biochim Biophys Acta 1772:989–1003

Lewis GN, Rice DA, McNair PJ (2012) Conditioned pain modulation in populations with chronic pain: a systematic review and meta-analysis. J Pain 13:936–944

Linley JE, Rose K, Ooi L, Gamper N (2010) Understanding inflammatory pain: ion channels contributing to acute and chronic nociception. Pflugers Arch 459:657–669

Malsch P, Andratsch M, Vogl C, Link AS, Alzheimer C, Brierley SM, Hughes PA, Kress M (2014) Deletion of interleukin-6 signal transducer gp130 in small sensory neurons attenuates mechanonociception and down-regulates mechanotransducer ion channel TRPA1. J Neurosci 34:9845–9856

McDougall JJ, Muley MM (2015) The role of proteases in pain. In: Schaible H-G (ed) Pain control. Springer, Berlin, pp 239–260

McMahon SB, Malcangio M (2009) Current challenges in glia-pain biology. Neuron 64:46–54

Mizumura K, Murase S (2015) Role of nerve growth factor in pain. In: Schaible H-G (ed) Pain control. Springer, Berlin, pp 57–77

Natura G, Bär K-J, Eitner A, Böttger M, Richter F, Hensellek S, Ebersberger A, Leuchtweis J, Maruyama T, Hofmann GO, Halbhuber K-J, Schaible H-G (2013) Neuronal prostaglandin E2 receptor subtype EP3 mediates antinociception during inflammation. Proc Natl Acad Sci U S A 110:13648–13653

Neugebauer V (2015) Amygdala pain mechanisms. In: Schaible H-G (ed) Pain control. Springer, Berlin, pp 261–284

Old EA, Clark AK, Malcangio M (2015) The role of glia in the spinal cord in neuropathic and inflammatory pain. In: Schaible H-G (ed) Pain control. Springer, Berlin, pp 145–170

Ossipov MH, Dussor GO, Porreca F (2010) Central modulation of pain. J Clin Invest 120:3779–3787

Phillips K, Clauw DJ (2013) Central pain mechanisms in the rheumatic diseases. Arthritis Rheum 65:291–302

Richter F, Natura G, Loeser S, Schmidt K, Viisanen H, Schaible H-G (2010) Tumor necrosis factor-α (TNF-α) causes persistent sensitization of joint nociceptors for mechanical stimuli. Arthritis Rheum 62:3806–3814

Richter F, Natura G, Ebbinghaus M, Segond von Banchet G, Hensellek S, König C, Bräuer R, Schaible H-G (2012) Interleukin-17 sensitizes joint nociceptors for mechanical stimuli and contributes to arthritic pain through neuronal IL-17 receptors in rodents. Arthritis Rheum 64:4125–4134

Rodriguez-Raecke R, Niemeier A, Ihle K, Ruether W, May A (2009) Brain gray matter decrease in chronic pain is the consequence and not the cause of pain. J Neurosci 29:13746–13750

Sandkühler J (2000) Learning and memory in pain pathways. Pain 88:113–118

Schaible H-G, Richter F (2004) Pathophysiology of pain. Langenbecks Arch Surg 389:237–243

Schaible H-G, Straub RH (2014) Function of the sympathetic supply in acute and chronic experimental joint inflammation. Auton Neurosci 182:55–64

Schaible H-G, Richter F, Ebersberger A, Boettger MK, Vanegas H, Natura G, Vazquez E, Segond von Banchet G (2009) Joint pain. Exp Brain Res 196:153–162

Schaible H-G, Segond von Banchet G, Boettger MK, Bräuer R, Gajda M, Richter F, Hensellek S, Brenn D, Natura G (2010) The role of proinflammatory cytokines in the generation and maintenance of joint pain. Ann N Y Acad Sci 1193:60–69

Schaible H-G, Ebersberger A, Natura G (2011) Update on peripheral mechanisms of pain: beyond prostaglandins and cytokines. Arthritis Res Ther 13:21

Schmelz M (2015) Itch and pain differences and commonalities. In: Schaible H-G (ed) Pain control. Springer, Berlin, pp 285–300

Schmidtko A (2015) Nitric oxide mediated pain processing in the spinal cord. In: Schaible H-G (ed) Pain control. Springer, Berlin, pp 103–117

Segond von Banchet G, Boettger MK, König C, Iwakura Y, Bräuer R, Schaible H-G (2013) Neuronal IL-17 receptor upregulates TRPV4 but not TRPV1 receptors in DRG neurons and mediates mechanical but not thermal hyperalgesia. Mol Cell Neurosci 52:152–160

Sommer C, Kress M (2004) Recent findings on how proinflammatory cytokines cause pain: peripheral mechanisms in inflammatory and neuropathic hyperalgesia. Neurosci Lett 361:184–187

Sorkin LS (2015) Modulation of peripheral inflammation by the spinal cord. In: Schaible H-G (ed) Pain control. Springer, Berlin, pp 191–206

Stein C, Clark JD, Oh U, Vasko MR, Wilcox GL, Overland AC, Vanderah TW, Spencer RH (2009) Peripheral mechanisms of pain and analgesia. Brain Res Rev 60:90–113

Thomas J, Mustafa S, Johnson J, Nicotra L, Hutchinson M (2015) The relationship between opioids and immune signalling in the spinal cord. In: Schaible H-G (ed) Pain control. Springer, Berlin, pp 207–238

Todd AJ (2015) Plasticity of inhibition in the spinal cord. In: Schaible H-G (ed) Pain control. Springer, Berlin, pp 171–190

Treede RD, Kenshalo DR, Gracely RH, Jones A (1999) The cortical representation of pain. Pain 79:105–111

Üceyler N, Schäfers M, Sommer C (2009) Mode of action of cytokines on nociceptive neurons. Exp Brain Res 196:67–78

Vanegas H, Schaible H-G (2004) Descending control of persistent pain: inhibitory or facilitatory? Brain Res Rev 46:295–309

Vogt BA (2005) Pain and emotion. Interactions in subregions of the cingulate cortex. Nat Rev Neurosci 6:533–544

Waldburger JM, Firestein GS (2010) Regulation of peripheral inflammation by the central nervous system. Curr Rheumatol Rep 12:370–378

Waxman SG, Zamponi GW (2014) Regulating excitability of peripheral afferents: emerging ion channel targets. Nat Neurosci 17:153–163

Woodhams SG, Sagar DR, Burston JJ, Chapman V (2015) The role of the endocannabinoid system in pain. In: Schaible H-G (ed) Pain control. Springer, Berlin, pp 119–143

Woolf CJ, Salter MW (2000) Neuronal plasticity: increasing the gain in pain. Science 288:1765–1768

The Pharmacology of Nociceptor Priming

Ram Kandasamy and Theodore J. Price

Contents

1 Introduction 16
2 Why Use Hyperalgesic Priming Models? 17
3 Mechanisms of Priming in the Periphery: A Model for Sustained Nociceptor Plasticity .. 18
   3.1 PKCε as a Crucial Mechanism of Nociceptor Priming 19
   3.2 Local Translation Is a Key Mediator of Nociceptor Priming 20
4 CNS Regulation of Hyperalgesic Priming 24
   4.1 Atypical PKCs and Brain-Derived Neurotropic Factor 24
   4.2 Endogenous Opioids, μ-Opioid Receptor Constitutive Activity, and Hyperalgesic Priming 27
   4.3 Surgery as a Priming Stimulus and the Effects of Opioids 29
5 Therapeutic Opportunities and Conclusions 31
References 33

Abstract

Nociceptors and neurons in the central nervous system (CNS) that receive nociceptive input show remarkable plasticity in response to injury. This plasticity is thought to underlie the development of chronic pain states. Hence, further understanding of the molecular mechanisms driving and maintaining this

---

R. Kandasamy
Department of Pharmacology, The University of Arizona, Tucson, AZ 85721, USA

T.J. Price (✉)
Department of Pharmacology, The University of Arizona, Tucson, AZ 85721, USA

Bio5 Institute, The University of Arizona, Tucson, AZ 85721, USA

Graduate Interdisciplinary Program in Neuroscience, The University of Arizona, Tucson, AZ 85721, USA

School of Brain and Behavioral Sciences, The University of Texas at Dallas, Richardson, TX 75080, USA
e-mail: theodore.price@utdallas.edu

plasticity has the potential to lead to novel therapeutic approaches for the treatment of chronic pain states. An important concept in pain plasticity is the presence and persistence of "hyperalgesic priming." This priming arises from an initial injury and results in a remarkable susceptibility to normally subthreshold noxious inputs causing a prolonged pain state in primed animals. Here we describe our current understanding of how this priming is manifested through changes in signaling in the primary nociceptor as well as through memory like alterations at CNS synapses. Moreover, we discuss how commonly utilized analgesics, such as opioids, enhance priming therefore potentially contributing to the development of persistent pain states. Finally we highlight where these priming models draw parallels to common human chronic pain conditions. Collectively, these advances in our understanding of pain plasticity reveal a variety of targets for therapeutic intervention with the potential to reverse rather than palliate chronic pain states.

Keywords

Atypical PKC • AMPA • NMDA • mTORC1 • PKC • Epac • Hyperalgesic priming • Prostaglandins • NGF • Interleukin 6

1 Introduction

A fundamental principle underlying our current understanding of pathological pain states is plasticity in the nociceptive system. While research into pathological pain states has long recognized this idea, it is only relatively recently that we have started to gain insight into mechanisms that cause this plasticity. On the most general level, plasticity in the pain system occurs at two locations, at the primary afferent nociceptor and at synapses receiving nociceptive input throughout the central nervous system (CNS). Preclinical models of acute and chronic inflammatory pain as well as models of neuropathic pain have revealed a plethora of molecular targets that have developed our understanding of how chronic pain develops as well as revealing important potential therapeutic intervention points. In the late 1990s and early 2000s, Jon Levine and colleagues developed "hyperalgesic priming" models (for review see Reichling and Levine 2009). These models provide unique insight into plasticity in the nociceptive system because they allow for molecular dissection of pain states in two distinct phases. These models involve a priming stimulus, aimed at causing an acute sensitization of peripheral nociceptors and their central inputs, albeit with some notable exceptions which will be discussed later. Next, in opposition to most other preclinical models, the initial sensitization is allowed to resolve and a second, normally subthreshold, stimulus is delivered. Importantly, this second stimulus, which has only a transient effect in naïve animals, leads to a prolonged state of pain hypersensitivity that allows for investigation of molecular mechanisms that define the primed nociceptor and/or the primed nociceptive system. Here we will argue that models of hyperalgesic priming have led to unique insight into how relatively brief pain states lead to reorganization of molecular machinery throughout the pain system rendering animals, and potentially humans, susceptible to prolonged pain states provoked by insults that would have little effect in unprimed individuals. This primed state, therefore, represents a kind of "pain memory" that, if reversed, has the potential to permanently remove the presence of a chronic pain state. Hence, our goals in this chapter will be to highlight (1) mechanisms underlying the priming in peripheral nociceptors, (2) mechanisms controlling priming in the CNS, and (3) potential therapeutic interventions elucidated by these findings with a view toward future pharmacological means to reverse chronic pain states.

2 Why Use Hyperalgesic Priming Models?

In order to fully grasp the importance of the research findings discussed herein, it is critical to first reflect on the utility of using hyperalgesic priming models to study pain plasticity. First, the experimental framework of the hyperalgesic priming model provides important insight into clinical chronic pain because it captures the recurrent nature of some of the most common pathological pain conditions (Reichling and Levine 2009). In 1921, Wilfred Harris described his clinical experience treating patients with presumed injuries to peripheral nerves. He described pain in these patients as episodic with pain episodes provoked by acute exacerbation (Harris 1921). Hence, from some of the earliest descriptions of pain as a disease, the notion of priming followed by subthreshold provocation of long-lived pain episodes has been apparent.

Population-based studies in several prevalent chronic pain conditions have directly demonstrated the episodic yet progressive nature of these disease states. Perhaps the best-known episodic pain condition is headache and, in the case of migraine, frequency of attacks is the best predictor of a transition to chronic migraine (Lipton 2009). In fact, the vast majority of migraineurs move from a low-frequency episodic headache stage to a high-frequency stage and eventually into chronic migraine (Bigal and Lipton 2008), highlighting the progressive worsening of this disorder. Moreover, migraines can frequently be provoked by what are often called migraine triggers. These are, by definition, subthreshold stimuli because they fail to provoke migraines in the non-migraineur population. This situation is not unique to migraine. Acute episodes of low back pain generally resolve (Bartleson 2001; Cassidy et al. 2005), but recurrence rates over 5 years are as high as 70 % (Von Korff and Saunders 1996; Carey et al. 1999; Cassidy et al. 2005; Kolb et al. 2011) and lifetime recurrence is estimated at 85 % (Andersson 1999; Tamcan et al. 2010). Moreover, the probability of low back pain recurrence increases with previous episodes of low back pain (Kolb et al. 2011). A similar clinical picture has been found for chronic neck pain (Croft et al. 2001; Nolet et al. 2010). Finally, in the case of surgery and chronic postsurgical pain, there is evidence that preexisting pain is a major risk factor for chronic post-incision pain suggesting that the preexisting pain can act as a priming stimulus causing a very long-lasting pain state induced by incision (Althaus et al. 2012; Pinto et al. 2013). Hence, we take the viewpoint, which is shared by others (Reichling and Levine 2009; Reichling et al. 2013), that the "priming" event in the hyperalgesic priming model may be viewed as an induction of the transition to chronic pain with important clinical implications for understanding molecular mechanisms involved in maintaining this disease state.

3 Mechanisms of Priming in the Periphery: A Model for Sustained Nociceptor Plasticity

Tissue injury, inflammation, and nerve injury are all examples where primary sensory nociceptors become sensitized. This sensitization leads to prolonged hyperalgesia, which sometimes outlasts or is disproportionate to the initial stimulus. A variety of chronic pain conditions have readily identified pathologies (e.g., rheumatoid arthritis, postherpetic neuropathic pain); however, some chronic pain conditions are defined by intermittent yet progressive periods of pain (e.g., low back pain, migraine, fibromyalgia) sometimes with no readily identified injury. However, one crucial question remains. What are the mechanisms responsible for pain chronification? In other words, how does acute pain ultimately transition to chronic pain?

The careful delineation of the mechanisms underlying the transition from acute to chronic pain can be difficult to examine, in part, due to the lack of a reliable animal model that could capture the complexity of this phenomenon. Although animal models of persistent, or "chronic" pain, exist (e.g., hindpaw CFA injection, spinal nerve ligation), it is not always straightforward to assess how an initial injury leads to persistent changes in signaling following that initial injury or stimulus. To address this, Jon Levine and colleagues developed the hyperalgesic priming model to isolate the mechanisms responsible for the persistence of the primed state, analogous to chronic pain. In this model, a hyperalgesic state is first evoked by a hindpaw injection of stimuli (e.g., carrageenan) that have traditionally been utilized to study acute pain in rodents. The subsequent hyperalgesia is transient and resolves within 4 days (Reichling and Levine 2009). After the resolution of the original hyperalgesia, a second stimulus is given to the same hindpaw. In naïve rodents, this stimulus (e.g., prostaglandin E_2; PGE_2) either fails to evoke hyperalgesia or is transient. In primed animals, this second stimulus causes a hyperalgesic state that lasts for at least 24 h and can persist for weeks (Fig. 1a) (Reichling and Levine 2009). Therefore, the model allows for the clear dissociation between the initiation phase, or priming, and the maintenance phase, which lacks any signs of hyperalgesia until another administration of an inflammatory mediator reestablishes hypersensitivity.

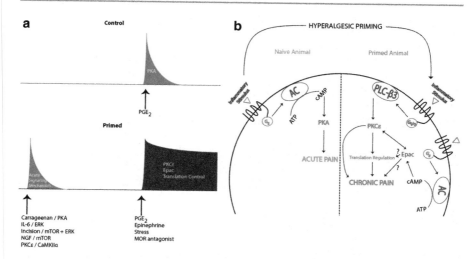

Fig. 1 Hyperalgesic priming induces a switch in nociceptor second messenger signaling pathways. In a naïve animal, administration (hindpaw injection) of an inflammatory stimulus (e.g., PGE_2) results in transient G_s/PKA-mediated acute pain (*green curve* in (**a**) and *left side* in (**b**)). In the primed animal, the same inflammatory stimulus recruits an additional $G_{i/o}$/PKCε and/or an Epac/PKCε-mediated pathway, which contributes to prolonged hyperalgesia (*red curve* in (**a**) and *right side* in (**b**)) in previously primed animals. Adapted from Reichling and Levine (2009)

3.1 PKCε as a Crucial Mechanism of Nociceptor Priming

How then is this exaggerated response to PGE_2, and other mediators like serotonin or A_2 adenosine receptor agonists (Aley et al. 2000), which can also precipitate hyperalgesia in primed animals, generated? Extensive studies have demonstrated that this priming effect is dependent on switches in signaling mechanisms in nociceptors. In naïve animals, the hyperalgesic effects of PGE_2 injection are mediated by adenylyl cyclase (AC) activation downstream of PGE_2 receptors causing protein kinase A (PKA) activation (Fig. 1b) (Aley and Levine 1999). This effect can be attenuated via injection of PKA antagonists (Aley and Levine 1999). Although the second messenger pathway underlying the early phase of PGE_2-induced hyperalgesia is PKA mediated, even in primed animals, the same cannot be said for the pathway responsible for the long-lasting hyperalgesia that is uniquely present in primed rodents. While PGE_2-induced hyperalgesia in primed animals is still cyclic AMP (cAMP) dependent, it now bypasses PKA to activate exchange proteins directly activated by cAMP (Epac) which can activate protein kinase Cε (PKCε, Fig. 1b, Hucho et al. 2005). Importantly, inflammatory stimulation of nociceptors leads to a decrease in G-protein receptor kinase 2 (GRK2) which results in enhanced Epac activity (Eijkelkamp et al. 2010; Wang et al. 2013). These changes occur in IB_4-positive nociceptors, and decreases in GRK2 and increases in Epac expression are correlated with the persistence of priming (Wang et al. 2013). Moreover, in primed animals, PGE_2 results in an activation of pertussis toxin-sensitive G-protein α_i subunits (Dina et al. 2009) and phospholipase Cβ (PLCβ) leading to a downstream engagement of PKCϵ (Joseph et al. 2007); hence, multiple pathways for PKCϵ engagement may exist in primed nociceptors (Fig. 1b). Critically, in primed animals, the long-lasting hyperalgesia arising from exposure to compounds that can precipitate priming is blocked by selective inhibition of PKCϵ (Aley et al. 2000) and by intrathecal delivery of antisense oligonucleotides knocking down PKCϵ expression (Parada et al. 2003). Additionally, injection of a PKCϵ agonist alone results in a prolonged hyperalgesic state and hyperalgesic priming, pointing to a key role for nociceptor PKCϵ in hyperalgesic priming (Reichling and Levine 2009). Importantly, this PKCϵ-dependent primed state does not require an initial bout of hyperalgesia as subthreshold doses of PKCϵ activators (Parada et al. 2003), previous exposure to unpredictable sound stress (Khasar et al. 2008), and even repeated administration of opioid agonists into the paw (Joseph et al. 2010) are capable of causing an emergence of a primed state (Fig. 1a).

Is there a distinct subset of nociceptors required for PKCϵ-dependent priming? To elucidate the population of nociceptors involved in hyperalgesic priming, Joseph and colleagues lesioned IB$_4$(+) nociceptors via intrathecal administration of IB$_4$-saporin (Joseph and Levine 2010). In these animals, the PKCϵ activator, $\psi\epsilon$RACK, produces an initial hyperalgesia; however, the PGE$_2$-induced prolonged hyperalgesia is absent in animals treated with toxin suggesting that priming is localized to IB$_4$(+) neurons. Furthermore, select agents known to act on peptidergic vs. IB$_4$(+) cells stimulate an initial hyperalgesic state in IB$_4$-lesioned animals but fail to establish priming indicating that a switch in signaling to PKCϵ selectively occurs in IB$_4$(+) nociceptors (Joseph and Levine 2010). As mentioned above, this may be linked to a pronounced loss of GRK2 expression and an increase in Epac activity in IB$_4$(+) nociceptors of primed animals (Wang et al. 2013).

Collectively these findings support a key role for PKCϵ in the initiation and maintenance of nociceptor priming therefore making this PKC isoform a key therapeutic target. However, PKCϵ-dependent forms of priming show a marked sexual dimorphism in rats with female rats failing to show priming, an effect that is apparently due to a protective effect of estrogen (Joseph et al. 2003). Importantly, this does not appear to be the case with mice where both carrageenan and PKC-ϵ-activating peptides cause robust priming to subsequent PGE$_2$ injection in females (Wang et al. 2013). At this point, it remains unclear as to whether there is sexual dimorphism in PKCϵ-mediated effects in humans. Despite these potential discrepancies, it is clear that one mechanism of priming is shared across species and sexes, a requirement for changes in gene expression at the level of protein synthesis.

3.2 Local Translation Is a Key Mediator of Nociceptor Priming

Due to the continuous turnover of cellular proteins, a simple switch in G-protein coupling is likely not sufficient to explain such a long-lasting maladaptive change

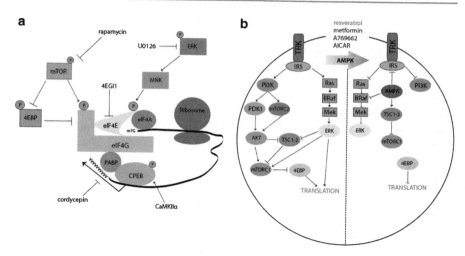

Fig. 2 Translational control pathways involved in hyperalgesic priming. (**a**) mTORC1 phosphorylates 4EBPs, negative regulators of eIF4F formation. This results in its dissociation from eIF4E, allowing the binding of eIF4E to eIF4G. Phosphorylation of eIF4E (via ERK/MNK1/2) or eIF4G (via mTORC1) enhances the formation of the eIF4F complex, promoting translation. Phosphorylation of CPEB by CamKIIα enhances translation efficiency by increasing the length of the poly A tail in mRNAs containing a CPE sequence. Taken together, eIF4F cap complex formation enhances cap-dependent translation, which is necessary for the induction of priming via translational control of gene expression in sensory afferents. (**b**) Pharmacological activation of AMPK results in the inhibition of IRS-1, dampening Trk-mediated signaling. AMPK activation also results in TSC2 phosphorylation, thereby inhibiting mTORC1. Moreover, AMPK phosphorylates BRaf leading to inhibition of ERK signaling. Taken together, AMPK activation decreases activity-dependent translation by turning off both the ERK and mTORC1 pathways, presenting a novel opportunity to prevent or reverse hyperalgesic priming. Adapted from Melemedjian et al. (2010) and Price and Dussor (2013)

in peripheral nociceptors (Bogen et al. 2012). It is likely that this effect is coupled to changes in gene expression and a possible solution to this problem is a localized change occurring at the level of protein translation. Translation can be controlled by extracellular factors signaling via kinase cascades offering a rapid, locally mediated control of gene expression. Two important kinases for translation control are the mechanistic target of rapamycin complex 1 (mTORC1) and extracellular signal-regulated kinase (ERK) (Fig. 2a). Both of these kinases signal to proteins that bind to the 5′ cap structure of mRNAs. mTORC1 phosphorylates 4E-binding proteins (4EBPs) leading to a release of inhibition on eukaryotic initiation factor (eIF) proteins allowing for the eIF4F complex (composed of eIF4E, the 5′ cap-binding protein; eIF4G, a scaffolding protein; and eIF4A, a helicase that unwinds secondary structure in mRNAs) to form (Fig. 2a) (Sonenberg and Hinnebusch 2009). ERK, on the other hand, activates the mitogen-activated kinase-interacting kinases (MNKs), which then phosphorylate eIF4E (Fig. 2a) (Wang et al. 1998). Phosphorylation of eIF4E, which is mediated specifically by MNK1/2, is thought to enhance eIF4F complex formation leading to an increase in translation (Ueda et al. 2010; Herdy et al. 2012). In sensory neurons, nerve growth factor (NGF) and interleukin 6 (IL6), two factors known to induce priming, induce an increase in ERK and mTORC1 signaling leading to a local, axonal increase in protein synthesis (Melemedjian et al. 2010, 2014). Blockade of these kinases, or blockade of eIF4F complex formation with the eIF4F inhibitor compound 4EGI1 (Fig. 2a), inhibits mechanical hypersensitivity induced by these factors and abrogates priming to subsequent PGE_2 exposure (Melemedjian et al. 2010; Asiedu et al. 2011). Hence, local translation is required for the induction of priming downstream of NGF and IL6 signaling.

An alternative mechanism to decrease ERK and mTORC1 signaling is via stimulation of adenosine monophosphate-activated protein kinase (AMPK) (Fig. 2b). AMPK is a ubiquitously expressed energy sensing kinase well known to inhibit mTORC1 signaling through multiple phosphorylation events (Carling et al. 2012). AMPK also abrogates ERK signaling in many cell types, and this effect has recently been linked to negative regulation of the upstream ERK activator BRaf (Fig. 2b) (Shen et al. 2013). In sensory neurons, AMPK activation with pharmacological stimulators (for review see Price and Dussor 2013) leads to decreased ERK and mTORC1 activity (Melemedjian et al. 2011; Tillu et al. 2012), decreased eIF4F complex formation (Melemedjian et al. 2011; Tillu et al. 2012), and inhibition of axonal protein synthesis as measured by enhanced processing body (P body) formation (Melemedjian et al. 2014). AMPK activators decrease peripheral nerve injury- and inflammation-induced mechanical hyperalgesia (Melemedjian et al. 2011; Russe et al. 2013) suggesting an important role for this kinase in pain plasticity. In the context of hyperalgesic priming, AMPK activation with the natural product resveratrol (Fig. 2b) decreases mechanical hypersensitivity caused by incision and completely blocks the development of priming when given locally around the time of incision (Tillu et al. 2012). These findings further suggest a role for local translation in the initiation of plasticity leading to priming of nociceptors.

Epac signaling, which was described above as an important mediator of priming induced by inflammatory mediators (Wang et al. 2013), may also play an important role in regulating translation in sensory neurons. Decreased GRK2 leads to enhanced activation of Epac causing an increase in ERK activity (Eijkelkamp et al. 2010). While this signaling event has not been linked to translation control, based on findings involving IL6-induced priming and its dependence of ERK/ eIF4E-mediated changes in translation control, it is conceivable that this is an important mediator for priming where ERK activity is increased. Furthermore, Epac activation causes an increase in mTORC1 activity in some transformed cells (Misra and Pizzo 2009) suggesting that enhanced Epac signaling in primed nociceptors may lead to convergent signaling onto the eIF4F complex in a manner analogous to that observed with NGF and IL6 (Melemedjian et al. 2010). It remains to be seen if these Epac-mediated events occur in sensory neurons and their axons, but this may be a fruitful area for further work and therapeutic intervention.

The regulation of translation via 5' cap-binding proteins and their upstream kinases clearly comprise an important mechanism for the priming of nociceptors. However, translation is likewise regulated by RNA-binding proteins that bind to either 5' or 3' untranslated regions (UTRs). The fragile X mental retardation protein

(FMRP) is a key RNA-binding protein regulating plasticity in the PNS and CNS (Bassell and Warren 2008). Mice lacking FMRP fail to sensitize in several preclinical pain models (Price et al. 2007; Price and Melemedjian 2012), and these mice also have deficits in priming induced by NGF and IL6 (Asiedu et al. 2011). Another important RNA-binding protein in hyperalgesic priming is the cytoplasmic polyadenylation element-binding protein (CPEB). CPEB binds preferentially to mRNAs containing a CPE sequence in their 3′ UTR near the polyadenylation sequence. These mRNAs contain short poly A tails, and CPEB, upon phosphorylation, can enhance the poly A tail length leading to enhanced translation efficiency (Fig. 2a) (Richter 2007). This process is linked to long-term potentiation (LTP) in the CNS (Udagawa et al. 2012) and has recently been shown to play an important role in nociceptor priming (Bogen et al. 2012; Ferrari et al. 2012, 2013a). CPEB is phosphorylated by the aurora family kinases (Mendez et al. 2000) and, likely more importantly for neurons, by Ca^{2+}/calmodulin-activated protein kinase IIα (Fig. 2a) (CaMKIIα, Atkins et al. 2005). In the PNS, CPEB is primarily expressed by $IB_4(+)$ nociceptors (Bogen et al. 2012). Knockdown of CPEB with intrathecally injected antisense oligonucleotides leads to inhibition of the initiation of priming induced by PKCε activators linking translation events to PKCε-dependent priming (Bogen et al. 2012). Importantly, in priming induced by peripheral inflammation, CPEB may act downstream of PKCε and CaMKIIα to initiate and maintain a primed state (Ferrari et al. 2013a). Since CPEB is thought to have prion-like properties that are crucial for its role in memory maintenance (Si et al. 2003a, b), these findings point to a potential role for CPEB in creating a permanently primed state in peripheral nociceptors.

While it is clear that translation regulation is required to initiate a primed state in the periphery, a crucial question is whether disruption of local translation is capable of interrupting priming once it has been fully established. This is an important question because it gives insight into devising therapeutic strategies to reverse priming in the clinical state. While there are certain situations (e.g., surgery) where inhibition of priming mechanisms at the time of injury is a viable strategy, the majority of clinical situations are likely to require intervention following the full establishment of priming. An experimental paradigm to test this translation dependency is to induce priming with a locally administered stimulus and then allow the initial hyperalgesia to resolve. Then, prior to injection of the stimulus to precipitate hyperalgesia in primed animals, translation inhibitors can be administered locally to test whether continuous translation is required to maintain a primed state (Asiedu et al. 2011; Ferrari et al. 2013a). In this regard, following injection of IL6 and resolution of mechanical hyperalgesia in mice, injection of anisomycin or rapamycin (at doses that block the initiation of priming) fail to reverse a primed state (Asiedu et al. 2011). In contrast, in rats, injection of carrageenan causes priming that is disrupted both at the time of carrageenan injection and during the maintenance phase by either the mTORC1 inhibitor rapamycin or the polyadenylation inhibitor cordycepin (Ferrari et al. 2013b). Interestingly, in addition to cordycepin action on polyadenylation, this compound has recently been identified as an AMPK activator (Wu et al. 2014). Similar effects with rapamycin and cordycepin are observed in rats primed with paw injection of CaMKIIα. Since CaMKIIα phosphorylates CPEB and CPEB regulates CaMKIIα translation, this raises the intriguing possibility that CaMKIIα/CPEB signaling could represent a positive feedback mechanism to maintain pain memory in the peripheral nociceptor (Ferrari et al. 2013a). Importantly, in rats where PKCε-induced priming is sexually dimorphic, priming that is dependent on local translation occurs both in male and female rats (Ferrari et al. 2013a). Hence, while there are conflicting results in different models, it is formally possible that brief disruption of local translation in primed nociceptors is capable of resolving a primed state. Further work is clearly needed to further interrogate mechanisms involved in these effects; however, they point to the tantalizing possibility that therapeutics targeting translation regulation might have disease-modifying effects in chronic pain conditions. This hypothesis is consistent with the observation that AMPK activators, which have a strong effect on translation regulation pathways, have disease-modifying effects in neuropathic pain models (Melemedjian et al. 2011, 2013b).

4 CNS Regulation of Hyperalgesic Priming

4.1 Atypical PKCs and Brain-Derived Neurotropic Factor

While there is strong evidence for memory-like mechanisms controlling hyperalgesic priming at the level of the peripheral nociceptor, plasticity in the CNS, especially in the dorsal horn of the spinal cord, also plays a central role in hyperalgesic priming. Here, analogies to mechanisms of memory formation and maintenance in other CNS regions, such as the hippocampus and cortex, have played a central role in directing research in the area. LTP, long recognized as a neurophysiological correlate of learning and memory at central synapses, can be divided into an early and late phase. Early LTP requires the activation of CaMKIIα, PKA, and conventional PKC leading to the phosphorylation of AMPA receptors (Huganir and Nicoll 2013). Early LTP also leads to changes in gene expression which occur both on the level of transcription and translation. These changes in gene expression are needed for the consolidation of early LTP into late LTP (Abraham and Williams 2008). Mechanisms involved in the maintenance of late LTP have been more difficult to clearly elucidate but are thought to involve brain-derived neurotropic factor (BDNF) and an atypical PKC (aPKC) isoform called PKMζ (Fig. 3) (Sacktor 2011). While LTP has been extensively studied in the cortex and hippocampus, LTP can also be induced at central synapses in the dorsal horn of the spinal cord, and many mechanisms elucidated in other CNS regions are shared at these spinal synapses (Sandkuhler 2007). Spinal LTP is a potential mechanism for primary hyperalgesia (Sandkuhler 2007), and heterosynaptic LTP may explain other aspects of pain plasticity that occur in chronic pain disorders (Klein et al. 2008).

Does spinal LTP explain features of the maintenance of hyperalgesic priming? While no direct measurements of spinal LTP as a correlate of hyperalgesic priming

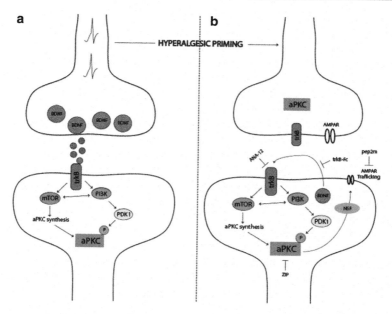

Fig. 3 The role of aPKCs in hyperalgesic priming initiation and maintenance. (**a**) Nociceptor activation leads to spinal BDNF release and a postsynaptic mTORC1-dependent translation of aPKC protein following trkB activation. These newly synthesized aPKCs are then phosphorylated by PDK1. Increased levels and phosphorylation of aPKCs are thought to be involved in initiating priming. (**b**) Once priming is established, increased aPKC protein and phosphorylation leads to a constitutive increase in AMPAR trafficking to the postsynaptic membrane. This appears to be regulated by BDNF signaling via trkB with BDNF potentially being released from postsynaptic dendrites in the maintenance stage of priming. Presynaptic trkB may also be activated by increased BDNF action in primed animals. Once established, hyperalgesic priming can be permanently reversed by inhibition of aPKCs with ZIP, by disruption of AMPAR trafficking with pep2M, or via inhibition of trkB/BDNF signaling with ANA-12 or trkB-Fc, respectively. Adapted from Price and Ghosh (2013)

have been made, pharmacological similarities abound. Inhibition of translation and/or CaMKIIα during LTP induction blocks consolidation of late LTP (Nicoll and Roche 2013). Likewise, intrathecal injection of translation inhibitors or CaMKIIα inhibitors at the time of priming induction in mice inhibits initial hyperalgesia and prevents priming. In contrast, intrathecal injection of these compounds following the resolution of initial hyperalgesia fails to resolve priming consistent with a lack of effect of these mechanisms on the maintenance of late LTP (Asiedu et al. 2011; Melemedjian et al. 2013a). Late LTP can be reversed by inhibition of aPKCs with a pseudo substrate inhibitor called ZIP (Pastalkova et al. 2006). Intrathecal injection of ZIP either at the time of priming induction or following the resolution of the initial hyperalgesia leads to a complete reversal of hyperalgesic priming (Asiedu et al. 2011; Melemedjian et al. 2013a). ZIP also reverses established pain states that have become dependent on central plasticity following sustained afferent input (Laferriere et al. 2011). These findings are consistent with a role for PKMζ in the maintenance of late LTP, memory retention, and the maintenance of a chronic pain state. However, recent experiments using genetic methods to dissect the role of PKMζ in late LTP and memory maintenance have called the specificity of ZIP and the role of PKMζ in these effects into question (Lee et al. 2013; Volk et al. 2013). It remains to be seen if PKMζ plays a specific role in the maintenance of hyperalgesic priming in the dorsal horn of the spinal cord (for review on this topic see Price and Ghosh 2013).

An important component of the proposed role of PKMζ in LTP and memory is the trafficking of AMPA receptors to synaptic sites leading to a persistent augmentation of postsynaptic glutamate-mediated signaling (Fig. 3) (Sacktor 2011). This trafficking can be disrupted with a peptide called pep2m (Migues et al. 2010). Similar to experiments in other CNS regions, intrathecal injection of pep2m disrupts the maintenance of hyperalgesic priming (Asiedu et al. 2011) suggesting that aPKC-mediated regulation of AMPA receptor trafficking may play a central role in chronic pain states (Fig. 3). This is consistent with a wide variety of experimental findings indicating that AMPA receptor trafficking plays a central role in mediating pain plasticity induced by peripheral injury (Tao 2012).

As mentioned above, while it is clear that ZIP is capable of permanently reversing a primed state in a variety of experimental models (Asiedu et al. 2011; Laferriere et al. 2011; Melemedjian et al. 2013a), the target of ZIP is less clear based on recent evidence from transgenic mice (Lee et al. 2013; Volk et al. 2013). One possibility is that aPKC isoforms play a redundant role in synaptic plasticity, and therefore PKCλ may be involved in maintenance mechanisms of hyperalgesic priming (Price and Ghosh 2013). Since this isoform is also inhibited by ZIP (Melemedjian et al. 2013a; Volk et al. 2013), this is a parsimonious explanation for the discrepancy between pharmacological effects of ZIP and findings from mice lacking aPKCs derived from the *Prckz* locus (PKMζ and PKCζ). If this were the case, upstream mechanisms that regulate all aPKCs isoforms, either via phosphorylation or through their translation at synapses, would represent potential alternative targets to reverse hyperalgesic priming. A candidate molecule fitting this description is BDNF.

BDNF has long been recognized as an important mediator of pain plasticity. BDNF is expressed by DRG neurons and released in the spinal dorsal horn (Balkowiec and Katz 2000), where it can act on pre- and postsynaptic trkB receptors to regulate plasticity of presynaptic afferent fibers (Matayoshi et al. 2005) and postsynaptic dorsal horn neurons (Kerr et al. 1999; Pezet et al. 2002; Garraway et al. 2003). BDNF expression shows considerable plasticity following peripheral injury (Mannion et al. 1999) and nociceptor-specific knockout of BDNF leads to abrogation of many forms of injury-induced pain plasticity (Zhao et al. 2006). BDNF is also a key factor in LTP. In the hippocampus, BDNF is required for the induction of LTP, and during late LTP, dendritic-expressed BDNF appears to play an autocrine role in the maintenance of late-phase LTP (Fig. 3) (Lu et al. 2008). Likewise, BDNF is capable of inducing LTP in dorsal horn neurons (Zhou et al. 2008) linking BDNF-induced pain plasticity to memory-like mechanisms that may be involved in the maintenance of hyperalgesic priming.

Indeed, intrathecal injection of a BDNF scavenging agent, trkB-fc, or systemic injection of a trkB antagonist, ANA-12 (Cazorla et al. 2011), blocks hyperalgesia induced by priming agents and prevents the precipitation of a primed state by subsequent PGE_2 injection. Significantly, interruption of BDNF/trkB signaling with either trkB-fc or ANA-12 after the establishment of a primed state leads to a resolution of priming precipitated by PGE_2 injection (Melemedjian et al. 2013a). This suggests a key role of BDNF/trkB signaling in the maintenance of a primed state. At spinal synapses, BDNF induces phosphorylation and translation of the two major aPKC isoforms found in the CNS, PKMζ and PKCλ (Fig. 3), suggesting a potential link between BDNF/trkB and aPKCs in the regulation of the maintenance of hyperalgesic priming (Melemedjian et al. 2013a). Because precise targets of ZIP remain unknown, these findings point to BDNF/trkB signaling as a therapeutic target for the reversal of established chronic pain states. Importantly, the precise location of BDNF action and trkB signaling in the maintenance of hyperalgesic priming has not been established. While it is tempting to speculate that an autocrine role for BDNF release from dendrites of dorsal horn neurons in the primed state plays a role in maintaining priming, especially based on the role of this mechanism in the maintenance of late LTP in other CNS structures (Lu et al. 2008), this hypothesis requires further experimental exploration.

4.2 Endogenous Opioids, μ-Opioid Receptor Constitutive Activity, and Hyperalgesic Priming

An intriguing question in the neurobiology of hyperalgesic priming is why does the initial hyperalgesia resolve if a primed state can persist for many weeks to months after the initial injury? This is especially interesting when one considers the strong pharmacological parallels between the maintenance of hyperalgesic priming and the maintenance of late LTP (e.g., aPKCs, AMPA receptor trafficking and BDNF/trkB signaling). In other words, if LTP persists and consolidates into late LTP to maintain priming, why does the initial hyperalgesia resolve? One possibility is that endogenous analgesic mechanisms mask the hyperalgesic state. In such a scenario, the precipitation of hyperalgesia in primed animals would be able to override this endogenous mechanism leading to a very long-lasting hyperalgesic state to a normally subthreshold stimulus. As described above, this is precisely what occurs in hyperalgesic priming models. One candidate for endogenous analgesia overriding hyperalgesia is the endogenous opioid system. This system is robust in the dorsal horn with interneurons capable of releasing peptides that act on μ-opioid receptors (MORs) expressed throughout the dorsal horn (Ribeiro-da-Silva et al. 1992; Ma et al. 1997), including on presynaptic nociceptor nerve endings (Schroeder et al. 1991; Schroeder and McCleskey 1993). Exciting new evidence suggests that this mechanism may be at play in hyperalgesic priming.

Animals exposed to a strong inflammatory stimulus demonstrate hyperalgesia that normally resolves within 10–14 days; however, when these same animals are infused with MOR inverse agonists, the initial hyperalgesia is prolonged for the

Fig. 4 Signaling pathways regulating endogenous opioid control of hyperalgesic priming. Prior to injury, MORs do not have constitutive activity and nociceptive input to the CNS is absent. However, after injury, nociceptor firing induces activity in CNS circuits leading to release of endogenous opioid peptides in the dorsal horn and the eventual development of MOR constitutive activity resulting in agonist-independent modulation of pain responses. Upon infusion of the MOR inverse agonist naltrexone, AC1 is disinhibited, likely in an NMDA receptor-dependent fashion, allowing for superactivation of cAMP and the reinstatement of hyperalgesia. Adapted from Corder et al. (2013)

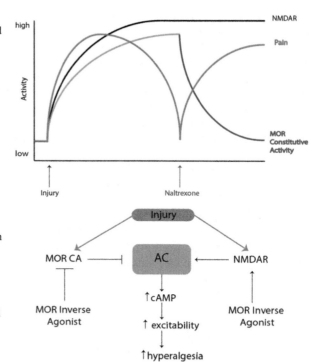

duration of inverse agonist infusion (Corder et al. 2013). Importantly, even when the initial hyperalgesia is allowed to resolve, infusion of MOR inverse agonists immediately precipitates a reinstatement of hyperalgesia, an effect that is absent in sham animals and an effect that is analogous to precipitation of hyperalgesia in primed animals with a subthreshold peripheral stimulus. What governs this effect? Peripheral inflammation, and presumably other nociceptive stimuli, induces a change in spinal MORs such that they now acquire constitutive activity (signaling through G-proteins in the absence of agonist). This MOR constitutive activity causes a tonic inhibition of pain signaling presumably masking a hyperalgesic state that would otherwise persist following the initial insult. When MOR inverse agonists remove this acquired MOR constitutive activity, a cAMP overshoot occurs (a classical cellular sign of opioid dependence) in a Ca^{2+}-dependent adenylyl cyclase 1 (AC1)-dependent fashion that also involves the engagement of NMDA receptors (Fig. 4) (Corder et al. 2013). This leads to the reinstatement of pain in animals that have a previous history of strong nociceptive input (Fig. 4).

These findings have several important implications for understanding central mechanisms governing hyperalgesic priming. First, they provide an elegant solution to the question of why initial hyperalgesia resolves despite the persistence of a primed state. This occurs, at least in part, because tonic opioid signaling, including MOR constitutive activity, masks the presence of mechanisms that would otherwise drive hyperalgesia (Corder et al. 2013). Second, this study provides some possible links to priming and late LTP maintenance that potentially solve questions stated above. Opioid-dependent mechanisms play an important role in governing spinal LTP. While there is evidence that high-dose opioids can stimulate LTP at certain synapses after their abrupt removal (Drdla et al. 2009), there is likewise evidence that MOR activation can resolve even late LTP at spinal synapses (Drdla-Schutting et al. 2012). Therefore, it is formally possible that the initial priming stimulus leads to late LTP consolidation, but this is subsequently resolved by endogenous opioid-mediated mechanisms. A key question then is: does the previous establishment of late LTP at central synapses lead to a drop in threshold for establishment of subsequent LTP? If the mechanisms governing the MOR-dependent reversal of spinal late LTP are constitutively expressed, as appears to be the case (Corder et al. 2013), then this may lead to a tonic reversal of late LTP with underlying mechanisms (e.g., aPKC and BDNF/trkB signaling) still in place. While this idea obviously requires extensive experimental work, it could represent an important mechanism linking changes in peripheral sensitivity to CNS plasticity responsible for the maintenance of priming. Reversing these mechanisms could lead to revolutionary new therapeutics with disease-modifying effects on chronic pain.

4.3 Surgery as a Priming Stimulus and the Effects of Opioids

Opioids have been the first line of therapy for moderate to severe acute pain for decades, if not centuries. However, there is abundant preclinical and some clinical evidence suggesting that opioid administration, designed to alleviate acute pain, paradoxically primes the patient and renders them more susceptible to a transition to chronic pain. Studies using rodents have demonstrated a steady reduction in withdrawal thresholds with intrathecal morphine administration, fentanyl boluses, and repeated systemic morphine (Vanderah et al. 2001; Gardell et al. 2002, 2006) or heroin administration (Mao et al. 1995; Celerier et al. 2000, 2001), generating a sensitized state that is presumably independent of noxious stimulation from the periphery. These preclinical studies highlight the neuroplastic changes induced by opioids and provide another mechanism for the induction of a sensitized and/or primed state in animals.

First, it is clear that surgical incision can produce priming in rodents and that these mechanisms are largely shared with other models involving the use of inflammatory mediators (Asiedu et al. 2011). Do the use of common postsurgical pain analgesics alleviate this priming? Here the answer appears to be no as opioid administration in the perioperative period can exacerbate a primed state revealed up to weeks later by injection of an inflammatory mediator (Fig. 5, Rivat et al. 2007). Specifically, in mouse models of postsurgical pain, remifentanil administration leads to the enhancement of the initial hyperalgesic state and the development of enhanced and prolonged response upon the precipitation of a second phase of hyperalgesia in animals primed with incision + opioids (Cabanero et al. 2009a, b). Additionally, in rodents that received incision and were simultaneously treated with

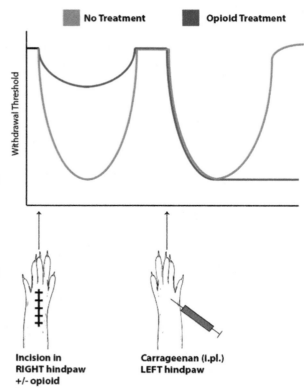

Fig. 5 Opioids given as postsurgical analgesics can exacerbate priming in animal models. Upon incision of the hindpaw, withdrawal thresholds in animals not receiving analgesics drop significantly compared to animals treated with opioids at the time of incision. Following resolution of the initial hyperalgesia, when carrageenan is subsequently injected into the *contralateral* hindpaw, withdrawal thresholds drop in both groups; however, animals previously treated with opioids demonstrate an exaggerated hyperalgesia to the same stimulus, suggesting that opioids, despite their analgesic effect, exacerbate nociceptive priming induced by injury

opioids, these animals were primed to develop a long-lasting hyperalgesic state in response to subsequent contralateral inflammation, environmental stress, or subsequent opioid administration (Fig. 5, Rivat et al. 2002, 2007; Cabanero et al. 2009a). Thus, opioid administration as an analgesic for postsurgery pain might be viewed as a catalyst that contributes to pain chronification after incision providing a model system to study the effect of opioids on exacerbation of the transition from acute to chronic pain. Significantly, this process appears to involve spinal NMDA receptors because the effect of opioids on enhancement of priming is prevented, in most cases, by concomitant treatment with intrathecal NMDA receptor antagonists (Rivat et al. 2013). Hence, despite their obviously beneficial postsurgical analgesic effect, opioids induce long-lasting sensitization after initial analgesia at least in animal models. Therefore, it is formally possible that high rates of pain chronification following surgery (Macrae 2001) may be at least partially due to the nearly universal use of opioids as postsurgical pain medications. Hence, continued work into mechanisms that might mitigate these effects (e.g., NMDA receptor blockers) or non-opioid analgesics is warranted.

5 Therapeutic Opportunities and Conclusions

We have made the case above for hyperalgesic priming as a model system to understand plasticity in the PNS and CNS underlying the maintenance of chronic pain states. While human experimental pain models have, as of yet, not given clear evidence of priming in our species, the chronic intermittent nature of many chronic pain states makes a strong case for priming as a key feature of clinical pain disorders. What then are the key opportunities for therapeutic intervention?

Targeting the peripheral nervous system is advantageous because it provides an opportunity to avoid CNS side effects. In our view, key targets here are PKCε, Epac, and translation regulation including AMPK. PKCε inhibitors have been the focus of intense research for some time now and may eventually enter pain clinical trials (Reichling and Levine 2009). Because PKCε is involved in both the initiation and maintenance of priming, this mechanism might be engaged for the reversal of a primed state; however, there is now evidence that several other key mechanisms are downstream of this kinase making them more attractive targets (Ferrari et al. 2013a; Wang et al. 2013). One of these is Epac. As noted above, Epac signaling may play a key role in signaling switches occurring exclusively in primed nociceptors (Hucho et al. 2005; Eijkelkamp et al. 2010; Ferrari et al. 2012; Wang et al. 2013). Hence, inhibition of Epac activity may likewise provide an opportunity for interruption of priming. Having said this, there is strong evidence that translation control is a final common denominator in all of these signaling mechanisms indicating that targeting gene expression at the level of local translation in the primed nociceptor may provide the greatest therapeutic opportunity. Additional preclinical work is needed to identify key signaling mechanisms involved in regulating translation regulation in primed nociceptors, but the currently available evidence points strongly to mTORC1, ERK (Melemedjian et al. 2010, 2013a; Tillu et al. 2012), and CPEB (Bogen et al. 2012; Ferrari et al. 2013a) as targets. It is currently not clear if CPEB can be targeted pharmacologically and its upstream kinases in priming have not been elucidated although CaMKIIα is a strong candidate. Targeting mTORC1 and ERK is likely a more viable approach because compounds that target these pathways are already in clinical use. While rapamycin clearly inhibits mTORC1, this approach also leads to feedback activation of ERK over the longer term, an action that may counteract rapamycin efficacy (Melemedjian et al. 2013b). Having said that, as detailed above, AMPK activation inhibits both mTORC1 and ERK pathways, and AMPK activators have already been shown to block the initiation of priming. Here, the common antidiabetic drug metformin is in wide clinical use for other indications and acts via AMPK activation (Shaw et al. 2005). While further work is needed to investigate preclinical effects of AMPK activators in hyperalgesic priming maintenance, clinical trials can be designed to test the hypothesis that AMPK activators have disease-modifying effects in chronic pain states (Price and Dussor 2013).

An alternative approach would be to target mechanisms in the CNS responsible for the maintenance of plasticity resulting in priming. These mechanisms share similarities to molecules involved in learning and memory; hence, they carry inherent risks that deserve appropriate caution (Price and Ghosh 2013). Having said this, innovative approaches to therapeutic intervention may lead to disease-modifying effects in chronic pain patients resulting from a permanent reversal of plasticity in dorsal horn circuitry. From this perspective, we propose that BDNF/trkB signaling is an attractive candidate. As mentioned above, short-term disruption of BDNF action at trkB receptors leads to a permanent reversal of a primed state (Melemedjian et al. 2013a). This suggests that in the clinical arena, therapeutics targeting this pathway could be used for a brief period of time to achieve long-lasting effects on chronic pain. TrkB-based therapeutics are under investigation for a wide variety of neurological disorders (Cazorla et al. 2011); therefore, clinical opportunities along this front may arise in the near future. In the longer term, continued research into the role of aPKCs in pain chronification may lead to important insight into a specific role for certain aPKC isoforms in the maintenance of priming. Because it now appears to be the case that PKMζ is not required for late LTP or learning and memory rostral to the spinal cord, an intriguing possibility is that this aPKC isoform plays a specific role in the pain pathway (Lee et al. 2013; Price and Ghosh 2013; Volk et al. 2013). This scenario could lead to therapeutic opportunities that would have decreased liability for disruption of plasticity in brain regions involved in learning and memory.

A final opportunity worth noting concerns the complex interrelationship of opioids with chronic pain. As discussed at length above, there is evidence that opioid analgesics can augment priming when given during the time of surgery (Rivat et al. 2013). On the other hand, endogenous opioid mechanisms may be crucial for masking hyperalgesia that would otherwise persist following injury. How can these mechanisms be resolved for better therapeutics? A clear opportunity exists in the development of non-opioid analgesics for the treatment of pain. This is a long-standing goal of research in the pain arena and continues to be the impetus for most target-based drug discovery in the field. From the perspective of MOR constitutive activity, which likely serves a beneficial function, targeting AC1 to avoid cAMP superactivation when this mechanism is disengaged could avoid the adverse consequences resulting from this MOR plasticity (Corder et al. 2013).

In closing, we have summarized research into hyperalgesic priming highlighting our current understanding of plasticity mechanisms in the peripheral nociceptor and in the CNS that govern this preclinical model of the transition from acute to chronic pain. The advent and subsequent proliferation of research into this model has led to a great expansion of our understanding of plasticity in the nociceptive system. Our view is that these findings provide great insight into therapeutic opportunities not only for the treatment of chronic pain but also for its potential reversal. Continued work in this area holds great promise for the development of revolutionary therapeutics for the permanent alleviation of chronic pain states.

Acknowledgments This work was supported by NIH grants NS065926 and GM102575 to T.J.P.

References

Abraham WC, Williams JM (2008) LTP maintenance and its protein synthesis-dependence. Neurobiol Learn Mem 89:260–268

Aley KO, Levine JD (1999) Role of protein kinase A in the maintenance of inflammatory pain. J Neurosci 19:2181–2186

Aley KO, Messing RO, Mochly-Rosen D, Levine JD (2000) Chronic hypersensitivity for inflammatory nociceptor sensitization mediated by the epsilon isozyme of protein kinase C. J Neurosci 20:4680–4685

Althaus A, Hinrichs-Rocker A, Chapman R, Arranz Becker O, Lefering R, Simanski C, Weber F, Moser KH, Joppich R, Trojan S, Gutzeit N, Neugebauer E (2012) Development of a risk index for the prediction of chronic post-surgical pain. Eur J Pain 16:901–910

Andersson GB (1999) Epidemiological features of chronic low-back pain. Lancet 354:581–585

Asiedu MN, Tillu DV, Melemedjian OK, Shy A, Sanoja R, Bodell B, Ghosh S, Porreca F, Price TJ (2011) Spinal protein kinase M zeta underlies the maintenance mechanism of persistent nociceptive sensitization. J Neurosci 31:6646–6653

Atkins CM, Davare MA, Oh MC, Derkach V, Soderling TR (2005) Bidirectional regulation of cytoplasmic polyadenylation element-binding protein phosphorylation by Ca2+/calmodulin-dependent protein kinase II and protein phosphatase 1 during hippocampal long-term potentiation. J Neurosci 25:5604–5610

Balkowiec A, Katz DM (2000) Activity-dependent release of endogenous brain-derived neurotrophic factor from primary sensory neurons detected by ELISA in situ. J Neurosci 20:7417–7423

Bartleson JD (2001) Low back pain. Curr Treat Options Neurol 3:159–168

Bassell GJ, Warren ST (2008) Fragile X syndrome: loss of local mRNA regulation alters synaptic development and function. Neuron 60:201–214

Bigal ME, Lipton RB (2008) Clinical course in migraine: conceptualizing migraine transformation. Neurology 71:848–855

Bogen O, Alessandri-Haber N, Chu C, Gear RW, Levine JD (2012) Generation of a pain memory in the primary afferent nociceptor triggered by PKCε activation of CPEB. J Neurosci 32:2018–2026

Cabanero D, Campillo A, Celerier E, Romero A, Puig MM (2009a) Pronociceptive effects of remifentanil in a mouse model of postsurgical pain: effect of a second surgery. Anesthesiology 111:1334–1345

Cabanero D, Celerier E, Garcia-Nogales P, Mata M, Roques BP, Maldonado R, Puig MM (2009b) The pro-nociceptive effects of remifentanil or surgical injury in mice are associated with a decrease in delta-opioid receptor mRNA levels: prevention of the nociceptive response by on-site delivery of enkephalins. Pain 141:88–96

Carey TS, Garrett JM, Jackman A, Hadler N (1999) Recurrence and care seeking after acute back pain: results of a long-term follow-up study. North Carolina Back Pain Project. Med Care 37:157–164

Carling D, Thornton C, Woods A, Sanders MJ (2012) AMP-activated protein kinase: new regulation, new roles? Biochem J 445:11–27

Cassidy JD, Cote P, Carroll LJ, Kristman V (2005) Incidence and course of low back pain episodes in the general population. Spine (Phila Pa 1976) 30:2817–2823

Cazorla M, Premont J, Mann A, Girard N, Kellendonk C, Rognan D (2011) Identification of a low-molecular weight TrkB antagonist with anxiolytic and antidepressant activity in mice. J Clin Invest 121:1846–1857

Celerier E, Rivat C, Jun Y, Laulin JP, Larcher A, Reynier P, Simonnet G (2000) Long-lasting hyperalgesia induced by fentanyl in rats: preventive effect of ketamine. Anesthesiology 92:465–472

Celerier E, Laulin JP, Corcuff JB, Le Moal M, Simonnet G (2001) Progressive enhancement of delayed hyperalgesia induced by repeated heroin administration: a sensitization process. J Neurosci 21:4074–4080

Corder G, Doolen S, Donahue RR, Winter MK, Jutras BL, He Y, Hu X, Wieskopf JS, Mogil JS, Storm DR, Wang ZJ, McCarson KE, Taylor BK (2013) Constitutive mu-opioid receptor activity leads to long-term endogenous analgesia and dependence. Science 341:1394–1399

Croft PR, Lewis M, Papageorgiou AC, Thomas E, Jayson MI, Macfarlane GJ, Silman AJ (2001) Risk factors for neck pain: a longitudinal study in the general population. Pain 93:317–325

Dina OA, Khasar SG, Gear RW, Levine JD (2009) Activation of Gi induces mechanical hyperalgesia poststress or inflammation. Neuroscience 160:501–507

Drdla R, Gassner M, Gingl E, Sandkuhler J (2009) Induction of synaptic long-term potentiation after opioid withdrawal. Science 325:207–210

Drdla-Schutting R, Benrath J, Wunderbaldinger G, Sandkuhler J (2012) Erasure of a spinal memory trace of pain by a brief, high-dose opioid administration. Science 335:235–238

Eijkelkamp N, Wang H, Garza-Carbajal A, Willemen HL, Zwartkruis FJ, Wood JN, Dantzer R, Kelley KW, Heijnen CJ, Kavelaars A (2010) Low nociceptor GRK2 prolongs prostaglandin E2 hyperalgesia via biased cAMP signaling to Epac/Rap1, protein kinase Cepsilon, and MEK/ERK. J Neurosci 30:12806–12815

Ferrari LF, Bogen O, Alessandri-Haber N, Levine E, Gear RW, Levine JD (2012) Transient decrease in nociceptor GRK2 expression produces long-term enhancement in inflammatory pain. Neuroscience 222:392–403

Ferrari LF, Bogen O, Levine JD (2013a) Role of nociceptor alphaCaMKII in transition from acute to chronic pain (hyperalgesic priming) in male and female rats. J Neurosci 33:11002–11011

Ferrari LF, Bogen O, Chu C, Levine JD (2013b) Peripheral administration of translation inhibitors reverses increased hyperalgesia in a model of chronic pain in the rat. J Pain 14:731–738

Gardell LR, Wang R, Burgess SE, Ossipov MH, Vanderah TW, Malan TP Jr, Lai J, Porreca F (2002) Sustained morphine exposure induces a spinal dynorphin-dependent enhancement of excitatory transmitter release from primary afferent fibers. J Neurosci 22:6747–6755

Gardell LR, King T, Ossipov MH, Rice KC, Lai J, Vanderah TW, Porreca F (2006) Opioid receptor-mediated hyperalgesia and antinociceptive tolerance induced by sustained opiate delivery. Neurosci Lett 396:44–49

Garraway SM, Petruska JC, Mendell LM (2003) BDNF sensitizes the response of lamina II neurons to high threshold primary afferent inputs. Eur J Neurosci 18:2467–2476

Harris W (1921) Persistent pain in lesions of the peripheral and central nervous system. Br Med J 2:896–900

Herdy B et al (2012) Translational control of the activation of transcription factor NF-kappaB and production of type I interferon by phosphorylation of the translation factor eIF4E. Nat Immunol 13:543–550

Hucho TB, Dina OA, Levine JD (2005) Epac mediates a cAMP-to-PKC signaling in inflammatory pain: an isolectin B4(+) neuron-specific mechanism. J Neurosci 25:6119–6126

Huganir RL, Nicoll RA (2013) AMPARs and synaptic plasticity: the last 25 years. Neuron 80:704–717

Joseph EK, Levine JD (2010) Hyperalgesic priming is restricted to isolectin B4-positive nociceptors. Neuroscience 169:431–435

Joseph EK, Parada CA, Levine JD (2003) Hyperalgesic priming in the rat demonstrates marked sexual dimorphism. Pain 105:143–150

Joseph EK, Bogen O, Alessandri-Haber N, Levine JD (2007) PLC-beta 3 signals upstream of PKC epsilon in acute and chronic inflammatory hyperalgesia. Pain 132:67–73

Joseph EK, Reichling DB, Levine JD (2010) Shared mechanisms for opioid tolerance and a transition to chronic pain. J Neurosci 30:4660–4666

Kerr BJ, Bradbury EJ, Bennett DL, Trivedi PM, Dassan P, French J, Shelton DB, McMahon SB, Thompson SW (1999) Brain-derived neurotrophic factor modulates nociceptive sensory inputs and NMDA-evoked responses in the rat spinal cord. J Neurosci 19:5138–5148

Khasar SG, Burkham J, Dina OA, Brown AS, Bogen O, Alessandri-Haber N, Green PG, Reichling DB, Levine JD (2008) Stress induces a switch of intracellular signaling in sensory neurons in a model of generalized pain. J Neurosci 28:5721–5730

Klein T, Stahn S, Magerl W, Treede RD (2008) The role of heterosynaptic facilitation in long-term potentiation (LTP) of human pain sensation. Pain 139:507–519

Kolb E, Canjuga M, Bauer GF, Laubli T (2011) Course of back pain across 5 years: a retrospective cohort study in the general population of Switzerland. Spine (Phila Pa 1976) 36:E268–E273

Laferriere A, Pitcher MH, Haldane A, Huang Y, Cornea V, Kumar N, Sacktor TC, Cervero F, Coderre TJ (2011) PKMzeta is essential for spinal plasticity underlying the maintenance of persistent pain. Mol Pain 7:99

Lee AM, Kanter BR, Wang D, Lim JP, Zou ME, Qiu C, McMahon T, Dadgar J, Fischbach-Weiss SC, Messing RO (2013) Prkcz null mice show normal learning and memory. Nature 493:416–419

Lipton RB (2009) Tracing transformation: chronic migraine classification, progression, and epidemiology. Neurology 72:S3–S7

Lu Y, Christian K, Lu B (2008) BDNF: a key regulator for protein synthesis-dependent LTP and long-term memory? Neurobiol Learn Mem 89:312–323

Ma W, Ribeiro-da-Silva A, De Koninck Y, Radhakrishnan V, Cuello AC, Henry JL (1997) Substance P and enkephalin immunoreactivities in axonal boutons presynaptic to physiologically identified dorsal horn neurons. An ultrastructural multiple-labelling study in the cat. Neuroscience 77:793–811

Macrae WA (2001) Chronic pain after surgery. Br J Anaesth 87:88–98

Mannion RJ, Costigan M, Decosterd I, Amaya F, Ma QP, Holstege JC, Ji RR, Acheson A, Lindsay RM, Wilkinson GA, Woolf CJ (1999) Neurotrophins: peripherally and centrally acting modulators of tactile stimulus-induced inflammatory pain hypersensitivity. Proc Natl Acad Sci U S A 96:9385–9390

Mao J, Price DD, Mayer DJ (1995) Mechanisms of hyperalgesia and morphine tolerance: a current view of their possible interactions. Pain 62:259–274

Matayoshi S, Jiang N, Katafuchi T, Koga K, Furue H, Yasaka T, Nakatsuka T, Zhou XF, Kawasaki Y, Tanaka N, Yoshimura M (2005) Actions of brain-derived neurotrophic factor on spinal nociceptive transmission during inflammation in the rat. J Physiol 569:685–695

Melemedjian OK, Asiedu MN, Tillu DV, Peebles KA, Yan J, Ertz N, Dussor GO, Price TJ (2010) IL-6- and NGF-induced rapid control of protein synthesis and nociceptive plasticity via convergent signaling to the eIF4F complex. J Neurosci 30:15113–15123

Melemedjian OK, Asiedu MN, Tillu DV, Sanoja R, Yan J, Lark A, Khoutorsky A, Johnson J, Peebles KA, Lepow T, Sonenberg N, Dussor G, Price TJ (2011) Targeting adenosine monophosphate-activated protein kinase (AMPK) in preclinical models reveals a potential mechanism for the treatment of neuropathic pain. Mol Pain 7:70

Melemedjian OK, Tillu DV, Asiedu MN, Mandell EK, Moy JK, Blute VM, Taylor CJ, Ghosh S, Price TJ (2013a) BDNF regulates atypical PKC at spinal synapses to initiate and maintain a centralized chronic pain state. Mol Pain 9:12

Melemedjian OK, Khoutorsky A, Sorge RE, Yan J, Asiedu MN, Valdez A, Ghosh S, Dussor G, Mogil JS, Sonenberg N, Price TJ (2013b) mTORC1 inhibition induces pain via IRS-1-dependent feedback activation of ERK. Pain 154(7):1080–1091

Melemedjian OK, Mejia GL, Lepow TS, Zoph OK, Price TJ (2014) Bidirectional regulation of P body formation mediated by eIF4F complex formation in sensory neurons. Neurosci Lett 563:169–174

Mendez R, Hake LE, Andresson T, Littlepage LE, Ruderman JV, Richter JD (2000) Phosphorylation of CPE binding factor by Eg2 regulates translation of c-mos mRNA. Nature 404:302–307

Migues PV, Hardt O, Wu DC, Gamache K, Sacktor TC, Wang YT, Nader K (2010) PKMzeta maintains memories by regulating GluR2-dependent AMPA receptor trafficking. Nat Neurosci 13:630–634

Misra UK, Pizzo SV (2009) Epac1-induced cellular proliferation in prostate cancer cells is mediated by B-Raf/ERK and mTOR signaling cascades. J Cell Biochem 108:998–1011

Nicoll RA, Roche KW (2013) Long-term potentiation: peeling the onion. Neuropharmacology 74:18–22

Nolet PS, Cote P, Cassidy JD, Carroll LJ (2010) The association between a lifetime history of a neck injury in a motor vehicle collision and future neck pain: a population-based cohort study. Eur Spine J 19:972–981

Parada CA, Yeh JJ, Reichling DB, Levine JD (2003) Transient attenuation of protein kinase Cepsilon can terminate a chronic hyperalgesic state in the rat. Neuroscience 120:219–226

Pastalkova E, Serrano P, Pinkhasova D, Wallace E, Fenton AA, Sacktor TC (2006) Storage of spatial information by the maintenance mechanism of LTP. Science 313:1141–1144

Pezet S, Malcangio M, Lever IJ, Perkinton MS, Thompson SW, Williams RJ, McMahon SB (2002) Noxious stimulation induces Trk receptor and downstream ERK phosphorylation in spinal dorsal horn. Mol Cell Neurosci 21:684–695

Pinto PR, McIntyre T, Ferrero R, Almeida A, Araujo-Soares V (2013) Risk factors for moderate and severe persistent pain in patients undergoing total knee and hip arthroplasty: a prospective predictive study. PLoS One 8:e73917

Price TJ, Dussor G (2013) AMPK: an emerging target for modification of injury-induced pain plasticity. Neurosci Lett 557(Pt A):9–18

Price TJ, Ghosh S (2013) ZIPping to pain relief: the role (or not) of PKMzeta in chronic pain. Mol Pain 9:6

Price TJ, Melemedjian OK (2012) Fragile X mental retardation protein (FMRP) and the spinal sensory system. Results Probl Cell Differ 54:41–59

Price TJ, Rashid MH, Millecamps M, Sanoja R, Entrena JM, Cervero F (2007) Decreased nociceptive sensitization in mice lacking the fragile X mental retardation protein: role of mGluR1/5 and mTOR. J Neurosci 27:13958–13967

Reichling DB, Levine JD (2009) Critical role of nociceptor plasticity in chronic pain. Trends Neurosci 32:611–618

Reichling DB, Green PG, Levine JD (2013) The fundamental unit of pain is the cell. Pain 154 (Suppl 1):S2–S9

Ribeiro-da-Silva A, De Koninck Y, Cuello AC, Henry JL (1992) Enkephalin-immunoreactive nociceptive neurons in the cat spinal cord. Neuroreport 3:25–28

Richter JD (2007) CPEB: a life in translation. Trends Biochem Sci 32:279–285

Rivat C, Laulin JP, Corcuff JB, Celerier E, Pain L, Simonnet G (2002) Fentanyl enhancement of carrageenan-induced long-lasting hyperalgesia in rats: prevention by the N-methyl-D-aspartate receptor antagonist ketamine. Anesthesiology 96:381–391

Rivat C, Laboureyras E, Laulin J-P, Le Roy C, Richebé P, Simonnet G (2007) Non-nociceptive environmental stress induces hyperalgesia, not analgesia, in pain and opioid-experienced rats. Neuropsychopharmacology 32:2217–2228

Rivat C, Bollag L, Richebe P (2013) Mechanisms of regional anaesthesia protection against hyperalgesia and pain chronicization. Curr Opin Anaesthesiol 26(5):621–625

Russe OQ, Moser CV, Kynast KL, King TS, Stephan H, Geisslinger G, Niederberger E (2013) Activation of the AMP-activated protein kinase reduces inflammatory nociception. J Pain 14:1330–1340

Sacktor TC (2011) How does PKMzeta maintain long-term memory? Nat Rev Neurosci 12:9–15

Sandkuhler J (2007) Understanding LTP in pain pathways. Mol Pain 3:9

Schroeder JE, McCleskey EW (1993) Inhibition of Ca^{2+} currents by a mu-opioid in a defined subset of rat sensory neurons. J Neurosci 13:867–873

Schroeder JE, Fischbach PS, Zheng D, McCleskey EW (1991) Activation of mu opioid receptors inhibits transient high- and low-threshold Ca^{2+} currents, but spares a sustained current. Neuron 6:13–20

Shaw RJ, Lamia KA, Vasquez D, Koo SH, Bardeesy N, Depinho RA, Montminy M, Cantley LC (2005) The kinase LKB1 mediates glucose homeostasis in liver and therapeutic effects of metformin. Science 310:1642–1646

Shen CH, Yuan P, Perez-Lorenzo R, Zhang Y, Lee SX, Ou Y, Asara JM, Cantley LC, Zheng B (2013) Phosphorylation of BRAF by AMPK impairs BRAF-KSR1 association and cell proliferation. Mol Cell 52:161–172

Si K, Lindquist S, Kandel ER (2003a) A neuronal isoform of the aplysia CPEB has prion-like properties. Cell 115:879–891

Si K, Giustetto M, Etkin A, Hsu R, Janisiewicz AM, Miniaci MC, Kim JH, Zhu H, Kandel ER (2003b) A neuronal isoform of CPEB regulates local protein synthesis and stabilizes synapse-specific long-term facilitation in aplysia. Cell 115:893–904

Sonenberg N, Hinnebusch AG (2009) Regulation of translation initiation in eukaryotes: mechanisms and biological targets. Cell 136:731–745

Tamcan O, Mannion AF, Eisenring C, Horisberger B, Elfering A, Muller U (2010) The course of chronic and recurrent low back pain in the general population. Pain 150:451–457

Tao YX (2012) AMPA receptor trafficking in inflammation-induced dorsal horn central sensitization. Neurosci Bull 28:111–120

Tillu DV, Melemedjian OK, Asiedu MN, Qu N, De Felice M, Dussor G, Price TJ (2012) Resveratrol engages AMPK to attenuate ERK and mTOR signaling in sensory neurons and inhibits incision-induced acute and chronic pain. Mol Pain 8:5

Udagawa T, Swanger SA, Takeuchi K, Kim JH, Nalavadi V, Shin J, Lorenz LJ, Zukin RS, Bassell GJ, Richter JD (2012) Bidirectional control of mRNA translation and synaptic plasticity by the cytoplasmic polyadenylation complex. Mol Cell 47:253–266

Ueda T, Sasaki M, Elia AJ, Chio II, Hamada K, Fukunaga R, Mak TW (2010) Combined deficiency for MAP kinase-interacting kinase 1 and 2 (Mnk1 and Mnk2) delays tumor development. Proc Natl Acad Sci U S A 107:13984–13990

Vanderah TW, Suenaga NM, Ossipov MH, Malan TP Jr, Lai J, Porreca F (2001) Tonic descending facilitation from the rostral ventromedial medulla mediates opioid-induced abnormal pain and antinociceptive tolerance. J Neurosci 21:279–286

Volk LJ, Bachman JL, Johnson R, Yu Y, Huganir RL (2013) PKM-zeta is not required for hippocampal synaptic plasticity, learning and memory. Nature 493:420–423

Von Korff M, Saunders K (1996) The course of back pain in primary care. Spine (Phila Pa 1976) 21:2833–2837, discussion 2838–2839

Wang X, Flynn A, Waskiewicz AJ, Webb BL, Vries RG, Baines IA, Cooper JA, Proud CG (1998) The phosphorylation of eukaryotic initiation factor eIF4E in response to phorbol esters, cell stresses, and cytokines is mediated by distinct MAP kinase pathways. J Biol Chem 273:9373–9377

Wang H, Heijnen CJ, van Velthoven CT, Willemen HL, Ishikawa Y, Zhang X, Sood AK, Vroon A, Eijkelkamp N, Kavelaars A (2013) Balancing GRK2 and EPAC1 levels prevents and relieves chronic pain. J Clin Invest 123(12):5023–5034

Wu C, Guo Y, Su Y, Zhang X, Luan H, Zhang X, Zhu H, He H, Wang X, Sun G, Sun X, Guo P, Zhu P (2014) Cordycepin activates AMP-activated protein kinase (AMPK) via interaction with the gamma1 subunit. J Cell Mol Med 18(2):293–304

Zhao J, Seereeram A, Nassar MA, Levato A, Pezet S, Hathaway G, Morenilla-Palao C, Stirling C, Fitzgerald M, McMahon SB, Rios M, Wood JN (2006) Nociceptor-derived brain-derived neurotrophic factor regulates acute and inflammatory but not neuropathic pain. Mol Cell Neurosci 31:539–548

Zhou LJ, Zhong Y, Ren WJ, Li YY, Zhang T, Liu XG (2008) BDNF induces late-phase LTP of C-fiber evoked field potentials in rat spinal dorsal horn. Exp Neurol 212:507–514

Sodium Channels and Pain

Abdella M. Habib, John N. Wood, and James J. Cox

Contents

1 Voltage-Gated Sodium Channel (Na_v) Family 40
2 Sodium Channels and Pain: Insights from Rodent Studies 41
    2.1 $Na_v1.7$ 41
    2.2 $Na_v1.8$ 45
    2.3 $Na_v1.9$ 46
    2.4 $Na_v1.3$ 46
3 Human Heritable Sodium Channelopathies 47
    3.1 Inherited Primary Erythromelalgia ($Na_v1.7$) 47
    3.2 Paroxysmal Extreme Pain Disorder ($Na_v1.7$) 49
    3.3 Small Fibre Neuropathy ($Na_v1.7$ and $Na_v1.8$) 49
    3.4 Familial Episodic Pain ($Na_v1.9$) 50
    3.5 Pain Insensitivity ($Na_v1.7$ and $Na_v1.9$) 51
4 Prospects for New $Na_v1.7$ Selective Analgesics 52
5 Summary 53
References 54

Abstract

Human and mouse genetic studies have led to significant advances in our understanding of the role of voltage-gated sodium channels in pain pathways. In this chapter, we focus on $Na_v1.7$, $Na_v1.8$, $Na_v1.9$ and $Na_v1.3$ and describe the insights gained from the detailed analyses of global and conditional transgenic Na_v knockout mice in terms of pain behaviour. The spectrum of human disorders caused by mutations in these channels is also outlined, concluding with a summary of recent progress in the development of selective $Na_v1.7$ inhibitors for the treatment of pain.

A.M. Habib • J.N. Wood • J.J. Cox (✉)
Molecular Nociception Group, Wolfson Institute for Biomedical Research, University College London, Gower Street, London WC1E 6BT, UK
e-mail: j.j.cox@ucl.ac.uk

© Springer-Verlag Berlin Heidelberg 2015
H.-G. Schaible (ed.), *Pain Control*, Handbook of Experimental Pharmacology 227,
DOI 10.1007/978-3-662-46450-2_3

Keywords

Voltage-gated sodium channels • Channelopathy • Transgenic mice • Analgesia • Chronic pain • Sensory neurons

1 Voltage-Gated Sodium Channel (Na$_v$) Family

Voltage-gated sodium channels (VGSCs) are integral transmembrane proteins that are critical for the electrical activity of cells (Eijkelkamp et al. 2012). The human genome contains ten structurally related sodium channel genes that are expressed in distinct spatial and temporal patterns (Table 1). Nine of the genes encode alpha subunits that are activated by membrane depolarisation, with a tenth atypical channel (Nax) activated by altered sodium concentrations (Goldin et al. 2000). The alpha subunits alone are capable of forming functional channels and have a size of ~260 kDa. However, beta subunits, encoded by the *SCN1B-SCN4B* genes, associate with alpha subunits and modulate channel biophysics and trafficking (Patino and Isom 2010). The alpha subunit is comprised of four repeated similar domains (I–IV), with each domain containing six transmembrane segments (S1–S6) (Fig. 1). The S5–S6 transmembrane regions fold to form the sodium ion selective pore of the channel. The S4 segment is comprised of arrays of positively charged amino acids and is involved in voltage sensing. At resting membrane potentials, VGSCs are closed, requiring depolarisation to be activated. Upon depolarisation of a cell, the S4 regions in domains I–IV move rapidly and induce a conformational change in the protein which opens the ion channel pore. The entry of sodium ions through the pore leads to the upstroke of the action potential in excitable cells. Inactivation of the channel then follows as a highly conserved trio of amino acids (IFM) located in the intracellular loop between domains III and IV moves into and occludes the channel pore, leading to the downstroke of the action potential. VGSCs open and close over a millisecond timescale in response to the voltage difference between the inside and outside of the cell. They therefore enable the

Table 1 Mammalian voltage-gated sodium channels

Protein	Gene	Human chromosome	Distribution	Tetrodotoxin sensitive
Na$_v$1.1	*SCN1A*	2q24.3	CNS; heart	+
Na$_v$1.2	*SCN2A*	2q24.3	CNS	+
Na$_v$1.3	*SCN3A*	2q24.3	Foetal DRG	+
Na$_v$1.4	*SCN4A*	17q23.3	Muscle	+
Na$_v$1.5	*SCN5A*	3p22.2	Heart	−
Na$_v$1.6	*SCN8A*	12q13.13	DRG; CNS	+
Na$_v$1.7	*SCN9A*	2q24.3	DRG; SCG	+
Na$_v$1.8	*SCN10A*	3p22.2	DRG	−
Na$_v$1.9	*SCN11A*	3p22.2	DRG	−
Nax	*SCN7A*	2q24.3	Lung nerve	+

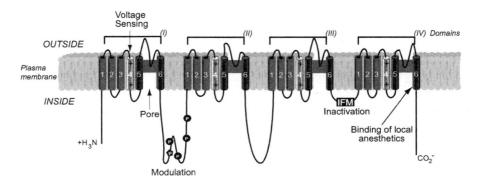

Fig. 1 Primary structure of the alpha subunit of a generic voltage-gated sodium channel within the plasma membrane lipid bilayer (in *grey*). Alpha helical transmembrane segments are shown as cylinders; *bold lines* represent polypeptide chains. P denotes sites of demonstrated protein phosphorylation by PKA (*circles*) and PKC (*pentagon*). Residues in the inner cavity of the channel pore involving the S6 segment of domains I, III and IV form the binding site for some local anaesthetic, antiepileptic and antiarrhythmic drugs, such as lidocaine, mexiletine and carbamazepine (Ragsdale et al. 1994, 1996)

rapid generation and propagation of electrical impulses. In terms of pain signalling, this can be critically important for survival. For a recent comprehensive review of the general structure and properties of the VGSC family, see Catterall (2014).

2 Sodium Channels and Pain: Insights from Rodent Studies

Detection of noxious stimuli and signalling of tissue damage are physiological processes that promote animal survival. Peripheral sensory neurons, with cell bodies located within the dorsal root ganglia (DRG), express receptors that are sensitive to noxious thermal, chemical or mechanical stimuli and transduce these damaging stimuli into electrical activity through the activation of VGSCs. Within DRG, several VGSCs can be detected with differing cellular expression profiles and biophysical properties, where they work to fine-tune the signalling of noxious information to the central nervous system. $Na_v1.7$, $Na_v1.8$ and $Na_v1.9$ are preferentially expressed within DRG neurons, whereas $Na_v1.3$ expression is upregulated following nervous system injuries. Through detailed studies of Na_v transgenic mice, the contribution of each channel to pain signalling is being elucidated (Table 2).

2.1 $Na_v1.7$

$Na_v1.7$ has seen special interest in recent years as several inherited human pain disorders, including congenital pain insensitivity, have been shown to result from mutations in this channel (see Sect. 3). The $Na_v1.7$ channel is characterised

Table 2 A summary of acute, inflammatory and neuropathic pain behaviour in Na_v^X transgenic mice (X denotes Cre line used to tissue-specifically ablate Na_v expression)

Transgenic mouse	Acute pain		Inflammatory pain		Neuropathic pain	
	No change	Reduced	No change	Reduced	No change	Reduced
$Na_v1.7$ flox × $Na_v1.8$ Cre	Hot plate[a,b], Cold plate[b], Hargreaves[b], Von Frey[a,b]	Randall-Selitto[a,b], Foot licking/biting acetone[b], Hargreaves[a]		Foot licking/biting[a] *formalin*, Hargreaves[a] *CFA, NGF, Carrageenan*, Von Frey[a] *CFA*		
$Na_v1.7$ flox × Advillin Cre	Hot plate[b], Von Frey[b], Cold plate[b]	Randall-Selitto[b], Foot licking/biting acetone[b], Hargreaves[b]			Von Frey[b] *SNT*	
$Na_v1.7$ flox × Wnt1 Cre	Cold plate[b], Von Frey[b]	Randall-Selitto[b], Foot licking/biting acetone[b], Hargreaves[b], Hot plate[b]				Von Frey[b] *SNT*

Genotype	Baseline tests					Inflammatory			Neuropathic
$Na_v1.8$ KO	Hargreaves[c,d]	Hot plate[c,d,e]	Von Frey[c,e]	Warm water[d]	Foot licking and biting/lifting after acetone[b]	Hargreaves[d] Carrageenan	Hargreaves[e] Carrageenan	Hargreaves[c,d] CFA	Hargreaves[c] after SNT
	Hargreaves[e]	Randall-Selitto[c,e]	Cold plate[b,c,f]			Von Frey[d] CFA		Von Frey[c] FCA	Von Frey[c,d] after SNT, SNI, CCI
						Foot licking and biting/lifting[d] formalin		Foot licking and biting/lifting[c] formalin, FCA	Weight-bearing[d] after SNI, CCI
									Flinching/licking/lifting[d] acetone after CCI, SNI
Floxed stop DTA × $Na_v1.8$ Cre[c]									Foot licking/flinching[d] acetone after CCI
$Na_v1.9$ KO	Von Frey[g,h]	Hargreaves[d,g]	Hot plate[d,g,h]	Cold plate[h]	Warm water[d]	Von Frey[d,h] CFA	Hargreaves[g] Carrageenan, CFA, PGE2		Von Frey[d,g,h] after SNI, CCI
						Hargreaves[d] CFA, Carrageenan	Foot licking/flinching[d,g,h,i] formalin, bradykinin, αβmet-ATP, UTP		Foot licking/flinching[d,h] acetone after SNI
							Von Frey[h,i] bradykinin, PGE2, IL-1β, NGF, Carrageenan		Weight-bearing[d] after SNI, CCI
							Hot plate[h] bradykinin, PGE2, IL-1β, NGF, CFA		
							Warm water[i] Carrageenan, CFA		
							Dynamic weight-bearing[i] Carrageenan, CFA		

(continued)

Table 2 (continued)

Transgenic mouse	Acute pain		Inflammatory pain		Neuropathic pain	
	No change	Reduced	No change	Reduced	No change	Reduced
Na$_v$1.3 KO	Hargreaves[j]		Foot licking/biting[j] *formalin*		Von Frey after spinal nerve ligation[j]	
	Hot plate[j]					
	Von Frey[j]		Hargreaves[j] *CFA*			
	Randall-Selitto[j]		Von Frey[j] *CFA*			

Hargreaves assay measures response to radiant heat; Randall-Selitto assay measures response to noxious mechanical stimuli; von Frey filaments are used to measure mechanical sensitivity; *SNI* sciatic nerve injury, *SNT* sciatic nerve transection, *CCI* chronic constriction injury, *CFA/FCA* complete Freund's adjuvant, *PGE2* prostaglandin E2, *IL-1β* interleukin-1β. For a comprehensive review of mouse pain behaviour see Minett et al. (2011)

[a]Nassar et al. (2004)
[b]Minett et al. (2012)
[c]Abrahamsen et al. (2008)
[d]Leo et al. (2010)
[e]Akopian et al. (1999)
[f]Zimmermann et al. (2007)
[g]Priest et al. (2005)
[h]Amaya et al. (2006)
[i]Lolignier et al. (2011)
[j]Nassar et al. (2006)

biophysically as a tetrodotoxin-sensitive, fast-activating and a fast-inactivating channel that recovers slowly from fast inactivation (Klugbauer et al. 1995). $Na_v1.7$ also has slow closed-state inactivation properties, allowing the channel to generate a ramp current in response to small, slow depolarisations (Cummins et al. 1998; Herzog et al. 2003). $Na_v1.7$-positive neurons are therefore able to amplify slowly developing subthreshold depolarising inputs, such as generator potentials arising in peripheral terminals of nociceptors. $Na_v1.7$ is expressed within DRG and trigeminal ganglia peripheral sensory neurons, as well as sympathetic neurons and olfactory epithelia (Toledo-Aral et al. 1997; Weiss et al. 2011). $Na_v1.7$ shows particularly high expression within the soma of small-diameter DRG neurons and along the peripherally and centrally directed C fibres of these cells (Black et al. 2012). Expression can also be detected within the peripheral terminals of DRG neurons in the skin and also in the preterminal central branches and terminals in the dorsal horn, as well as at nodes of Ranvier in a subpopulation of small-diameter myelinated fibres.

In 2004, conditional deletion of $Na_v1.7$ within $Na_v1.8$-positive nociceptors highlighted the importance of this channel to pain behaviour, with knockout animals losing acute noxious mechanosensation and inflammatory pain (Nassar et al. 2004). In 2012, an Advillin Cre line was used to ablate $Na_v1.7$ expression in all DRG sensory neurons, with knockout mice showing an additional loss of noxious thermosensation (Minett et al. 2012). This suggests that $Na_v1.7$ expressed within $Na_v1.8$-positive sensory neurons is important for acute noxious mechanosensation, whilst $Na_v1.7$ expressed within $Na_v1.8$-negative DRG neurons is essential for acute noxious thermosensation. The contribution of $Na_v1.7$ to the development of neuropathic pain behaviour has also been assessed using different Cre mice to delete $Na_v1.7$ in subgroups of cell populations. Neuropathic pain behaviour develops normally in mice where $Na_v1.7$ has been deleted in $Na_v1.8$-positive sensory neurons and in Advillin-positive DRG neurons (Nassar et al. 2005; Minett et al. 2012). In contrast, mice in which $Na_v1.7$ is deleted from all sensory neurons as well as sympathetic neurons (using a Wnt1-Cre) show a dramatic reduction in mechanical hypersensitivity following a surgical model of neuropathic pain, demonstrating an important role for $Na_v1.7$ in sympathetic neurons in the development of neuropathic pain (Minett et al. 2012).

2.2 $Na_v1.8$

The TTX-resistant $Na_v1.8$ sodium channel is expressed in the majority of small-diameter unmyelinated nociceptive DRG neurons (Akopian et al. 1996). $Na_v1.8$ displays ~10-fold slower kinetics with a depolarised voltage dependence of activation and inactivation than the fast and rapidly inactivating TTX-S DRG channels (Akopian et al. 1996; Cummins and Waxman 1997). Analysis of $Na_v1.8$ null mice shows that $Na_v1.8$ carries the majority of the current underlying the upstroke of the action potential in nociceptive neurons (Akopian et al. 1999; Renganathan et al. 2001). Furthermore, $Na_v1.8$ is cold resistant, meaning that it is essential in maintaining the excitability of nociceptors at low temperatures (Zimmermann et al. 2007).

In contrast to $Na_v1.7$ global knockout mice, $Na_v1.8$ null mice are viable, fertile and apparently normal (Akopian et al. 1999). However, pain behaviour analyses show that $Na_v1.8$ null mice have reduced sensitivity to noxious mechanical stimuli (tail pressure), noxious thermal stimuli (radiant heat) and are insensitive to noxious cold stimuli (Akopian et al. 1999; Zimmermann et al. 2007) (Table 2). Furthermore, $Na_v1.8$ is important in nerve growth factor-induced thermal hyperalgesia, although neuropathic pain behaviour is normal in these mice following peripheral nerve injury (Kerr et al. 2001). Transgenic mice in which $Na_v1.8$-expressing neurons are ablated by the targeted expression of diphtheria toxin A are resistant to noxious cold and noxious mechanical stimuli but show a normal hot plate response (Abrahamsen et al. 2008). Furthermore, mechanical and thermal hyperalgesia induced by inflammatory insults is attenuated in these mice, although neuropathic pain behaviour is normal.

2.3 $Na_v1.9$

$Na_v1.9$ was cloned in 1998 and has a similar expression pattern within DRG to $Na_v1.8$ (Dib-Hajj et al. 1998; Fang et al. 2002). $Na_v1.9$ generates extremely slow persistent TTX-resistant currents and can be activated at potentials close to the resting membrane potential (Cummins et al. 1999; Dib-Hajj et al. 2002). The activation kinetics are too slow to contribute to the upstroke of an action potential, although $Na_v1.9$ may amplify subthreshold depolarisations and lower the threshold for action potential induction (Cummins et al. 1999; Herzog et al. 2001; Baker et al. 2003). Mice lacking the $Na_v1.9$ channel have normal sensitivity to acute mechanical and thermal noxious stimuli suggesting that $Na_v1.9$ is not essential for setting basal pain thresholds; however, under inflammatory conditions, the channel plays a predominant role in mechanical and thermal hyperalgesia (Table 2) (Priest et al. 2005; Amaya et al. 2006; Lolignier et al. 2011).

2.4 $Na_v1.3$

$Na_v1.3$, like $Na_v1.7$, is a TTX-sensitive channel with fast kinetics and a rapid recovery from fast inactivation. Unlike $Na_v1.7$, $Na_v1.3$ has very low expression levels in the DRG in adults, instead having a predominant expression pattern in the central and peripheral nervous system during embryogenesis (Beckh et al. 1989; Waxman et al. 1994). In humans, $Na_v1.3$ is widely expressed in adult brain and is thought to play a role in propagating synaptic signals from dendrites to the soma and integration of electrical signals within the soma prior to the initiation of axonal action potentials (Whitaker et al. 2001). $Na_v1.3$ levels increase in peripheral sensory neurons following nerve injury or inflammation, suggesting that this channel could play a role in pain (Waxman et al. 1994; Dib-Hajj et al. 1999).

However, $Na_v1.3$ global and nociceptor-specific knockout mice show normal neuropathic pain behaviour (Table 2) (Nassar et al. 2006).

3   Human Heritable Sodium Channelopathies

Genetic analyses of sporadic patients and families with rare inherited pain disorders have given important insights into the role of VGSCs in human pain pathways (Goldberg et al. 2012a). In recent years, the mutations that underlie several human pain disorders have been identified, and with the introduction of exome and whole genome sequencing, the pace of disease gene identification has significantly increased and will continue to do so (Table 3). With the obvious importance of the Na_v family in neuronal signalling, it is perhaps unsurprising that several human disorders, including epilepsy ($Na_v1.1$, $Na_v1.2$ and $Na_v1.3$), hyperkalaemic periodic paralysis ($Na_v1.4$), Brugada syndrome ($Na_v1.5$) and learning disability with cerebellar ataxia ($Na_v1.6$), are caused by mutations in these channels. Likewise, inherited mutations in three of the sodium channel genes expressed in damage-sensing neurons ($Na_v1.7$, $Na_v1.8$ and $Na_v1.9$) give rise to distinct human disorders with phenotypes ranging from lacerating chronic pain to complete pain insensitivity. By understanding how the gene mutations specifically affect sodium channel function, we can learn more about the aetiology of each pain disorder and also flag potential new human-validated analgesic drug targets.

3.1   Inherited Primary Erythromelalgia ($Na_v1.7$)

Inherited primary erythromelalgia (IEM) is a chronic and debilitating pain disorder affecting approximately 1 in 100,000 individuals (Reed and Davis 2009). Symptoms usually, but not always, appear at a young age and include excruciating symmetrical burning pain in the hands and feet, increased skin temperature, oedema

Table 3 Human pain-related channelopathies

Protein	Gene	Loss of function phenotype	Gain-of-function phenotype
$Na_v1.7$	SCN9A	Channelopathy-associated insensitivity to pain (2006)	Primary erythromelalgia (2004)
		Hereditary sensory and autonomic neuropathy type IID (2013)	Paroxysmal extreme pain disorder (2006)
			Painful small-fibre neuropathy (2011)
$Na_v1.8$	SCN10A	None reported in humans	Painful small-fibre neuropathy (2012)
$Na_v1.9$	SCN11A	None reported in humans	Congenital inability to experience pain (2013)
			Familial episodic pain (2013)

Year of first-published disease-causing mutations given in parentheses and erythema of affected areas. The phenotype can range from mild to severe, even within the same family, and is related to an increased rate of suicide due to its devastating effects on the lifestyle of severely affected patients. Painful attacks are often triggered by warmth or moderate exercise, and the pain can be reduced by immersing the affected areas in ice-cold water. Analgesics such as opioids, anticonvulsants and antidepressants are often prescribed but frequently with minimal positive effects on controlling symptoms. Patients therefore often try to modify their environment (e.g. cooling their houses, wearing open-toed shoes) and try to limit their physical exertion, but often with limited success in reducing the painful attacks.

IEM is an autosomal dominant disorder, and in 2004, the first disease-causing mutations were identified in a Chinese family and a sporadic patient using a positional cloning strategy (Yang et al. 2004). Both mutations were located within the *SCN9A* gene, which encodes $Na_v1.7$, and replaced highly conserved amino acids (Fig. 2). The L858H mutation is located within the domain II/S5 segment whereas the I848T mutation is positioned within the cytoplasmic loop region preceding domain II/S5. Subsequent electrophysiological analysis of the mutations showed that they led to a gain of function of the sodium channel (Cummins et al. 2004). Specifically, the mutations led to a significant hyperpolarising shift in voltage dependence of activation (facilitating channel opening), slowed channel deactivation (keeping the channel open for longer once activated) and an increased ramp current (i.e. causing an increase in amplitude of the current produced by $Na_v1.7$ in response to slow, small depolarisations). Numerous more IEM *SCN9A* mutations have since been identified and characterised electrophysiologically, with

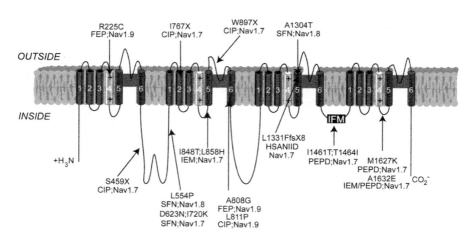

Fig. 2 Examples of the first-published human pain Na_v channelopathy mutations. *CIP* congenital insensitivity to pain (Cox et al. 2006; Leipold et al. 2013), *HSANIID* hereditary and sensory autonomic neuropathy type IID (Yuan et al. 2013), *IEM* inherited primary erythromelalgia (Yang et al. 2004), *PEPD* paroxysmal extreme pain disorder (Fertleman et al. 2006), *FEP* familial episodic pain (Zhang et al. 2013), *SFN* painful small-fibre neuropathy (Faber et al. 2012a, b)

similar changes in channel kinetics recorded (Dib-Hajj et al. 2010). Curiously, even in the presence of a family history, only approximately 10 % of IEM families are proven to have an *SCN9A* mutation, suggesting that other genetic causes underlie this devastating and debilitating disorder (Goldberg et al. 2012a).

3.2 Paroxysmal Extreme Pain Disorder (Na$_v$1.7)

Paroxysmal extreme pain disorder (PEPD), formerly known as familial rectal pain syndrome, is typically an early-onset dominantly inherited condition characterised by severe burning pain in the rectal, submandibular and ocular areas (Hayden and Grossman 1959). Pain attacks are often triggered by bowel movements and can also be accompanied by tonic nonepileptic seizures, syncopes, bradycardia and occasionally asystole. In 2006, the first disease-causing mutations were mapped to the *SCN9A* gene in 11 families and 2 sporadic cases (Fig. 2) (Fertleman et al. 2006). As with the IEM mutations, electrophysiological analysis of the PEPD mutations shows a gain of function in Na$_v$1.7, rendering the channel hyperexcitable. However, some important biophysical differences exist between IEM- and PEPD-mutated sodium channels. For example, PEPD mutations shift voltage dependency of steady-state fast inactivation in a depolarising direction and may make inactivation incomplete, resulting in a persistent current. In contrast to IEM, PEPD symptoms are often well controlled by anticonvulsant sodium channel blockers, such as carbamazepine.

A fascinating feature of IEM and PEPD is their distinct clinical symptoms, even though both disorders are caused by a gain of function in Na$_v$1.7. The reasons for the phenotypic differences are still unclear, although a patient has been reported with clinical symptoms reminiscent of both IEM and PEPD (Estacion et al. 2008). This patient was found to have a missense mutation in a highly conserved amino acid (A1632E) located in the cytoplasmic linker between transmembrane segments S4 and S5 of domain IV (Fig. 2). Electrophysiological analysis of the mutation showed biophysical properties common to both IEM and PEPD mutations.

3.3 Small Fibre Neuropathy (Na$_v$1.7 and Na$_v$1.8)

Small fibre neuropathy (SFN) is typically an adult-onset pain disorder characterised by burning pain, often throughout the body, together with autonomic symptoms such as orthostatic dizziness, palpitations, dry eyes and a dry mouth. Small-diameter unmyelinated and thinly myelinated fibres are affected with a reduced intraepidermal nerve fibre density detected by a skin biopsy. Large-diameter fibres are unaffected, resulting in normal strength, tendon reflexes and vibration sense. Underlying causes for SFN include Fabry disease, diabetes mellitus and impaired glucose tolerance, but in a significant proportion of individuals, the cause of the neuropathy remains unexplained.

In 2011, mutation screening of the *SCN9A* gene in 28 patients with idiopathic SFN identified heterozygous missense mutations in approximately a third of the cohort (Fig. 2) (Faber et al. 2012a). The phenotype of these *SCN9A* SFN patients differed from patients with IEM in several ways: (1) severe autonomic features were common in the SFN patients, whereas erythema is typically the only predominant autonomic symptom in IEM; (2) pain was found throughout the body in SFN, whereas it tends to localise to the extremities in IEM; and (3) the symptoms in SFN patients were not triggered by warmth or relieved by cold, as is characteristic of IEM. Electrophysiological analysis of the SFN mutant $Na_v1.7$ channels showed impaired slow inactivation, depolarised fast and slow inactivation or enhanced resurgent currents. As with the IEM and PEPD mutations, DRG neurons expressing the mutant SFN channels were rendered hyperexcitable. However, the SFN mutant channels did not show the incomplete fast inactivation that is found in PEPD, nor the hyperpolarised activation and enhanced ramp responses that are characteristic of IEM.

Recently, gain-of-function mutations in $Na_v1.8$ have also been identified in painful SFN patients (Fig. 2) (Faber et al. 2012b; Huang et al. 2013). Electrophysiological analysis of two of these mutations (L554P and A1304T) showed that they altered the gating properties of the channels in a proexcitatory manner and increased the excitability of DRG neurons. However, it is still unclear how these specific mutations result in axonal degeneration of small-diameter peripheral neurons.

3.4 Familial Episodic Pain ($Na_v1.9$)

In 2010, a family from Colombia was analysed in which 15 affected individuals suffered from episodes of debilitating upper body pain that was triggered by fasting, cold temperatures or fatigue (Kremeyer et al. 2010). This disorder was shown to be the result of a gain-of-function missense mutation in the S4 transmembrane segment of TRPA1, the chemosensitive transient receptor potential channel. In late 2013, a new familial episodic pain disorder was described in two large Chinese families, with affected individuals spanning five generations (Zhang et al. 2013). The phenotype seen in these patients comprised of (1) intense pain localised principally to the distal lower extremities and occasionally in the upper body, especially in the joints of fingers and arms, (2) episodic pain appearing late in the day and relapses once every 2–5 days for a total of >10 recurrences for one cycle, (3) pain exacerbated with fatigue, such as catching a cold or performing hard exercise and (4) intense pain usually being accompanied by sweating. The painful attacks could be relieved through the oral administration of anti-inflammatory medicines or a hot compress on the painful region.

A combination of a genome-wide linkage scan with microsatellite markers, exome sequencing and Sanger sequencing identified the disease-causing mutations to be located within the *SCN11A* gene on chromosome 3. Both mutations were predicted to change highly conserved amino acid residues within the $Na_v1.9$

channel (Fig. 2). The R225C mutation mapped to domain I/S4 whereas the A808G mutation mapped to domain II/S6. Electrophysiological analysis of these two mutations via overexpression in isolated DRG neurons showed an increase in peak current densities and higher electrical activities than for overexpressed wild-type channels. This promoted action potential firing in the DRG neurons, probably explaining the episodic pain symptoms.

3.5   Pain Insensitivity ($Na_v1.7$ and $Na_v1.9$)

In contrast to the excess pain disorders described above, mutations in VGSC genes can also result in pain insensitivity. In 2006, three consanguineous families originating from northern Pakistan were studied in which affected members had a complete inability to perceive pain (Cox et al. 2006). They had never felt any pain, at any time, in any part of their bodies and consequently had suffered serious but painless injuries such as fractures and biting of the lips and tongue. The patients had normal intelligence, normal sensations of touch, warm and cold temperatures, proprioception, tickle and pressure, no overt autonomic nervous system deficits, normal nerve conduction tests and a normal sural nerve biopsy. The disorder could therefore be differentiated from the hereditary sensory and autonomic neuropathies which also give a pain-insensitive phenotype but with additional syndromic features (Rotthier et al. 2012).

The pain-free phenotype in the three consanguineous families segregated in an autosomal recessive manner and by using autozygosity mapping, a region that was homozygous by descent was identified in the affected individuals. Each family mapped to a shared region on chromosome 2, with the minimum critical interval containing approximately 50 genes. One of those genes, *SCN9A*, was selected for mutation screening, and Sanger sequencing showed a different truncating mutation in each of the three pain-free families. Expression of the truncated sodium channels in HEK293A cells showed that they led to a complete loss of function. Several more individuals from a range of ethnic groups have since been found to have recessive mutations in *SCN9A*, including some with loss of function missense mutations (Goldberg et al. 2007; Cox et al. 2010). Furthermore, the phenotype has been refined to show that in addition to being pain-free, the only other significant clinical feature is the presence of anosmia or hyposmia (absence or reduction of smell) (Weiss et al. 2011).

In 2013, a new phenotype was ascribed to recessive loss of function mutations in *SCN9A*: hereditary sensory and autonomic neuropathy type IID (Yuan et al. 2013). Using next-generation sequencing of candidate genes, two unrelated Japanese families were found to have the same homozygous frameshift mutation within domain III, leading to premature truncation of the protein or activation of the nonsense-mediated mRNA decay machinery (Fig. 2). The phenotype was characterised by adolescent or congenital onset with loss of pain and temperature sensation, autonomic nervous system dysfunctions, hearing loss and hyposmia. Sural nerve biopsy analyses gave variable results in the patients, with a marked loss of myelinated fibres and axonal degeneration of unmyelinated fibres in one patient but only slight losses in myelinated fibres in a second patient. Why this particular mutation gives rise to a phenotype that is distinct from that seen in other pain-free patients with recessive loss of function mutations in *SCN9A* remains unclear.

Congenital inability to experience pain is not only restricted to individuals with recessive loss of function mutations in the *SCN9A* gene. Recently two patients have been identified with the same de novo mutation in *SCN11A*, encoding $Na_v1.9$ (Leipold et al. 2013). The phenotype seen in these patients was characterised by a congenital inability to experience pain since birth, resulting in self-mutilations, slow-healing wounds and multiple painless fractures. The pain-free phenotype was accompanied by mild muscular weakness, delayed motor development, slightly reduced motor and sensory nerve conduction velocities with normal amplitudes, a sural nerve biopsy without axonal loss, prominent hyperhidrosis, gastrointestinal dysfunction and no intellectual disability. Exome sequencing identified a heterozygous missense change (L811P) that changed a highly conserved amino acid within domain II/S6 (Fig. 2). To investigate the pathogenicity of this mutation, a heterozygous knock-in mouse was generated in which the orthologous leucine residue was mutated to proline. Behavioural analyses of these knock-in mice showed a reduced sensitivity to pain (although a limited loss of acute pain perception and self-inflicted tissue lesions were observed). Electrophysiological analysis of dorsal root ganglion neurons from the knock-in mice showed the mutant $Na_v1.9$ channels to have excessive activity at resting voltages, causing sustained depolarisation of nociceptors, impaired generation of action potentials and aberrant synaptic transmission. Therefore, in contrast to the $Na_v1.7$ loss of function pain-free mutations, these mutations result in pain insensitivity through a gain-of-function mechanism.

4 Prospects for New $Na_v1.7$ Selective Analgesics

The search for drugs that selectively target $Na_v1.7$ has been stimulated by the finding that humans with loss of function mutations in this channel are completely pain insensitive, but without associated cognitive, cardiac or motor deficits (Cox et al. 2006; Goldberg et al. 2007). Although complete analgesia can result in dangerous and life-threatening injuries due to the absence of protective pain signals, a dose-selective block of this channel could potentially prove beneficial in the treatment of a plethora of chronic pain conditions found in the general population. Furthermore, patients with rare inherited $Na_v1.7$ channelopathies, such as those with primary erythromelalgia, are beginning to directly benefit from the development of $Na_v1.7$ selective analgesics (Goldberg et al. 2012b).

Historically, the development of selective inhibitors of $Na_v1.7$ has been hampered by its high structural similarity to other neuronal sodium channels ($Na_v1.1$, $Na_v1.2$, $Na_v1.3$ and $Na_v1.6$) and to the skeletal muscle sodium channel $Na_v1.4$ (Theile and Cummins 2011). To overcome this problem, natural peptide toxins have been studied as they are often highly potent and efficacious with low nanomolar IC_{50}s. Binding of these toxins is regulated by multiple contact points, meaning that they are often very selective. One example is ProTx-II, a venom toxin from the tarantula *Thrixopelma prurient*, which blocks $Na_v1.7$ with an IC_{50} of 0.3 nM and with >100-fold selectivity over other Na_v isoforms (Middleton et al. 2002; Priest et al. 2007; Schmalhofer et al. 2008). ProTx-II inhibits $Na_v1.7$ activity by inducing a depolarising shift in the voltage dependence of activation. In an isolated skin nerve preparation in which the nerve was de-sheathed, ProTx-II was effective at reducing C-fibre action potential firing frequency (Schmalhofer et al. 2008). However, ProTx-II was ineffective in reducing inflammatory pain following intravenous or intrathecal delivery, possibly related to its inability to cross the blood-nerve barrier. Recently a selective inhibitor purified from centipede venom, Ssm6a, was reported that potently inhibits $Na_v1.7$ with an IC_{50} of ~25 nM (Yang et al. 2013). Ssm6a has more than 150-fold selectivity for $Na_v1.7$ over all other human Na_v subtypes, with the exception of $Na_v1.2$, for which the selectivity is 32-fold. In mouse pain behaviour experiments, Ssm6a was significantly more effective than morphine (on a molar basis) at reducing the paw-licking pain behaviour in both phases of the formalin test. Furthermore, Ssm6a was equally as effective as morphine in reducing abdominal writing induced by intraperitoneal injection of acid and was also similarly effective as morphine at reducing thermal pain. Intraperitoneal injection of Ssm6a at doses up to 1 μmol/kg produced no adverse effects on blood pressure, heart rate or motor function and was highly stable in human plasma. Ssm6a is therefore a promising lead molecule as a selective $Na_v1.7$ analgesic.

Numerous pharmaceutical companies have focused on the development of selective small molecule $Na_v1.7$ blockers, rather than purifying toxins as an alternative route to develop new analgesics (Theile and Cummins 2011; Goldberg et al. 2012b; McCormack et al. 2013). One example is XEN402, developed by Xenon Pharmaceuticals, which exhibits potent voltage-dependent block of $Na_v1.7$ (IC_{50} of 80 nM). In a pilot study in a small cohort of IEM patients, XEN402 was shown to significantly attenuate the ability to induce pain in these patients compared to placebo and also significantly reduced the amount of pain (42 % less) after pain was induced. XEN402 and other small molecule selective $Na_v1.7$ blockers are currently in phase II clinical trials.

5 Summary

The importance of VGSCs to pain perception is indisputable. However, much is still to be understood at the cellular and molecular level, for example, what is the full extent of Na_v interacting proteins and post-translational modifications? Have all tissue-specific splice isoforms been detected? Is RNA editing a significant phenomenon for mammalian Na_vs? How are these genes regulated at the transcriptional level? By understanding these and other related questions, we may be able to further understand specific aspects of the human Na_v channelopathies such as, why does the IEM age of onset vary from infancy to adult? Why do gain-of-function mutations in $Na_v1.7$ give such distinct phenotypes? Why have humans evolved without a compensation mechanism for the loss of function of $Na_v1.7$? Furthermore, by understanding the basic biology of $Na_v1.7$ function, we hope to give further impetus to the development of a new class of analgesics to better treat pain.

References

Abrahamsen B, Zhao J et al (2008) The cell and molecular basis of mechanical, cold, and inflammatory pain. Science 321(5889):702–705

Akopian AN, Sivilotti L et al (1996) A tetrodotoxin-resistant voltage-gated sodium channel expressed by sensory neurons. Nature 379(6562):257–262

Akopian AN, Souslova V et al (1999) The tetrodotoxin-resistant sodium channel SNS has a specialized function in pain pathways. Nat Neurosci 2(6):541–548

Amaya F, Wang H et al (2006) The voltage-gated sodium channel Na(v)1.9 is an effector of peripheral inflammatory pain hypersensitivity. J Neurosci 26(50):12852–12860

Baker MD, Chandra SY et al (2003) GTP-induced tetrodotoxin-resistant Na+ current regulates excitability in mouse and rat small diameter sensory neurones. J Physiol 548(Pt 2):373–382

Beckh S, Noda M et al (1989) Differential regulation of three sodium channel messenger RNAs in the rat central nervous system during development. EMBO J 8(12):3611–3616

Black JA, Frezel N et al (2012) Expression of Nav1.7 in DRG neurons extends from peripheral terminals in the skin to central preterminal branches and terminals in the dorsal horn. Mol Pain 8:82

Catterall WA (2014) 2013 Sharpey-Schafer Prize Lecture: structure and function of voltage-gated sodium channels at atomic resolution. Exp Physiol 99(1):35–51

Cox JJ, Reimann F et al (2006) An SCN9A channelopathy causes congenital inability to experience pain. Nature 444(7121):894–898

Cox JJ, Sheynin J et al (2010) Congenital insensitivity to pain: novel SCN9A missense and in-frame deletion mutations. Hum Mutat 31(9):E1670–E1686

Cummins TR, Waxman SG (1997) Downregulation of tetrodotoxin-resistant sodium currents and upregulation of a rapidly reprinting tetrodotoxin-sensitive sodium current in small spinal sensory neurons after nerve injury. J Neurosci 17(10):3503–3514

Cummins TR, Howe JR et al (1998) Slow closed-state inactivation: a novel mechanism underlying ramp currents in cells expressing the hNE/PN1 sodium channel. J Neurosci 18(23):9607–9619

Cummins TR, Dib-Hajj SD et al (1999) A novel persistent tetrodotoxin-resistant sodium current in SNS-null and wild-type small primary sensory neurons. J Neurosci 19(24):RC43

Cummins TR, Dib-Hajj SD et al (2004) Electrophysiological properties of mutant Nav1.7 sodium channels in a painful inherited neuropathy. J Neurosci 24(38):8232–8236

Dib-Hajj SD, Tyrrell L et al (1998) NaN, a novel voltage-gated Na channel, is expressed preferentially in peripheral sensory neurons and down-regulated after axotomy. Proc Natl Acad Sci U S A 95(15):8963–8968

Dib-Hajj SD, Fjell J et al (1999) Plasticity of sodium channel expression in DRG neurons in the chronic constriction injury model of neuropathic pain. Pain 83(3):591–600

Dib-Hajj S, Black JA et al (2002) NaN/Nav1.9: a sodium channel with unique properties. Trends Neurosci 25(5):253–259

Dib-Hajj SD, Cummins TR et al (2010) Sodium channels in normal and pathological pain. Annu Rev Neurosci 33:325–347

Eijkelkamp N, Linley JE et al (2012) Neurological perspectives on voltage-gated sodium channels. Brain 135(Pt 9):2585–2612

Estacion M, Dib-Hajj SD et al (2008) NaV1.7 gain-of-function mutations as a continuum: A1632E displays physiological changes associated with erythromelalgia and paroxysmal extreme pain disorder mutations and produces symptoms of both disorders. J Neurosci 28(43):11079–11088

Faber CG, Hoeijmakers JG et al (2012a) Gain of function Nanu1.7 mutations in idiopathic small fiber neuropathy. Ann Neurol 71(1):26–39

Faber CG, Lauria G et al (2012b) Gain-of-function Nav1.8 mutations in painful neuropathy. Proc Natl Acad Sci U S A 109(47):19444–19449

Fang X, Djouhri L et al (2002) The presence and role of the tetrodotoxin-resistant sodium channel Na(v)1.9 (NaN) in nociceptive primary afferent neurons. J Neurosci 22(17):7425–7433

Fertleman CR, Baker MD et al (2006) SCN9A mutations in paroxysmal extreme pain disorder: allelic variants underlie distinct channel defects and phenotypes. Neuron 52(5):767–774

Goldberg YP, MacFarlane J et al (2007) Loss-of-function mutations in the Nav1.7 gene underlie congenital indifference to pain in multiple human populations. Clin Genet 71(4):311–319

Goldberg Y, Pimstone S et al (2012a) Human Mendelian pain disorders: a key to discovery and validation of novel analgesics. Clin Genet 82(4):367–373

Goldberg YP, Price N et al (2012b) Treatment of Na(v)1.7-mediated pain in inherited erythromelalgia using a novel sodium channel blocker. Pain 153(1):80–85

Goldin AL, Barchi RL et al (2000) Nomenclature of voltage-gated sodium channels. Neuron 28(2):365–368

Hayden R, Grossman M (1959) Rectal, ocular, and submaxillary pain; a familial autonomic disorder related to proctalgia fugax: report of a family. AMA J Dis Child 97(4):479–482

Herzog RI, Cummins TR et al (2001) Persistent TTX-resistant Na+ current affects resting potential and response to depolarization in simulated spinal sensory neurons. J Neurophysiol 86(3):1351–1364

Herzog RI, Cummins TR et al (2003) Distinct repriming and closed-state inactivation kinetics of Nav1.6 and Nav1.7 sodium channels in mouse spinal sensory neurons. J Physiol 551 (Pt 3):741–750

Huang J, Yang Y et al (2013) Small-fiber neuropathy Nav1.8 mutation shifts activation to hyperpolarized potentials and increases excitability of dorsal root ganglion neurons. J Neurosci 33(35):14087–14097

Kerr BJ, Souslova V et al (2001) A role for the TTX-resistant sodium channel Nav 1.8 in NGF-induced hyperalgesia, but not neuropathic pain. Neuroreport 12(14):3077–3080

Klugbauer N, Lacinova L et al (1995) Structure and functional expression of a new member of the tetrodotoxin-sensitive voltage-activated sodium channel family from human neuroendocrine cells. EMBO J 14(6):1084–1090

Kremeyer B, Lopera F et al (2010) A gain-of-function mutation in TRPA1 causes familial episodic pain syndrome. Neuron 66(5):671–680

Leipold E, Liebmann L et al (2013) A de novo gain-of-function mutation in SCN11A causes loss of pain perception. Nat Genet 45(11):1399–1404

Leo S, D'Hooge R et al (2010) Exploring the role of nociceptor-specific sodium channels in pain transmission using Nav1.8 and Nav1.9 knockout mice. Behav Brain Res 208(1):149–157

Lolignier S, Amsalem M et al (2011) Nav1.9 channel contributes to mechanical and heat pain hypersensitivity induced by subacute and chronic inflammation. PLoS One 6(8):e23083

McCormack K, Santos S et al (2013) Voltage sensor interaction site for selective small molecule inhibitors of voltage-gated sodium channels. Proc Natl Acad Sci U S A 110(29):E2724–E2732

Middleton RE, Warren VA et al (2002) Two tarantula peptides inhibit activation of multiple sodium channels. Biochemistry 41(50):14734–14747

Minett MS, Quick K, Wood JN (2011) Behavioral measures of pain thresholds. In: Auwerx J, Brown SD, Justice M, Moore DD, Ackerman SL, Nadeau J (eds) Current protocols in mouse biology. Wiley, Hoboken

Minett MS, Nassar MA et al (2012) Distinct Nav1.7-dependent pain sensations require different sets of sensory and sympathetic neurons. Nat Commun 3:791

Nassar MA, Stirling LC et al (2004) Nociceptor-specific gene deletion reveals a major role for Nav1.7 (PN1) in acute and inflammatory pain. Proc Natl Acad Sci U S A 101(34):12706–12711

Nassar MA, Levato A et al (2005) Neuropathic pain develops normally in mice lacking both Nav1.7 and Nav1.8. Mol Pain 1:24

Nassar MA, Baker MD et al (2006) Nerve injury induces robust allodynia and ectopic discharges in Nav1.3 null mutant mice. Mol Pain 2:33

Patino GA, Isom LL (2010) Electrophysiology and beyond: multiple roles of Na+ channel beta subunits in development and disease. Neurosci Lett 486(2):53–59

Priest BT, Murphy BA et al (2005) Contribution of the tetrodotoxin-resistant voltage-gated sodium channel NaV1.9 to sensory transmission and nociceptive behavior. Proc Natl Acad Sci U S A 102(26):9382–9387

Priest BT, Blumenthal KM et al (2007) ProTx-I and ProTx-II: gating modifiers of voltage-gated sodium channels. Toxicon 49(2):194–201

Ragsdale DS, McPhee JC et al (1994) Molecular determinants of state-dependent block of Na+ channels by local anesthetics. Science 265(5179):1724–1728

Ragsdale DS, McPhee JC et al (1996) Common molecular determinants of local anesthetic, antiarrhythmic, and anticonvulsant block of voltage-gated Na+ channels. Proc Natl Acad Sci U S A 93(17):9270–9275

Reed KB, Davis MD (2009) Incidence of erythromelalgia: a population-based study in Olmsted County, Minnesota. J Eur Acad Dermatol Venereol 23(1):13–15

Renganathan M, Cummins TR et al (2001) Contribution of Na(v)1.8 sodium channels to action potential electrogenesis in DRG neurons. J Neurophysiol 86(2):629–640

Rotthier A, Baets J et al (2012) Mechanisms of disease in hereditary sensory and autonomic neuropathies. Nat Rev Neurol 8(2):73–85

Schmalhofer WA, Calhoun J et al (2008) ProTx-II, a selective inhibitor of NaV1.7 sodium channels, blocks action potential propagation in nociceptors. Mol Pharmacol 74(5):1476–1484

Theile JW, Cummins TR (2011) Recent developments regarding voltage-gated sodium channel blockers for the treatment of inherited and acquired neuropathic pain syndromes. Front Pharmacol 2:54

Toledo-Aral JJ, Moss BL et al (1997) Identification of PN1, a predominant voltage-dependent sodium channel expressed principally in peripheral neurons. Proc Natl Acad Sci U S A 94 (4):1527–1532

Waxman SG, Kocsis JD et al (1994) Type III sodium channel mRNA is expressed in embryonic but not adult spinal sensory neurons, and is reexpressed following axotomy. J Neurophysiol 72 (1):466–470

Weiss J, Pyrski M et al (2011) Loss-of-function mutations in sodium channel Nav1.7 cause anosmia. Nature 472(7342):186–190

Whitaker WR, Faull RL et al (2001) Comparative distribution of voltage-gated sodium channel proteins in human brain. Brain Res Mol Brain Res 88(1–2):37–53

Yang Y, Wang Y et al (2004) Mutations in SCN9A, encoding a sodium channel alpha subunit, in patients with primary erythermalgia. J Med Genet 41(3):171–174

Yang S, Xiao Y et al (2013) Discovery of a selective NaV1.7 inhibitor from centipede venom with analgesic efficacy exceeding morphine in rodent pain models. Proc Natl Acad Sci U S A 110 (43):17534–17539

Yuan J, Matsuura E et al (2013) Hereditary sensory and autonomic neuropathy type IID caused by an SCN9A mutation. Neurology 80(18):1641–1649

Zhang XY, Wen J et al (2013) Gain-of-function mutations in SCN11A cause familial episodic pain. Am J Hum Genet 93(5):957–966

Zimmermann K, Leffler A et al (2007) Sensory neuron sodium channel Nav1.8 is essential for pain at low temperatures. Nature 447(7146):855–858

Role of Nerve Growth Factor in Pain

Kazue Mizumura and Shiori Murase

Contents

1 Introduction 58
2 NGF and Its Receptor 58
3 Role of NGF in the Development of Nociceptive System 59
4 Pain and Mechanical/Thermal Hyperalgesia Induced by Exogenously Injected NGF 60
   4.1 Animals 60
   4.2 Humans 60
5 Role of NGF in Various Painful Conditions 62
   5.1 Role of NGF in Inflammatory Pain 62
   5.2 Role of NGF in Neuropathic Pain Resulting from Nerve Injury 62
   5.3 Role of NGF in Musculoskeletal Pain 63
   5.4 Role of NGF in Visceral Painful Conditions 66
   5.5 Role of NGF in Cancer Pain (and Cachexia) and Other Conditions 67
6 Effect of NGF on Nociceptor Activities and Their Axonal Properties 67
7 Action Mechanism of NGF in Modulating the Nociceptive System 68
   7.1 Mechanism of NGF-Induced Acute Sensitization of Nociceptors to Heat 68
   7.2 Mechanism of NGF-Induced Long-Lasting Sensitization to Heat 69
   7.3 Mechanism of NGF-Induced Sensitization to Mechanical Stimuli 70
8 Therapeutic Perspective 71
References 71

Abstract

Nerve growth factor (NGF) was first identified as a substance that is essential for the development of nociceptive primary neurons and later found to have a role in inflammatory hyperalgesia in adults. Involvement of NGF in conditions with no apparent inflammatory signs has also been demonstrated. In this review we look at the hyperalgesic effects of exogenously injected NGF into different tissues,

K. Mizumura (✉) • S. Murase
Department of Physical Therapy, College of Life and Health Sciences, Chubu University, 1200 Matsumoto-cho, Kasugai 487-8501, Japan
e-mail: mizu@isc.chubu.ac.jp

both human and animal, with special emphasis on the time course of these effects. The roles of NGF in inflammatory and neuropathic conditions as well as cancer pain are then reviewed. The role of NGF in delayed onset muscle soreness is described in more detail than its other roles based on the authors' recent observations. Acute effects are considered to be peripherally mediated, and accordingly, sensitization of nociceptors by NGF to heat and mechanical stimulation has been reported. Changes in the conductive properties of axons have also been reported. The intracellular mechanisms so far proposed for heat sensitization are direct phosphorylation and membrane trafficking of TRPV1 by TrkA. Little investigation has been done on the mechanism of mechanical sensitization, and it is still unclear whether mechanisms similar to those for heat sensitization work in mechanical sensitization. Long-lasting sensitizing effects are mediated both by changed expression of neuropeptides and ion channels (Na channels, ASIC, TRPV1) in primary afferents and by spinal NMDA receptors. Therapeutic perspectives are briefly discussed at the end of the chapter.

Keywords
Neuropathic pain • Cancer pain • Noninflammatory pain • Delayed onset muscle soreness • Mechanical hyperalgesia • Heat hyperalgesia

1 Introduction

Nerve growth factor (NGF) was the first identified neurotrophic factor. Its role in the development of nociceptive primary neurons/fibers was discovered first, after which its role in inflammatory hyperalgesia in adults was found. Recently, the involvement of NGF in painful conditions with no apparent inflammatory signs has been demonstrated, broadening its value in therapy. This chapter reviews the hyperalgesic effects of exogenously injected NGF and then introduces the roles of NGF in hyperalgesia in different physiological and pathological conditions, with special emphasis on its roles in delayed onset muscle soreness (DOMS). Finally, proposed action mechanisms of NGF will be presented, and the therapeutic perspective will be briefly discussed.

2 NGF and Its Receptor

NGF is a peptide composed of three subunits, α, β, and γ, in an $\alpha_2\beta\gamma_2$ structure. The β-subunit itself is a homodimer of peptides composed of 118 amino acids. The β-subunit alone accounts for the full biological activity of NGF. Its amino acid composition and sequence are well preserved among different species, with 90 % homology existing between mouse and human NGF.

NGF was first found as a substance that is essential for the survival, outgrowth, and development of peripheral sensory and sympathetic neurons (see Sect. 3). Later, it turned out that in adults NGF plays an important role in pain/hyperalgesia, a point that will be discussed throughout this chapter. NGF also has a central nervous function (Berry et al. 2012 for review), but that is beyond the scope of this chapter.

ProNGF is a precursor of NGF and its predominant form. It binds to a low affinity receptor, p75NTR (p75 neurotrophin receptor). NGF can also bind to p75NTR.

NGF exerts its activity by binding to a specific receptor, TrkA (receptor tyrosine kinase A, tropomyosin-related kinase A), which is composed of one transmembrane structure, an extracellular domain that binds with NGF, and an intracellular domain with tyrosine kinase activity. When a molecule of NGF binds to TrkA, another molecule of TrkA joins, making a TrkA dimer. Tyrosine kinase is then activated, and information is transmitted intracellularly at the site of binding (see Sect. 7.1). In addition, NGF that binds to TrkA is internalized and transported in the axon to the neuronal cell body (signaling endosome), where it changes the expression of neuropeptides, channels, and receptors. Increased neuropeptides, channels, and receptors are then transported both to the central and peripheral terminals, changing their sensitivities. This takes a longer time than the local action (see Schmieg et al. 2014 for review).

3 Role of NGF in the Development of Nociceptive System

The importance of NGF in the development of the nociceptive system was first shown by Ritter et al. (1991) with repetitive administration of anti-NGF antibody to neonatal rats up to 5 weeks after birth. In these rats, high threshold mechanonociceptors in the Aδ range (2–13 m/s) were completely abolished. Instead, the number of D-hairs increased. These authors also showed that administration of anti-NGF antibody to adult rats had no such effect. They did not mention anything about C-fibers. Later experiments with NGF knockout mice showed lack of sympathetic neurons and small primary afferent neurons, which would be cell bodies of nociceptors. These animals showed no response to tail pinch and had an elongated response latency in the tail flick test (Crowley et al. 1994). These results clearly show that NGF is essential for the development of nociceptors.

The validity of these observations was confirmed in humans by Indo et al. in 1996 (Indo et al. 1996) with the finding that patients who have congenital insensitivity to pain with anhidrosis (CIPA) have altered trkA genes. In this condition, sufferers cannot perceive any pain, itch, or heat, and at the same time they cannot sweat at all. Patients with a mutation in NGF-β gene are also reported to suffer from pain loss (Einarsdottir et al. 2004).

In adult animals NGF is no longer required for the survival of sensory neurons; instead, it works as a sensitizer of nociceptors as follows.

4 Pain and Mechanical/Thermal Hyperalgesia Induced by Exogenously Injected NGF

4.1 Animals

Lewin et al. (1993) reported that *thermal* hyperalgesia induced by injecting βNGF to adult rats began 15 min after injection and lasted more than 2 h. Amann et al. (1996) also checked thermal hyperalgesia by intraplantar injection of 4 μg mouse βNGF, and similar to Lewin's group, they observed significant thermal hyperalgesia in 10 min that disappeared in 3 h.

Cutaneous mechanical hyperalgesia was also reported by Lewin et al. (1993), who first observed it 6 h after injection. Therefore, they thought that the mechanical hyperalgesia may be due to central changes (in DRGs and in the spinal cord, see Lewin et al. 1992), whereas heat hyperalgesia is likely to result at least in part from the sensitization of peripheral nociceptors to heat. In contrast, mechanical hyperalgesia that appeared as early as 10 min after NGF (1 μg) injection and lasted 120 h (measured by Randall–Selitto apparatus; the measured tissue was not specified but was probably deep tissue) was reported by Malik-Hall et al. (2005). They also revealed that the TrkA receptor was involved in NGF-induced mechanical hyperalgesia. However, Khodorova et al. (2013) recently reported that mechanical hyperalgesia is mediated by binding to p75NTR.

Muscular mechanical hyperalgesia was also reported by intramuscular injection of rat recombinant NGF in rats. Mechanical hyperalgesia induced by 0.2 μM (20 μL) NGF appeared 2 h after the injection and lasted up to 5 h after the injection, while 0.8 μM NGF-induced hyperalgesia appeared 3 h after the injection and lasted up to 2 days after the injection (Murase et al. 2010).

4.2 Humans

The hyperalgesic effect of human recombinant NGF in humans was first reported by Petty et al. (1994). At that time NGF was proposed as a candidate for the treatment or prevention of peripheral or central nervous diseases, and that experiment was performed to assess its safety, tolerance, and other aspects. They reported that *intravenous* injection of human recombinant NGF induced, interestingly, mild to moderate myalgia (pain in swallowing, chewing, eye movements) starting 60–90 min after injection and lasting up to 7 weeks. Abdominal and limb muscles were also sometimes involved.

Using *subcutaneous* injection, Dyck et al. (1997) reported localized tenderness of the injected site from 3 h to 21 days after injections. Tenderness of deep structures to palpation was also reported, whereas no tactile allodynia and no abnormality in the vibratory or cooling detection threshold were observed. The heat-pain threshold was significantly lowered 1, 3, and 7 days (and in some cases at 3 h and 14 and 21 days) after NGF injection. These authors regarded the time course (appearance in 3 h) of tenderness to pressure and heat hyperalgesia to be too rapid to be explained by uptake of NGF by nociceptor terminals, retrograde transport, and upregulation of pain modulators in DRGs. Thus, local tissue mechanisms appear to be implicated.

The effects of *intradermal* injection of NGF into the human skin were somewhat different in their time course from the effects of subcutaneous injection. Intradermal injection induced heat hyperalgesia from 1 day up to 21 days after injection. In contrast, mechanical hyperalgesia to impact stimulation was observed first as late as day 3 and lasted up to 49 days after injection (Rukwied et al. 2010). It must be noted that the same group reported an increased proportion of mechanoresponsive C-fibers and their lowered response threshold in pigs 3 weeks after injection (see Sect. 6) (Hirth et al. 2013). They also reported that this change was not followed by an increase in the density of intraepidermal nerve fibers. Thus, sensitization in peripheral nociceptors remains for a long time.

Muscular mechanical hyperalgesia: Myalgia was the first reported sensory change in a clinical trial with intravenous injection of NGF in humans, as introduced above (Petty et al. 1994). Experimental injection of 5 μg human recombinant NGF into the masseter muscle induced no pain by itself and induced no changes in either the pressure pain threshold (PPT) or pressure tolerance threshold 1 h after injection. It significantly reduced the PPT from 1 day to 7 days after injection and the pressure tolerance threshold for 1 day ($p < 0.001$). The pain rating for chewing and yawning was significantly increased for 7 days following NGF injection (Svensson et al. 2003). The same amount of NGF injected into the tibialis anterior muscle lowered the PPT at the injection site as early as 3 h after injection and lasted up to 1 day after injection (Andersen et al. 2008). The authors attributed this difference in the time course to the relatively small amount of NGF with respect to the larger tibialis anterior muscle compared with the masseter muscle. No change in the hardness of the muscle was induced.

Long-lasting and widespread muscular mechanical hyperalgesia was reported after repetitive intramuscular injection of 5 μg human recombinant NGF for 3 days (day 0 to day 2) (Hayashi et al. 2013). In that work, increases in soreness scores (modified Likert scale) started 3 h after the first injection and continued until day 2, after which they remained at the increased level until day 16. Decreased PPTs were observed on days 1 to 3 and facilitated temporal summation of pressure pain at days 1 to 10. In addition, referred pain areas distant from the injection site appeared. Facilitated temporal summation and appearance of distant referred pain areas suggest involvement of central mechanisms.

Fascia has long been considered not only to be an important nociceptive organ but also, clinically, to be a target for the treatment of musculoskeletal pain (Hoheisel et al. 2012). Despite this, its neurophysiological properties have not so far been examined. The innervation characteristics and nociceptive properties of thoracolumbar fascia (Taguchi et al. 2008a, b; Hoheisel et al. 2011) and crural fascia have been clarified only recently (Taguchi et al. 2013). The effect of NGF on this structure in humans has also been examined: 1 μg NGF injected under ultrasound guidance to the fascia of the musculus erector spinae muscle evoked a lasting (days 1–7) and significant reduction of PPTs, exercise-evoked muscle pain, and hyperalgesia to impact stimuli (12 m/s) (Deising et al. 2012).

5 Role of NGF in Various Painful Conditions

5.1 Role of NGF in Inflammatory Pain

Increased NGF in tissue has been shown in a number of experimental inflammatory conditions, including inflammation induced by carrageenan (Mcmahon et al. 1995; Woolf et al. 1996), Freund's adjuvant (Woolf et al. 1994), and turpentine (Oddiah et al. 1998). Involvement of this upregulated NGF in hyperalgesia has been suggested by the finding that hyperalgesia is relieved with the use of TrkA–IgG fusion molecule (Mcmahon et al. 1995; Koltzenburg et al. 1999).

Increased NGF content was found in synovium of patients with osteoarthritis or rheumatoid arthritis (Aloe et al. 1992; Halliday et al. 1998). NGF immunoreactivity was found in lymphocyte-like cells in the perivascular region (Aloe et al. 1992). In Crohn's disease and ulcerative colitis, both NGF mRNA and trkA mRNA were increased (di Mola et al. 2000). In these conditions NGF mRNA was expressed in polymorphonuclear-like cells of the lamina propria and mast cells.

In conditions with no apparent involvement of inflammation, such as interstitial cystitis/bladder pain syndrome (IC/BPS) (Liu et al. 2013), painful intervertebral disk (Freemont et al. 2002), and osteoarthritis (Halliday et al. 1998; Iannone et al. 2002), increased NGF was observed in patients' serum or tissues. Upregulation of NGF has been reported in DOMS (Murase et al. 2010) and hyperalgesia after immobilization induced by casting (Sekino et al. 2014). In these two conditions, there is also no apparent sign of inflammation. In DOMS, muscle-contracting activity induces upregulation of NGF in the muscle (described in Sect. 5.3 in more detail).

5.2 Role of NGF in Neuropathic Pain Resulting from Nerve Injury

In a mouse model of neuropathic pain by sciatic nerve **chronic constriction injury (CCI)**, mouse monoclonal anti-NGF antibody (i.p., 2 weeks after surgery) produced recovery of the withdrawal threshold (Wild et al. 2007). In an L5 spinal nerve ligation model, NGF protein content in the ipsilateral L4 DRG significantly increased 14 days after surgery. In contrast, NGF mRNA was increased in the ipsilateral L5 DRG and sciatic nerve, but not in the ipsilateral L4 DRG or L4 spinal nerve. These authors speculated that NGF may be synthesized in the sciatic nerve (perhaps in the L5 spinal nerve distal to the ligation site) and diffused into the L4 spinal nerve, which retrogradely transports NGF to the L4 DRG neurons. Local application of anti-NGF antibody to the L4 spinal nerve beside the L5 spinal nerve ligation site prevented the development of thermal hyperalgesia for 5 days after ligation (Fukuoka et al. 2001).

Our preliminary experiment showed a neuropathic condition in which L5 spinal ligation induced muscular mechanical hyperalgesia and upregulation of NGF in the innervating muscle (medial gastrocnemius muscle). Muscular hyperalgesia was partially reversed by injection of anti-NGF antibody into the muscle. In all these experiments on neuropathic pain models, it was not clear which cell types (muscle cells, endothelial cells, and/or Schwann cells) produce NGF.

5.3 Role of NGF in Musculoskeletal Pain

5.3.1 Role of NGF in DOMS

DOMS appears after unaccustomed, strenuous exercise, typically after some delay (ca. 24 h). Characteristic symptoms are tenderness and movement-related pain. In addition, the existence of muscle hardening and trigger point-like sensitive spots, which are characteristic symptoms of myofascial pain syndrome, has been reported (Itoh et al. 2004; Kawakita et al. 2008). Thus, DOMS is often used as a model for myofascial pain syndrome (Andersen et al. 2006). Several causes have been proposed (Armstrong 1984; Smith 1991), of which the most widely accepted has been inflammation following micro-injury of the muscle fibers (Armstrong 1984). However, anti-inflammatory drugs were not effective in reducing the established mechanical hyperalgesia (Cheung et al. 2003). To find the real cause for DOMS, we developed a rat model of DOMS by applying lengthening contraction (LC) to the lower hind leg. Muscular mechanical hyperalgesia developed 1 day after the application of LC and lasted for 3 days (Fig. 1a) (Taguchi et al. 2005a). In our model, inflammatory signs and damage to muscle fibers were almost absent in exercised muscle (Fujii et al. 2008), yet mechanical hyperalgesia developed. This difference from previous experiments might have originated from differences in the strength of exercise and the method of inducing muscle contraction (Crameri et al. 2007). This is currently being studied.

Increased NGF mRNA in exercised muscle was first detected 12 h after LC; that is, up to that time point, NGF mRNA did not change (Fig. 1b). NGF protein also increased 12 h after LC up to 1 day after LC (Murase et al. 2010). In addition, anti-NGF antibody injected intramuscularly 2 days after LC, when the muscular mechanical hyperalgesia was the strongest, clearly reversed the mechanical hyperalgesia in 3 h (Fig. 1a). In situ hybridization showed that increased signals of NGF mRNA were seen around the nuclei of muscle/satellite cells 12 h after LC (Murase et al. 2012). Single-fiber recording from muscle C-fibers (sample recording in Fig. 2a, b) showed that NGF 0.8 μM (5 μL) decreased the mechanical threshold of these fibers (Fig. 2c) and increased the discharge number in response to ramp mechanical stimulation (Fig. 2d) (Murase et al. 2010). These changes were observed 10–20 min after injection and lasted as long as the recordings were continued (up to 2 h). This observation differs from that by Mense's group, which observed no sensitization induced by NGF (Hoheisel et al. 2005), possibly because they recorded only up to 15 min after injection.

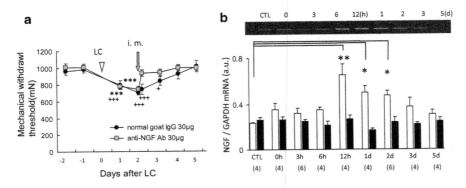

Fig. 1 Muscular mechanical hyperalgesia after exercise and changed expression of NGF mRNA in the muscle of rat. (**a**) Muscular mechanical hyperalgesia developed after lengthening contraction (LC) of the extensor digitorum muscle (*black circle*), and it was reversed by intramuscular injection of anti-NGF antibody (10 μg, *gray square*). ***, +++ $p < 0.001$ compared with day -1. (**b**) NGF mRNA was increased 12 h–3 days after LC (*white columns*, exercised side; *black columns*, contralateral side) (cited from Murase et al. 2010)

NGF also facilitates the heat response of muscle C-fibers (Queme et al. 2013). Notably, the heat sensitivity of muscle afferents 2 days after LC was not facilitated in comparison with that recorded from normal muscle (Taguchi et al. 2005b; Queme et al. 2013). This is puzzling because NGF is upregulated in the muscle after LC and NGF can sensitize muscle thin fibers to heat (Queme et al. 2013). This discrepancy might be due to a concentration of NGF in DOMS muscle that is too low to induce heat sensitization or to a substance other than NGF that is involved in the sensitization of nociceptors to mechanical stimulation but not to heat stimulation. We have reported that GDNF is also upregulated in the muscle cells/satellite cells of exercised muscle and implicated in DOMS (Murase et al. 2012, 2013). Our preliminary observation suggests that NGF and GDNF may collaborate in inducing mechanical hyperalgesia.

The upregulation of NGF in the muscle after LC was found to be blocked by HOE 140, a bradykinin B2 receptor antagonist, injected before LC but not after LC. The development of DOMS was also blocked by this procedure (Murase et al. 2010). This observation suggests that a B2 receptor agonist is released during exercise and stimulates NGF upregulation. This B2 agonist is probably Arg-bradykinin in rats, as Boix et al. (2002) have already shown. However, a B2 agonist alone would not seem to be sufficient to induce upregulation of NGF in the muscle, and so some unknown factor(s) is (are) also needed (Murase et al. 2012).

Involvement of TRPV1 in NGF-induced muscular mechanical hyperalgesia has also been shown using capsazepine and TRPV1 knockout mice (Ota et al. 2013). Involvement of ASIC in DOMS was also demonstrated (Fujii et al. 2008).

It is well known that when an LC bout is repeated after days or a few weeks, DOMS after the second bout is much less severe. This is called a repeated bout

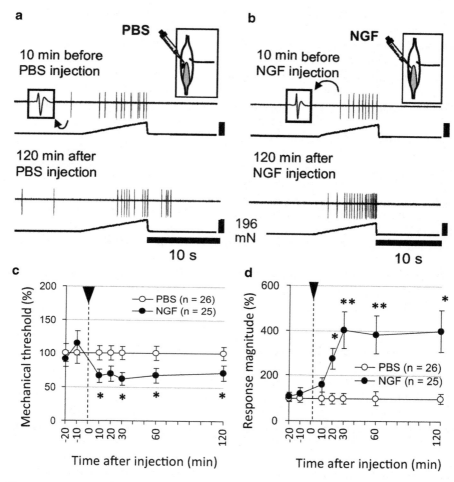

Fig. 2 Sensitization of muscular C-fiber afferents to mechanical stimulation by intramuscularly injected NGF (0.8 μM, 5 μL). (**a**, **b**) Sample recordings of C-fiber mechanosensitive afferents from the rat muscle–nerve preparation. (**c**) Change in the mechanical threshold; (**d**) number of discharges responding to mechanical stimulation. Values are presented as % of the averaged value of PBS group at each time point. *$p < 0.05$ and **$p < 0.01$ compared with PBS group at each time point (cited from Murase et al. 2010)

effect (Chen et al. 2007). NGF upregulation was also found to be reduced after the second LC (Urai et al. 2012).

5.3.2 Role in Cast Immobilization and Osteoarthritis Models

Cast immobilization is known to induce a CRPS 1-like painful condition (Guo et al. 2004; Ohmichi et al. 2012). After experimental cast immobilization of the hind leg, NGF is reported to increase in the hind paw skin (Sekino et al. 2014) and the gastrocnemius muscle (unpublished observation from our lab). NGF upregulation after joint immobilization was also observed in the DRGs (Nishigami et al. 2013). Whether hyperalgesia in these conditions is reduced by anti-NGF antibody or TrkA antagonist has not yet been examined. In addition, the cause of NGF upregulation has also not been explored.

Involvement of NGF in osteoarthritis has been suggested by the observation that NGF is upregulated in osteoarthritic chondrocytes (Iannone et al. 2002) and synovial fluid (Aloe et al. 1992) and pain is alleviated by injection of a soluble NGF receptor, TrkAd5, in a mouse model of osteoarthritis induced by surgical joint destabilization (McNamee et al. 2010). In the latter model, NGF was upregulated in the joints during both postoperative (day 3) and OA (16 weeks) pain phases, but not in the non-painful stage of disease (8 weeks post-surgery). The important role of NGF in OA pain has been indicated by the high effectiveness of TrkAd5 in suppressing pain in both painful phases (McNamee et al. 2010).

Inflammatory processes do not seem to be involved in the abovementioned hyperalgesic conditions; therefore, NGF upregulation might be induced by another mechanism, for example, muscle contraction in the case of DOMS.

In another painful condition of the musculoskeletal system, primary fibromyalgia, an increase of NGF in the cerebrospinal fluid has been reported (Giovengo et al. 1999; Sarchielli et al. 2007). However, the origin of this NGF is not clear.

5.4 Role of NGF in Visceral Painful Conditions

Upregulation of NGF and its high affinity receptor TrkA was shown in samples obtained from inflammatory bowel disease patients (Crohn's disease and ulcerative colitis) (di Mola et al. 2000). Cells expressing NGF and TrkA are reported to be polymorphonuclear-like cells of the lamina propria, mast cells, and some ganglionic cells.

There are also conditions where no apparent inflammation is the cause of NGF upregulation. Urinary (Kim et al. 2006) and serum (Jiang et al. 2013) levels of NGF were found to be higher in overactive bladder patients. Because NGF levels were decreased with antimuscarinic and botulinum toxin treatment (Liu et al. 2009), contraction of the bladder smooth muscles seems to be the cause of this upregulation of NGF. Urinary levels of NGF can be a marker for diagnosis of overactive bladder and evaluation of the effects of treatment for it (see Bhide et al. 2013 for review). In addition, when the vector encoding NGF was experimentally infected into the bladder, NGF was upregulated fourfold, and bladder overactivity was induced without any histological evidence of inflammation (Lamb et al. 2004). Collectively, contraction of the bladder and NGF upregulation seem to develop into a vicious cycle. The question of which is the initiating event will be an interesting issue for study.

5.5 Role of NGF in Cancer Pain (and Cachexia) and Other Conditions

All nociceptive, inflammatory, and neuropathic pains can be induced in cancer. Bone metastasis of breast cancer, prostate cancer, and lung cancer induces bone pain. Breakthrough pain induced by movement of involved bone is difficult to treat, and more effective treatments are being sought. A bone metastasis model made by injecting tumor cells into the bone marrow (Halvorson et al. 2005; Bloom et al. 2011) showed that tumors themselves and peripheral bone in the vicinity of the tumor lack detectable innervation, whereas the periosteum is densely innervated in areas where NGF is upregulated. Anti-NGF antibody alleviates pain and decreases the ectopic sprouting of nociceptors into the periosteum (Halvorson et al. 2005; Bloom et al. 2011). NGF blockade decreases tumor proliferation, nociception, and weight loss (Ye et al. 2011).

Administration of a blocking antibody to NGF produced a significant reduction in both early- and late-stage bone cancer pain-related behaviors (Halvorson et al. 2005). Thus, anti-NGF antibody therapy may be particularly effective in blocking bone cancer pain (Pantano et al. 2011). Clinical trials of anti-NGF antibody for the treatment of bone cancer pain still continue.

6 Effect of NGF on Nociceptor Activities and Their Axonal Properties

In normal skin, chronic (10–12 days) deprivation of NGF produced by continuous infusion of TrkA–IgG fusion molecule by an osmotic minipump decreased the percentage of cutaneous nociceptors that responded to heat and/or bradykinin in adult rats but not the percentage that responded to mechanical stimuli (Bennett et al. 1998). This report also showed that the innervation density was reduced in the epidermis with sequestration of NGF. This experiment demonstrates that endogenous NGF in normal adult animals modulates sensitivities to heat and bradykinin but not to mechanical stimuli and the innervation density of terminal axons.

In an inflammatory condition induced by carrageenan, the percentage of spontaneously active fibers recorded in vitro from skin–nerve preparations increased. Sensitivity to heat and bradykinin, but not to mechanical stimulation, also increased. When the TrkA–IgG fusion molecule was coadministered with carrageenan, sensitization to heat and bradykinin of cutaneous nociceptors did not occur, and again, their mechanical sensitivities were not changed (Koltzenburg et al. 1999).

The effects of NGF on muscle nociceptors seem somewhat different from those on skin. Mann et al. (2006) reported that intramuscular injection of human (not rat) NGF into the rat masseter muscle failed to evoke afferent discharges; however, it did decrease the mechanical threshold of masseter A-delta afferent fibers (Mann et al. 2006). Single-fiber recording from C-fibers innervating the extensor digitorum longus (EDL, sample recording in Fig. 2a, b) muscle showed that NGF 0.8 µM (5 µL) decreased their mechanical threshold (Fig. 2c) and increased the discharge number in response to ramp mechanical stimulation (Fig. 2d) (Murase et al. 2010) when compared with fibers recorded from normal rats that received PBS injection.

Axonal properties are also reported to be changed by NGF. Djouhri et al. (2001) reported that NGF sequestration by injecting NGF-binding domain (amino acids 285–413 of TrkAIg2) prevented the following CFA-induced changes in nociceptive neurons with A-delta or C-fibers: increased frequency that a fiber can follow, increased proportions of units with ongoing activity, and decreased action potential duration.

Hirth et al. (2013) also showed by single-fiber recordings 3 weeks after one-time intradermal injection of NGF to pig skin that NGF increased conduction velocity and decreased activity-dependent slowing of mechano-insensitive fibers. They also showed an increase in mechanosensitive fibers and decrease in median mechanical threshold. In contrast to the previous report using continuous infusion of NGF to the rat ankle skin (Bennett et al. 1998), they could not find any increase in the density of intraepidermal nerve fibers (Hirth et al. 2013). The abovementioned changes in axonal properties, especially activity-dependent slowing of conduction velocity, are reported to be related to the availability of Na channels (De Col et al. 2008). The tetrodotoxin-resistant (TTX-r) sodium channels Nav1.8 and Nav1.9 are predominantly expressed in small-/medium-sized nociceptive neurons that are cell bodies of thin-fiber afferents, and increased expression of these channels by NGF has been reported (Fjell et al. 1999; Bielefeldt et al. 2003).

7 Action Mechanism of NGF in Modulating the Nociceptive System

7.1 Mechanism of NGF-Induced Acute Sensitization of Nociceptors to Heat

7.1.1 Direct Phosphorylation by TrkA

TRPV1 is activated by heat (ca. 43 °C) and believed to be involved in heat transduction in nociceptors (Caterina et al. 1997), at least in inflammatory heat hyperalgesia (Caterina et al. 2000; Davis et al. 2000). It was reported that NGF was unable to induce heat hyperalgesia in TRPV1-deficient mice (Chuang et al. 2001). Therefore, the mechanism of the acute sensitizing effect of NGF on the heat response of nociceptors has been often studied using a TRPV1 stimulant, capsaicin, instead of heat.

Shu and Mendell (1999, 2001) first showed that a 10-min application of NGF facilitated capsaicin-induced currents in DRG neurons and later confirmed this observation (Zhu et al. 2004). Even though TrkA is connected with the PLC pathway and an earlier study showed that NGF-induced sensitization was blocked by PKA inhibition (Shu and Mendell 2001), this was not confirmed in later reports (Bonnington and McNaughton 2003; Zhu and Oxford 2007). Activation of protein kinase C by phorbol ester can sensitize the nociceptive neuron response to capsaicin (Bhave et al. 2003), while PKC inhibition abolishes or reduces NGF-induced TRPV1 sensitization (Bonnington and McNaughton 2003; Zhu and Oxford 2007). The effect of inhibiting CaMKII also differed among reports (Bonnington and McNaughton 2003; Zhu and Oxford 2007).

7.1.2 Membrane Trafficking of TRPV1 by TrkA

NGF promotes TRPV1 insertion into the plasma membrane (Zhang et al. 2005), for which involvement of PI3kinase (Stein et al. 2006) and its downstream Src kinase were reported. Src kinase reportedly phosphorylates Tyr200 of TRPV1 and translocates it to the cell membrane (Zhang et al. 2005). The early phase of heat hyperalgesia can be explained by this rapid sensitization (within several min) to heat by NGF.

7.1.3 Indirect Action of NGF Through Degradation of Mast Cells

Previous mast cell degranulation by compound 48/80 or pretreatment with antagonists of 5HT, contained in mast cell granules in rats, reduced the early phase of heat hyperalgesia (or delayed its onset) (Lewin et al. 1994; Amann et al. 1996; Woolf et al. 1996). These observations suggest that NGF also acts indirectly by activating mast cells and neutrophils, which in turn release additional inflammatory mediators causing hypersensitivity to heat.

7.1.4 Involvement of Sympathetic Nerve

The early phase of NGF-induced heat sensitization is partially dependent on sympathetic neurons, as sympathectomy partly reduced the effect of NGF in causing heat hyperalgesia (Andreev et al. 1995; Woolf et al. 1996).

7.2 Mechanism of NGF-Induced Long-Lasting Sensitization to Heat

The later phase (7 h–4 days after NGF) of heat hyperalgesia appeared to be centrally maintained, since it could be selectively blocked by the noncompetitive NMDA receptor antagonist MK-801 (Lewin et al. 1994). NGF binds to TrkA and is transported to DRG neurons to change the expression of neuropeptides (Donnerer et al. 1992, 1993; Leslie et al. 1995), sodium channels (Fjell et al. 1999), ASIC (Mamet et al. 2002), and other properties. In addition, NGF can increase TRPV1 expression (Donnerer et al. 2005; Xue et al. 2007), via the Ras–mitogen-activated protein kinase pathway (Ji et al. 2002) in DRG neurons. This increased expression of TRPV1 by NGF is implicated in maintaining the heat hyperalgesia in inflammation. Plastic changes in synaptic connections of muscle afferents in the spinal cord have been also reported after long-lasting injection of NGF to the muscle (Lewin et al. 1992).

7.3 Mechanism of NGF-Induced Sensitization to Mechanical Stimuli

The short latency action of NGF on heat sensitivity is well accepted; however, discrepancy exists in the time course of NGF-induced mechanical sensitization. An earlier study showed a latency of 7 h. The shortest latency of sensitization was 10–20 min in single-fiber recording in vitro (Murase et al. 2010) and also in nociceptive behavior although its peak was observed 3 h after injection (Malik-Hall et al. 2005). The longest latency so far reported is 3 days (Hirth et al. 2013). Medium latency of 1 h has been reported (sensation in humans by Svensson et al. 2003, afferent activities by Mann et al. 2006).

Mechanical hypersensitivity several hours after intraplantar injection of NGF was abolished in sympathectomized animals or delayed in mast cell degranulated animals by compound 48/80 (Woolf et al. 1996).

Malik-Hall et al. (2005) reported that acute mechanical hyperalgesia was reduced by inhibitors of the three major pathways for TrkA receptor signaling, extracellular signal-related kinase (ERK)/mitogen-activated protein kinase kinase (MEK), PI3K, and PLCγ. However, inhibitors of kinases downstream of PI3K and PLCγ (glycogen synthetase kinase 3, CAMII-K, or PKC) failed to reduce mechanical hyperalgesia. Thus, they could not clarify the downstream pathways.

Not much cell-based research has been done so far, possibly because the calcium imaging method cannot be applied for the mechanical response or because mechanotransducing channels of nociceptors have not been identified yet. Di Castro et al. (2006) showed that mechanically activated currents in cultured small and IB4(−) neurons was increased after application of NGF for 8 h (not 1 h) through a transcriptional mechanism. The augmented currents were further facilitated by activation of PKC by phorbol ester, and this effect was blocked by tetanus toxin, suggesting that the insertion of new channels into the cell membrane is involved in sensitization (Di Castro et al. 2006). In this report, no early sensitization was observed.

Involvement of TRPV1 was reported in mechanical hyperalgesia after lengthening contraction, where NGF plays a pivotal role (Fujii et al. 2008; Ota et al. 2013). Further research is needed to answer the question of whether mechanisms reported for heat hyperalgesia or augmented response to capsaicin (Bhave et al. 2003; Bonnington and McNaughton 2003; Zhang et al. 2005; Zhu and Oxford 2007) also work in NGF-induced mechanical hyperalgesia.

7.3.1 TrkA or p75NTR

A receptor for NGF that is believed to be involved in heat and mechanical hyperalgesia is TrkA. Recent reports also showed involvement of p75NTR in mechanical hyperalgesia (Iwakura et al. 2010; Khodorova et al. 2013; Matsuura et al. 2013). Downstream signaling cascades are different between these two NGF receptors, and the p75NTR cascade is a sphingomyelin signaling cascade that includes neutral sphingomyelinase(s) (nSMase), ceramide, and the atypical protein kinase C (aPKC) and protein kinase M zeta (PKMζ) (Zhang et al. 2012, also see

Nicol and Vasko 2007 for review). On the question of the relative importance of TrkA and p75NTR in NGF-induced hyperalgesia, controversial results have been reported, such as that NGF still produces hyperalgesia in p75 knockout mice (Bergmann et al. 1998, also see Lewin and Nykjaer 2014 for review).

8 Therapeutic Perspective

Efforts for the antagonization or reduction of NGF action have been directed toward the development of (1) humanized monoclonal antibodies (mAbs), (2) small molecules that bind NGF and change its molecular shape such that it can no longer bind to its receptor(s), (3) peptides that competitively bind TrkA or p75NTR receptors (Eibl et al. 2012), and (4) small molecules that block TrkA activities (Ghilardi et al. 2011). The specificity of mAbs is quite high, but it must be intravenously or intramuscularly injected. In addition, administration of mAbs entails the risk of immune reactions. In contrast, while small molecule inhibitors of kinase activity may not be as specific as mAbs or small molecules that bind to NGF and block binding to its receptor, they may have equal therapeutic potential. They can be orally administered, are less expensive to produce, and have greater flexibility in dosing. Except for mAbs, these agents are still in the preclinical stage. Clinical trials using a humanized anti-NGF antibody, tanezumab, have been conducted for low back pain (Kivitz et al. 2013), osteoarthritis (Sanga et al. 2013), and bone cancer pain, and outcomes have been good. However, because of a serious side effect (joint destruction), all clinical trials except for one on bone cancer pain were for a while suspended and now restarted. We hope that in the near future, some of these agents can be used for the treatment of pain.

References

Aloe L, Tuveri MA, Carcassi U, Levi-Montalcini R (1992) Nerve growth factor in the synovial fluid of patients with chronic arthritis. Arthritis Rheum 35:351–355

Amann R, Schuligoi R, Herzog G, Donnerer J (1996) Intraplantar injection of nerve growth factor into the rat hind paw: local edema and effects on thermal nociceptive threshold. Pain 64:323–329

Andersen H, Arendt-Nielsen L, Danneskiold-Samsoe B, Graven-Nielsen T (2006) Pressure pain sensitivity and hardness along human normal and sensitized muscle. Somatosens Mot Res 23:97–109

Andersen H, Arendt-Nielsen L, Svensson P, Danneskiold-Samsoe B, Graven-Nielsen T (2008) Spatial and temporal aspects of muscle hyperalgesia induced by nerve growth factor in humans. Exp Brain Res 191:371–382

Andreev NY, Dimitrieva N, Koltzenburg M, Mcmahon SB (1995) Peripheral administration of nerve growth factor in the adult rat produces a thermal hyperalgesia that requires the presence of sympathetic post-ganglionic neurones. Pain 63:109–115

Armstrong RB (1984) Mechanisms of exercise-induced delayed onset muscular soreness: a brief review. Med Sci Sports Exerc 16:529–538

Bennett DLH, Koltzenburg M, Priestley JV, Shelton DL, Mcmahon SB (1998) Endogenous nerve growth factor regulates the sensitivity of nociceptors in the adult rat. Eur J Neurosci 10:1282–1291

Bergmann I, Reiter R, Toyka KV, Koltzenburg M (1998) Nerve growth factor evokes hyperalgesia in mice lacking the low-affinity neurotrophin receptor p75. Neurosci Lett 255:87–90

Berry A, Bindocci E, Alleva E (2012) NGF, brain and behavioral plasticity. Neural Plast 2012:784040

Bhave G, Hu HJ, Glauner KS, Zhu W, Wang H, Brasier DJ, Oxford GS, Gereau RW (2003) Protein kinase C phosphorylation sensitizes but does not activate the capsaicin receptor transient receptor potential vanilloid 1 (TRPV1). Proc Natl Acad Sci U S A 100:12480–12485

Bhide AA, Cartwright R, Khullar V, Digesu GA (2013) Biomarkers in overactive bladder. Int Urogynecol J 24:1065–1072

Bielefeldt K, Ozaki N, Gebhart GF (2003) Role of nerve growth factor in modulation of gastric afferent neurons in the rat. Am J Physiol Gastrointest Liver Physiol 284:G499–G507

Bloom AP, Jimenez-Andrade JM, Taylor RN, Castaneda-Corral G, Kaczmarska MJ, Freeman KT, Coughlin KA, Ghilardi JR, Kuskowski MA, Mantyh PW (2011) Breast cancer-induced bone remodeling, skeletal pain, and sprouting of sensory nerve fibers. J Pain 12:698–711

Boix F, Rosenborg L, Hilgenfeldt U, Knardahl S (2002) Contraction-related factors affect the concentration of a kallidin-like peptide in rat muscle tissue. J Physiol 544:127–136

Bonnington JK, McNaughton PA (2003) Signalling pathways involved in the sensitisation of mouse nociceptive neurones by nerve growth factor. J Physiol 551:433–446

Caterina MJ, Schumacher MA, Tominaga M, Rosen TA, Levine JD, Julius D (1997) The capsaicin receptor: a heat-activated ion channel in the pain pathway. Nature 389:816–824

Caterina MJ, Leffler A, Malmberg AB, Martin WJ, Trafton J, Petersen-Zeitz KR, Kolzenburg M, Basbaum AI, Julius D (2000) Impaired nociception and pain sensation in mice lacking the capsaicin receptor. Science 288:306–313

Chen TC, Nosaka K, Sacco P (2007) Intensity of eccentric exercise, shift of optimum angle, and the magnitude of repeated-bout effect. J Appl Physiol 102:992–999

Cheung K, Hume P, Maxwell L (2003) Delayed onset muscle soreness: treatment strategies and performance factors. Sports Med 33:145–164

Chuang H-H, Prescott ED, Kong H, Shields S, Jordt SE, Basbaum AI, Chao MV, Julius D (2001) Bradykinin and nerve growth factor release the capsaicin receptor from PtdIns(4,5)P$_2$-mediated inhibition. Nature 411:957–962

Crameri RM, Aagaard P, Qvortrup K, Langberg H, Olesen J, Kjaer M (2007) Myofibre damage in human skeletal muscle: effects of electrical stimulation versus voluntary contraction. J Physiol 583:365–380

Crowley C, Spencer SD, Nishimura MC, Chen KS, Pittsmeek S, Armanini MP, Ling LH, Mcmahon SB, Shelton DL, Levinson AD, Phillips HS (1994) Mice lacking nerve growth factor display perinatal loss of sensory and sympathetic neurons yet develop basal forebrain cholinergic neurons. Cell 76:1001–1011

Davis JB, Gray J, Gunthorpe MJ, Hatcher JP, Davey PT, Overend P, Harries MH, Latcham J, Clapham C, Atkinson K, Hughes SA, Rance K, Grau E, Harper AJ, Pugh PL, Rogers DC, Bingham S, Randall A, Sheardown SA (2000) Vanilloid receptor-1 is essential for inflammatory thermal hyperalgesia. Nature 405:183–187

De Col R, Messlinger K, Carr RW (2008) Conduction velocity is regulated by sodium channel inactivation in unmyelinated axons innervating the rat cranial meninges. J Physiol 586:1089–1103

Deising S, Weinkauf B, Blunk J, Obreja O, Schmelz M, Rukwied R (2012) NGF-evoked sensitization of muscle fascia nociceptors in humans. Pain 153:1673–1679

Di Castro A, Drew LJ, Wood JN, Cesare P (2006) Modulation of sensory neuron mechanotransduction by PKC- and nerve growth factor-dependent pathways. Proc Natl Acad Sci U S A 103:4699–4704 di Mola FF, Friess H, Zhu ZW, Koliopanos A, Bley T, Di Sebastiano P, Innocenti P, Zimmermann A, Buchler MW (2000) Nerve growth factor and Trk high affinity receptor (TrkA) gene expression in inflammatory bowel disease. Gut 46:670–679

Djouhri L, Dawbarn D, Robertson A, Newton R, Lawson SN (2001) Time course and nerve growth factor dependence of inflammation-induced alterations in electrophysiological membrane properties in nociceptive primary afferent neurons. J Neurosci 21:8722–8733

Donnerer J, Schuligoi R, Stein C (1992) Increased content and transport of substance P and calcitonin gene-related peptide in sensory nerves innervating inflamed tissue: evidence for a regulatory function of nerve growth factor in vivo. Neuroscience 49:693–698

Donnerer J, Schuligoi R, Stein C, Amann R (1993) Upregulation, release and axonal transport of substance-P and calcitonin gene-related peptide in adjuvant inflammation and regulatory function of nerve growth factor. Regul Pept 46:150–154

Donnerer J, Liebmann I, Schicho R (2005) Differential regulation of 3-beta-hydroxysteroid dehydrogenase and vanilloid receptor TRPV1 mRNA in sensory neurons by capsaicin and NGF. Pharmacology 73:97–101

Dyck PJ, Peroutka S, Rask C, Burton E, Baker MK, Lehman KA, Gillen DA, Hokanson JL, O'Brien PC (1997) Intradermal recombinant human nerve growth factor induces pressure allodynia and lowered heat-pain threshold in humans. Neurology 48:501–505

Eibl JK, Strasser BC, Ross GM (2012) Structural, biological, and pharmacological strategies for the inhibition of nerve growth factor. Neurochem Int 61:1266–1275

Einarsdottir E, Carlsson A, Minde J, Toolanen G, Svensson O, Solders G, Holmgren G, Holmberg D, Holmberg M (2004) A mutation in the nerve growth factor beta gene (NGFB) causes loss of pain perception. Hum Mol Genet 13:799–805

Fjell J, Cummins TR, Dib-Hajj SD, Fried K, Black JA, Waxman SG (1999) Differential role of GDNF and NGF in the maintenance of two TTX-resistant sodium channels in adult DRG neurons. Mol Brain Res 67:267–282

Freemont AJ, Watkins A, Le MC, Baird P, Jeziorska M, Knight MT, Ross ER, O'Brien JP, Hoyland JA (2002) Nerve growth factor expression and innervation of the painful intervertebral disc. J Pathol 197:286–292

Fujii Y, Ozaki N, Taguchi T, Mizumura K, Sugiura Y (2008) TRP channels and ASICs mediate mechanical hyperalgesia in models of inflammatory muscle pain and delayed onset muscle soreness. Pain 140:292–304

Fukuoka T, Kondo E, Dai Y, Hashimoto N, Noguchi K (2001) Brain-derived neurotrophic factor increases in the uninjured dorsal root ganglion neurons in selective spinal nerve ligation model. J Neurosci 21:4891–4900

Ghilardi JR, Freeman KT, Jimenez-Andrade JM, Mantyh WG, Bloom AP, Bouhana KS, Trollinger D, Winkler J, Lee P, Andrews SW, Kuskowski MA, Mantyh PW (2011) Sustained blockade of neurotrophin receptors TrkA, TrkB and TrkC reduces non-malignant skeletal pain but not the maintenance of sensory and sympathetic nerve fibers. Bone 48:389–398

Giovengo SL, Russell IJ, Larson AA (1999) Increased concentrations of nerve growth factor in cerebrospinal fluid of patients with fibromyalgia. J Rheumatol 26:1564–1569

Guo TZ, Offley SC, Boyd EA, Jacobs CR, Kingery WS (2004) Substance P signaling contributes to the vascular and nociceptive abnormalities observed in a tibial fracture rat model of complex regional pain syndrome type I. Pain 108:95–107

Halliday DA, Zettler C, Rush RA, Scicchitano R, McNeil JD (1998) Elevated nerve growth factor levels in the synovial fluid of patients with inflammatory joint disease. Neurochem Res 23:919–922

Halvorson KG, Kubota K, Sevcik MA, Lindsay TH, Sotillo JE, Ghilardi JR, Rosol TJ, Boustany L, Shelton DL, Mantyh PW (2005) A blocking antibody to nerve growth factor attenuates skeletal pain induced by prostate tumor cells growing in bone. Cancer Res 65:9426–9435

Hayashi K, Shiozawa S, Ozaki N, Mizumura K, Graven-Nielsen T (2013) Repeated intramuscular injections of nerve growth factor induced progressive muscle hyperalgesia, facilitated temporal summation and expanded pain areas. Pain 154:2344–2352

Hirth M, Rukwied R, Gromann A, Turnquist B, Weinkauf B, Francke K, Albrecht P, Rice F, Hagglof B, Ringkamp M, Engelhardt M, Schultz C, Schmelz M, Obreja O (2013) Nerve growth factor induces sensitization of nociceptors without evidence for increased intraepidermal nerve fiber density. Pain 154:2500–2511

Hoheisel U, Unger T, Mense S (2005) Excitatory and modulatory effects of inflammatory cytokines and neurotrophins on mechanosensitive group IV muscle afferents in the rat. Pain 114:168–176

Hoheisel U, Taguchi T, Treede RD, Mense S (2011) Nociceptive input from the rat thoracolumbar fascia to lumbar dorsal horn neurones. Eur J Pain 15:810–815

Hoheisel U, Taguchi T, Mense S (2012) Nociception: the thoracolumbar fascia as a sensory organ. In: Schleip R, Findley TW, Chaitow L, Huijing PA (eds) Fascia: the tensional network of the human body: the science and clinical applications in manual and movement therapy. Churchill Livingstone/Elsevier, Edinburgh, pp 95–101

Iannone F, De BC, Dell'Accio F, Covelli M, Patella V, Lo BG, Lapadula G (2002) Increased expression of nerve growth factor (NGF) and high affinity NGF receptor (p140 TrkA) in human osteoarthritic chondrocytes. Rheumatology 41:1413–1418

Indo Y, Tsuruta M, Hayashida Y, Karim MA, Ohta K, Kawano T, Mitsubuchi H, Tonoki H, Awaya Y, Matsuda I (1996) Mutations in the TRKA/NGF receptor gene in patients with congenital insensitivity to pain with anhidrosis. Nat Genet 13:485–488

Itoh K, Okada K, Kawakita K (2004) A proposed experimental model of myofascial trigger points in human muscle after slow eccentric exercise. Acupunct Med 22:2–12

Iwakura N, Ohtori S, Orita S, Yamashita M, Takahashi K, Kuniyoshi K (2010) Role of low-affinity nerve growth factor receptor inhibitory antibody in reducing pain behavior and calcitonin gene-related Peptide expression in a rat model of wrist joint inflammatory pain. J Hand Surg Am 35:267–273

Ji R, Samad T, Jin S, Schmoll R, Woolf C (2002) p38 MAPK activation by NGF in primary sensory neurons after inflammation increases TRPV1 levels and maintains heat hyperalgesia. Neuron 36:57–68

Jiang YH, Peng CH, Liu HT, Kuo HC (2013) Increased pro-inflammatory cytokines, C-reactive protein and nerve growth factor expressions in serum of patients with interstitial cystitis/bladder pain syndrome. PLoS One 8:e76779

Kawakita K, Itoh K, Okada K (2008) Experimental model of trigger points using eccentric exercise. J Musculoskelet Pain 16:29–35

Khodorova A, Nicol GD, Strichartz G (2013) The p75(NTR) signaling cascade mediates mechanical hyperalgesia induced by nerve growth factor injected into the rat hind paw. Neuroscience 254:312–323

Kim JC, Park EY, Seo SI, Park YH, Hwang TK (2006) Nerve growth factor and prostaglandins in the urine of female patients with overactive bladder. J Urol 175:1773–1776

Kivitz AJ, Gimbel JS, Bramson C, Nemeth MA, Keller DS, Brown MT, West CR, Verburg KM (2013) Efficacy and safety of tanezumab versus naproxen in the treatment of chronic low back pain. Pain 154:1009–1021

Koltzenburg M, Bennett DL, Shelton DL, Mcmahon SB (1999) Neutralization of endogenous NGF prevents the sensitization of nociceptors supplying inflamed skin. Eur J Neurosci 11:1698–1704

Lamb K, Gebhart GF, Bielefeldt K (2004) Increased nerve growth factor expression triggers bladder overactivity. J Pain 5:150–156

Leslie TA, Emson PC, Dowd PM, Woolf CJ (1995) Nerve growth factor contributes to the up-regulation of growth- associated protein 43 and preprotachykinin A messenger RNAs in primary sensory neurons following peripheral inflammation. Neuroscience 67:753–761

Lewin GR, Nykjaer A (2014) Pro-neurotrophins, sortilin, and nociception. Eur J Neurosci 39:363–374

Lewin GR, Winter J, Mcmahon SB (1992) Regulation of afferent connectivity in the adult spinal cord by nerve growth factor. Eur J Neurosci 4:700–707

Lewin GR, Ritter AM, Mendell LM (1993) Nerve growth factor-induced hyperalgesia in the neonatal and adult rat. J Neurosci 13:2136–2148

Lewin GR, Rueff A, Mendell LM (1994) Peripheral and central mechanisms of NGF-induced hyperalgesia. Eur J Neurosci 6:1903–1912

Liu HT, Chancellor MB, Kuo HC (2009) Decrease of urinary nerve growth factor levels after antimuscarinic therapy in patients with overactive bladder. BJU Int 103:1668–1672

Liu HT, Jiang YH, Kuo HC (2013) Increased serum adipokines implicate chronic inflammation in the pathogenesis of overactive bladder syndrome refractory to antimuscarinic therapy. PLoS One 8:e76706

Malik-Hall M, Dina OA, Levine JD (2005) Primary afferent nociceptor mechanisms mediating NGF-induced mechanical hyperalgesia. Eur J Neurosci 21:3387–3394

Mamet J, Baron A, Lazdunski M, Voilley N (2002) Proinflammatory mediators, stimulators of sensory neuron excitability via the expression of acid-sensing ion channels. J Neurosci 22:10662–10670

Mann MK, Dong XD, Svensson P, Cairns BE (2006) Influence of intramuscular nerve growth factor injection on the response properties of rat masseter muscle afferent fibers. J Orofac Pain 20:325–336

Matsuura Y, Iwakura N, Ohtori S, Suzuki T, Kuniyoshi K, Murakami K, Hiwatari R, Hashimoto K, Okamoto S, Shibayama M, Kobayashi T, Ogawa Y, Sukegawa K, Takahashi K (2013) The effect of Anti-NGF receptor (p75 Neurotrophin Receptor) antibodies on nociceptive behavior and activation of spinal microglia in the rat brachial plexus avulsion model. Spine (Phila Pa 1976) 38:E332–E338

Mcmahon SB, Bennett DL, Priestley JV, Shelton DL (1995) The biological effects of endogenous nerve growth factor on adult sensory neurons revealed by a trkA-IgG fusion molecule. Nat Med 1:774–780

McNamee KE, Burleigh A, Gompels LL, Feldmann M, Allen SJ, Williams RO, Dawbarn D, Vincent TL, Inglis JJ (2010) Treatment of murine osteoarthritis with TrkAd5 reveals a pivotal role for nerve growth factor in non-inflammatory joint pain. Pain 149:386–392

Murase S, Terazawa E, Queme F, Ota H, Matsuda T, Hirate K, Kozaki Y, Katanosaka K, Taguchi T, Mizumura K (2010) Bradykinin and nerve growth factor play pivotal roles in muscular mechanical hyperalgesia after exercise (delayed onset muscle soreness). J Neurosci 30:3752–3761

Murase S, Yamanaka Y, Kanda H, Mizumura K (2012) COX-2, nerve growth factor (NGF) and glial cell-derived neurotrophic factor (GDNF), which play pivotal roles in delayed onset muscle soreness (DOMS), are produced by exercised skeletal muscle. J Physiol Sci 62(Suppl):S179

Murase S, Terazawa E, Hirate K, Yamanaka H, Kanda H, Noguchi K, Ota H, Queme F, Taguchi T, Mizumura K (2013) Upregulated glial cell line-derived neurotrophic factor through cyclooxygenase-2 activation in the muscle is required for mechanical hyperalgesia after exercise in rats. J Physiol 591:3035–3048

Nicol GD, Vasko MR (2007) Unraveling the story of NGF-mediated sensitization of nociceptive sensory neurons: ON or OFF the Trks? Mol Interv 7:26–41

Nishigami T, Osako Y, Ikeuchi M, Yuri K, Ushida T (2013) Development of heat hyperalgesia and changes of TRPV1 and NGF expression in rat dorsal root ganglion following joint immobilization. Physiol Res 62:215–219

Oddiah D, Anand P, Mcmahon SB, Rattray M (1998) Rapid increase of NGF, BDNF and NT-3 mRNAs in inflamed bladder. Neuroreport 9:1455–1458

Ohmichi Y, Sato J, Ohmichi M, Sakurai H, Yoshimoto T, Morimoto A, Hashimoto T, Eguchi K, Nishihara M, Arai YC, Ohishi H, Asamoto K, Ushida T, Nakano T, Kumazawa T (2012) Two-week cast immobilization induced chronic widespread hyperalgesia in rats. Eur J Pain 16:338–348

Ota H, Katanosaka K, Murase S, Kashio M, Tominaga M, Mizumura K (2013) TRPV1 and TRPV4 play pivotal roles in delayed onset muscle soreness. PLoS One 8:e65751

Pantano F, Zoccoli A, Iuliani M, Lanzetta G, Vincenzi B, Tonini G, Santini D (2011) New targets, new drugs for metastatic bone pain: a new philosophy. Expert Opin Emerg Drugs 16:403–405

Petty BG, Cornblath DR, Adornato BT, Chaudhry V, Flexner C, Wachsman M, Sinicropi D, Burton LE, Peroutka SJ (1994) The effect of systemically administered recombinant human nerve growth factor in healthy human subjects. Ann Neurol 36:244–246

Queme F, Taguchi T, Mizumura K, Graven-Nielsen T (2013) Muscular heat and mechanical pain sensitivity after lengthening contractions in humans and animals. J Pain 14:1425–1436

Ritter AM, Lewin GR, Kremer NE, Mendell LM (1991) Requirement for nerve growth factor in the development of myelinated nociceptors in vivo. Nature 350:500–502

Rukwied R, Mayer A, Kluschina O, Obreja O, Schley M, Schmelz M (2010) NGF induces non-inflammatory localized and lasting mechanical and thermal hypersensitivity in human skin. Pain 148:407–413

Sanga P, Katz N, Polverejan E, Wang S, Kelly KM, Haeussler J, Thipphawong J (2013) Efficacy, safety, and tolerability of fulranumab, an anti-nerve growth factor antibody, in the treatment of patients with moderate to severe osteoarthritis pain. Pain 154:1910–1919

Sarchielli P, Mancini ML, Floridi A, Coppola F, Rossi C, Nardi K, Acciaresi M, Pini LA, Calabresi P (2007) Increased levels of neurotrophins are not specific for chronic migraine: evidence from primary fibromyalgia syndrome. J Pain 8:737–745

Schmieg N, Menendez G, Schiavo G, Terenzio M (2014) Signalling endosomes in axonal transport: travel updates on the molecular highway. Semin Cell Dev Biol 27:32–43. doi:10.1016/j.semcdb.2013.10.004

Sekino Y, Nakano J, Hamaue Y, Chuganji S, Sakamoto J, Yoshimura T, Origuchi T, Okita M (2014) Sensory hyperinnervation and increase in NGF, TRPV1 and P2X3 expression in the epidermis following cast immobilization in rats. Eur J Pain 18:639–648

Shu X, Mendell LM (1999) Nerve growth factor acutely sensitizes the response of adult rat sensory neurons to capsaicin. Neurosci Lett 274:159–162

Shu X, Mendell LM (2001) Acute sensitization by NGF of the response of small-diameter sensory neurons to capsaicin. J Neurophysiol 86:2931–2938

Smith LL (1991) Acute inflammation—the underlying mechanism in delayed onset muscle soreness. Med Sci Sports Exerc 23:542–551

Stein AT, Ufret-Vincenty CA, Hua L, Santana LF, Gordon SE (2006) Phosphoinositide 3-kinase binds to TRPV1 and mediates NGF-stimulated TRPV1 trafficking to the plasma membrane. J Gen Physiol 128:509–522

Svensson P, Cairns BE, Wang K, Arendt-Nielsen L (2003) Injection of nerve growth factor into human masseter muscle evokes long-lasting mechanical allodynia and hyperalgesia. Pain 104:241–247

Taguchi T, Matsuda T, Tamura R, Sato J, Mizumura K (2005a) Muscular mechanical hyperalgesia revealed by behavioural pain test and c-Fos expression in the spinal dorsal horn after eccentric contraction in rats. J Physiol 564:259–268

Taguchi T, Sato J, Mizumura K (2005b) Augmented mechanical response of muscle thin-fiber sensory receptors recorded from rat muscle-nerve preparations *in vitro* after eccentric contraction. J Neurophysiol 94:2822–2831

Taguchi T, Hoheisel U, Mense S (2008a) Dorsal horn neurons having input from low back structures in rats. Pain 138:119–129

Taguchi T, Hoheisel U, Mense S (2008b) Neuroanatomy and electrophysiology of low back pain: experiments on rats. Aktuelle Urol 34:472–477

Taguchi T, Yasui M, Kubo A, Abe M, Kiyama H, Yamanaka A, Mizumura K (2013) Nociception originating from the crural fascia in rats. Pain 154:1103–1114

Urai H, Murase S, Mizumura K (2012) Decreased nerve growth factor upregulation is a mechanism for reduced mechanical hyperalgesia after the second bout of exercise in rats. Scand J Med Sci Sports 23:e96–e101

Wild KD, Bian D, Zhu D, Davis J, Bannon AW, Zhang TJ, Louis JC (2007) Antibodies to nerve growth factor reverse established tactile allodynia in rodent models of neuropathic pain without tolerance. J Pharmacol Exp Ther 322:282–287

Woolf CJ, Safieh-Garabedian B, Ma QP, Crilly P, Winter J (1994) Nerve growth factor contributes to the generation of inflammatory sensory hypersensitivity. Neuroscience 62:327–331

Woolf CJ, Ma QP, Allchorne A, Poole S (1996) Peripheral cell types contributing to the hyperalgesic action of nerve growth factor in inflammation. J Neurosci 16:2716–2723

Xue Q, Jong B, Chen T, Schumacher MA (2007) Transcription of rat TRPV1 utilizes a dual promoter system that is positively regulated by nerve growth factor. J Neurochem 101:212–222

Ye Y, Dang D, Zhang J, Viet CT, Lam DK, Dolan JC, Gibbs JL, Schmidt BL (2011) Nerve growth factor links oral cancer progression, pain, and cachexia. Mol Cancer Ther 10:1667–1676

Zhang X, Huang J, McNaughton PA (2005) NGF rapidly increases membrane expression of TRPV1 heat-gated ion channels. EMBO J 24:4211–4223

Zhang YH, Kays J, Hodgdon KE, Sacktor TC, Nicol GD (2012) Nerve growth factor enhances the excitability of rat sensory neurons through activation of the atypical protein kinase C isoform, PKMzeta. J Neurophysiol 107:315–335

Zhu W, Oxford GS (2007) Phosphoinositide-3-kinase and mitogen activated protein kinase signaling pathways mediate acute NGF sensitization of TRPV1. Mol Cell Neurosci 34:689–700

Zhu W, Galoyan SM, Petruska JC, Oxford GS, Mendell LM (2004) A developmental switch in acute sensitization of small dorsal root ganglion (DRG) neurons to capsaicin or noxious heating by NGF. J Neurophysiol 92:3148–3152

Central Sensitization in Humans: Assessment and Pharmacology

Lars Arendt-Nielsen

Contents

1 Introduction 80
2 Central Sensitization in Chronic Pain Patients 82
   2.1 Extraterritorial Manifestations of Sensitization 82
   2.2 Widespread Manifestations of Sensitization 83
3 QST for Assessing Central Sensitization 84
4 Conclusion and Future Perspectives 94
References 94

Abstract

It is evident that chronic pain can modify the excitability of central nervous system which imposes a specific challenge for the management and for the development of new analgesics. The central manifestations can be difficult to quantify using standard clinical examination procedures, but quantitative sensory testing (QST) may help to quantify the degree and extend of the central reorganization and effect of pharmacological interventions. Furthermore, QST may help in optimizing the development programs for new drugs.

Specific translational mechanistic QST tools have been developed to quantify different aspects of central sensitization in pain patients such as threshold ratios, provoked hyperalgesia/allodynia, temporal summation (wind-up like pain), after sensation, spatial summation, reflex receptive fields, descending pain modulation, offset analgesia, and referred pain areas. As most of the drug development programs in the area of pain management have not been very successful, the pharmaceutical industry has started to utilize the complementary knowledge obtained from QST profiling. Linking patients QST profile with drug efficacy

---

L. Arendt-Nielsen (✉)
Center for Sensory-Motor Interaction (SMI), Department of Health Science and Technology, School of Medicine, Aalborg University, Fredrik Bajers Vej 7-D3, 9220 Aalborg, Denmark
e-mail: LAN@HST.AAU.DK

© Springer-Verlag Berlin Heidelberg 2015
H.-G. Schaible (ed.), *Pain Control*, Handbook of Experimental Pharmacology 227,
DOI 10.1007/978-3-662-46450-2_5

profile may provide the fundamentals for developing individualized, targeted pain management programs in the future. Linking QST-assessed pain mechanisms with treatment outcome provides new valuable information in drug development and for optimizing the management regimes for chronic pain.

Keywords
Pain assessment • Quantitative sensory testing • Spreading sensitization

1 Introduction

Chronic pain is a disabling condition for the individual patient. In recent years, the many studies in animals, healthy volunteers, and pain patients have been successful in providing mechanistic knowledge leading to a better understanding of the clinical signs and symptoms associated with chronic pain. However, the treatment of chronic pain is still generally unsatisfactory due to the lack of adequate diagnostic tools and analgesics. Individualized and personalized treatment regimes are far from being implemented in clinical practice. Reorganization of the pain system, resulting in central sensitization, is a main hurdle for efficient pain management of chronic pain with the drugs currently available. The main treatment regimes still consist of opioids, nonsteroidal anti-inflammatory drugs (NSAIDs), analgesics, anticonvulsants, antidepressants, and combinations thereof. Some of those drug classes may interact with the central reorganization but with highly unpredictable efficacy in the individual patient.

This lack of efficient drugs has prompted pharmaceutical companies to invest substantially in the search for new compounds to treat chronic pain, and many new molecules and molecular mechanisms have been tested or are currently in clinical trials. The majority of the drugs are being investigated in conditions like neuropathic pain (post herpetic neuralgia, diabetic painful neuropathy), osteoarthritis (OA) pain, complex regional pain syndrome, trigeminal neuralgia, and chronic low back pain. Despite the substantial investments in developing new molecules, the success rate has been very low and the pharmaceutical industry is currently moving away from centrally acting compounds towards peripherally acting compounds (Nishimura et al. 2010) in order to minimize unwanted side effects, drugs interacting with the glia cells (Milligan and Watkins 2009), or even drugs interacting with genetic factors (Crow et al. 2013). Reducing the peripheral nociceptive drive may be a more efficient way of reducing the central consequences than attempting interacting directly with the central targets (Richards and McMahon 2013).

Unfortunately in many chronic pain conditions with associated central changes, the nociceptive driver is not fully understood, e.g., fibromyalgia, whiplash, endometriosis, and irritable bowel syndrome. To complicate the picture further, central sensitization may also be found in conditions where pain may not be part of the characteristic clinical picture as in, e.g., multiple sclerosis, Parkinson's disease, multiple chemical hypersensitivity, restless leg syndrome, chronic fatigue, postdeployment syndrome, posttraumatic stress, borderline personality disorders, major depression, and schizophrenia (Arendt-Nielsen et al. 2012a, b).

It is generally accepted that injury or inflammation can drive the nociceptive nerve endings or afferents in the somatic or visceral tissue. This peripheral drive may cause various adaptive changes in the central neuronal network and eventually result in pain hypersensitivity. This increase in central gain facilitates the spontaneous and/or ongoing pain intensity which is the main problem for the patient. In addition to the increased gain, the central changes and reorganization may also manifest as allodynia experienced by the patient as an additional factor contributing to the overall pain problem. Most animal studies on central sensitization focus specifically on the reorganization by assessing modality-specific hyperalgesia and/or allodynia although this is not considered as the main problem for the chronic pain patients. It has been found that generalized central hyperalgesia affects 17.5–35.3 % of the chronic pain population (Schliessbach et al. 2013).

The involvement of sensitization may also explain the lack of relationship between actual tissue damage in, e.g., osteoarthritis (OA) (Davis et al. 1992; Hannan et al. 2000; Neogi et al. 2009) and the pain intensity.

The central sensitization occurring within spinal and supraspinal networks has been summarized in many book chapters and reviews (Basbaum et al. 2009; Sandkühler 2009). In general, the central sensitization involves both presynaptic effects, i.e., changes in transmitter release, and postsynaptic effects. The latter are primarily due to increases in intracellular calcium, due to increased influx through ionotropic channels (e.g., NMDA channels) and voltage-dependent calcium channels, or due to release of calcium from intracellular stores produced by receptors (e.g., metabotropic receptors) or tyrosine kinases. In addition, there is a loss of GABA- and glycine-mediated inhibition within the spinal cord and a loss of descending monoaminergically mediated inhibitory controls from the medulla. Finally, immune cells are recruited to sites of inflammation and injury, and microglial activation occurs in the dorsal horn of the spinal cord, close to the central terminals of injured sensory neurons (Schomberg and Olson 2012). In recent years, this has led to an increasing interest in glia cells as a nonneuronal mediator of sensitization and a suggested molecular target for pain treatment. Human experimental pain models and quantitative sensory testing (QST) can be used to profile the efficacy of new drugs, optimize the early drug development programs, and help predicting which pain patient populations would benefit and hence be included in early clinical trials (Arendt-Nielsen and Hoeck 2011; Hayes et al. 2014; Arendt-Nielsen and Curatolo 2013; Oertel and Lötsch 2013).

This chapter will (1) introduce some aspects of central sensitization in chronic pain, (2) provide a summary on which QST techniques can be used for assessing central sensitization in pain patients, and (3) describe how pharmacological interventions in pain patients can be profiled using QST techniques.

2   Central Sensitization in Chronic Pain Patients

The term central sensitization is a too broad term to use from a mechanistic point of view as "central" may refer to (1) ipsilateral sensitization associated with the local nociceptive focus, (2) segmental sensitization contralateral to the local nociceptive focus, (3) extraterritorial spreading sensitization around local nociceptive focus, or (4) generalized widespread sensitization. The development from local ipsilateral sensitization to a widespread sensitization represents a progression in the development of the chronic pain manifestations as many chronic pain conditions start with a local injury/trauma which over time develops into chronic widespread pain associated with widespread sensitization (e.g., fibromyalgia) (Lee et al. 2014; Sarzi-Puttini et al. 2011; Staud 2011).

This transition is felt by the patient as a spreading of pain from localized to a more diffuse and larger presentation, and it is found that 10–20 % of individuals with regional pain subsequently develop widespread pain (Sarzi-Puttini et al. 2011; Atzeni et al. 2011; Mourão et al. 2010).

In patients with low back pain, 24.5 % of patients develop widespread pain over 18 years (Lapossy et al. 1995), and studies of subjects with chronic neck pain or whiplash injuries have found that 10–22 % develop chronic widespread pain (Holm et al. 2007; Macfarlane 1999).

The underlying mechanism and the nature of the predisposing risk factors for developing widespread pain and sensitization have yet to be discovered although both neurophysiological and biosocial (psychological) explanations have been suggested (Sarzi-Puttini et al. 2011). Furthermore, genetic factors (Smith et al. 2011, 2012) and cognitive, social, and lifestyle factors (Mourão et al. 2010) may be important.

The picture can be further complicated as somatic pain cannot only cause widespread sensitization but also visceral hyperalgesia (Miranda et al. 2004) and vice versa (Giamberardino 2003).

2.1   Extraterritorial Manifestations of Sensitization

Bilateral presentation of pain symptoms occurred in about 5–15 % of patients with CRPS (Veldman and Goris 1996; Kozin et al. 1976), and another study showed a contralateral pain pattern in 49 % and diagonal pattern in 14 % of the CRPS patients (van Rijn et al. 2011) indicating the possible role of extraterritorial spreading.

There is ample clinical evidence that in neuropathic conditions, the signs and symptoms extend into regions beyond those directly innervated by the injured nerve (Konopka et al. 2012; Malan et al. 2000).

On the contrary to many of the above findings for postherpetic neuralgia (PHN), the pain remains localized with no contralateral effects on neurogenic inflammation (Baron and Saguer 1994) or facilitated capsaicin provoked pain (Petersen et al. 2000) suggesting PHN as a specific class of neuropathic pain. This may as such be problematic to use PHN as a generalized neuropathic pain model in drug screening studies and for drug approval studies. In patients with temporomandibular pain and atypical facial pain, the pain reaction often occurs bilaterally, and trigeminal neuropathic and inflammatory pain often spreads beyond the area of injury to involve nearby areas or even remote regions in which there may be hypersensitivity (Fillingim et al. 1998; John et al. 2003; Fernández-de-las-Peñas et al. 2010; Zakrzewska 2013).

During migraine attacks, lowered thresholds to mechanical and thermal stimulation of cephalic and extracephalic areas also may reflect extraterritorial hyperalgesia and allodynia (Burstein et al. 2000).

2.2 Widespread Manifestations of Sensitization

In patients, the pain duration, intensity, and number of pain locations are important for the potency of the spreading sensitization. In knee OA, the pain intensity/duration code for the degree of extrasegmental pressure hyperalgesia (Arendt-Nielsen et al. 2010), similarly in TMD (Fernández-de-las-Peñas et al. 2009), similarly for chronic tension-type headache (Fernández-de-las-Peñas et al. 2008). The number of OA locations are important for how diffuse the OA knee pain is perceived (Thompson et al. 2010), and the more myofascial trigger points exist, the more generalized is sensitization (Xu et al. 2010). Widespread sensitization is found in patients with pain after revision total knee arthroplasty indicating the importance of ongoing nociceptive input for the chronification process generation of widespread sensitization (Skou et al. 2013).

Patients with chronic neck and shoulder pain (whiplash) have also shown mechanical hypersensitivity in lower body regions (Koelbaek Johansen et al. 1999; Scott et al. 2005). It should also be noted that there are many comorbidities between different pain syndromes with high prevalence of, e.g., temporomandibular disorder in fibromyalgia (Leblebici et al. 2007) and chronic low back pain (Balasubramaniam et al. 2007).

Laursen et al. (2005) showed widespread hypersensitivity to pressure stimulation across different chronic musculoskeletal (fibromyalgia, whiplash, rheumatoid arthritis) and urogenital (endometriosis) pain conditions. Similar findings are also found in irritable bowel syndrome (Piché et al. 2010).

Several studies have shown that if the ongoing pain in whiplash patients is changed by local blocks, the generalized hypersensitivity to various experimental stimuli changes accordingly: reduced pain causes reduced generalized sensitization (Schneider et al. 2010; Herren-Gerber et al. 2004).

Similarly, it has been shown that the widespread sensitization in painful knee OA can be normalized after successful total knee replacement with no residual pain (Graven-Nielsen et al. 2012). Another manifestation of widespread hypersensitivity may be the number of palpable trigger points, and a significantly higher number of these points were found in lower limb muscles in OA patients (Bajaj et al. 2001a, b). The more widespread a musculoskeletal pain problem becomes, the more abnormal the quantitative sensory findings are (Carli et al. 2002).

Generalized spreading sensitization in fibromyalgia and whiplash patients is further supported by increased nociceptive withdrawal reflexes (Banic et al. 2004).

In migraine, spreading of multimodal allodynia and hyperalgesia beyond the locus of migraine headache is mediated by sensitized thalamic neurons that process nociceptive information from the cranial meninges together with sensory information from the skin of the scalp, face, body, and limbs (Burstein et al. 2010). Pain thresholds to mechanical and thermal stimulation of cephalic and extracephalic areas show significantly sensitization (hyperalgesia and allodynia) during migraine attacks (Burstein et al. 2000).

For other craniofacial pain conditions such as temporomandibular disorders, clinical pain may occur outside the craniofacial region (Fillingim et al. 1998; John et al. 2003).

There are some human volunteer studies showing that widespread hypersensitivity can be provoked. Ingestion of monosodium glutamate (MSG) has been given to volunteers over 5 days and provokes generalized pain sensitization (Shimada et al. 2013).

During systemic inflammation elicited by the administration of endotoxin to volunteers generalized sensitization developed to pressure pain, electrical pain, and cold pressor pain (de Goeij et al. 2013). The same model has shown to induce visceral hypersensitivity (Benson et al. 2012) assessed by sensory and pain thresholds and subjective pain ratings rectal distension by a barostat. Furthermore, low-dose endotoxin potentiates capsaicin-induced pain and evoked hyperalgesia and allodynia in volunteers (Hutchinson et al. 2013).

3   QST for Assessing Central Sensitization

QST involves a large variety of stimulus modalities (thermal, mechanical, chemical, electrical), assessment methods (psychophysics, electrophysiology, imaging, microdialysis), and target structures (skin, musculoskeletal, and viscera). QST can provide an understanding of the mechanisms involved in pain transduction, transmission, modulation, and perception under normal and pathophysiological conditions and as such contribute to mechanism-based diagnosis, prevention, and management of pain in future (Jensen and Baron 2002).

Different QST protocols have been suggested for profiling patients, and the QST battery developed by the German Research Network on Neuropathic Pain is the one applied in many studies (Geber et al. 2011; Magerl et al. 2010; Maier et al. 2010). Briefly, the protocol assesses the function of small (thermal thresholds) and large (tactile and vibration thresholds) nerve fiber pathways and increased/decreased pain sensitivity (hyperalgesia, allodynia, hyperpathia, wind-up-like pain). The battery consists predominantly of cutaneous stimulus modalities and is therefore not adequate for profiling musculoskeletal or visceral pain conditions.

Some of the measures are specifically addressing that central reorganization is associated with sensitization such as facilitated wind-up-like pain and allodynia. However, the battery does not include aspects related to, e.g., pain modulation. In most clinical studies assessing central hyperexcitability, there is a focus on "gain of function," but based on the many studies using the above platform, it has been evident that often the pain patients (in particular neuropathic pain) experience also modality-specific "loss of function" (Haanpää et al. 2011; Jensen and Baron 2002). The platform has also been used for sensory profiling conditions like restless legs syndrome (Stiasny-Kolster et al. 2013).

The present chapter will not address surrogate pain models of sensitization where transient central sensitization is induced in healthy volunteers (e.g., intradermal capsaicin) (Werner et al. 2013; Cavallone et al. 2013).

The focus of this section is to highlight the opportunities to assess quantitatively central sensitization in pain patients by QST techniques and, when possible, refer to clinical studies where pharmacological interventions have been profiles using QST in pain patients.

The problem of localized vs. general hyperalgesia: It is evident that modality-specific analgesia/hyperalgesia can be detected by different phasic experimental stimuli (mechanical, thermal). In case of a localized pain focus (e.g., nerve trauma, muscle/tendon inflammation, and joint pathology), the local peripheral hyperalgesia is sometimes evident when comparing the thresholds or pain responses to the contralateral location. However, in many cases, the contralateral location is not a control area due to the central manifestations, and the separation between peripheral and central sensitization can be difficult. An example could be pressure pain threshold assessment over the knee joint in patients with osteoarthritis (OA) where localized joint sensitization is detected. In the same patient, the pressure pain thresholds are also reduced as compared with control subjects when assessed from the arm (Arendt-Nielsen et al. 2010). If a control group is not possible to include in a study, the ratio can be calculated between a pain threshold assessed from the painful site and a threshold assessed from a non-painful site. When comparing the ratios within a given group of patients, the relative contribution between localized and generalized sensitization can be estimated and used for phenotyping patients.

Widespread sensory manifestations of sensitization are also found in, e.g., unilateral epicondylitis (Fernández Carnero et al. 2009a, b). There is ample evidence that in neuropathic conditions, the signs and symptoms also extend into regions beyond those directly innervated by the injured nerve (Konopka et al. 2012; Malan et al. 2000). In most chronic pain conditions, true control site is therefore most often nonexisting (Konopka et al. 2012). The way to overcome this problem is to use normative databases for all the QST tests applied (Neziri et al. 2011) and to use statistical techniques such as Z-scores to judge when an individual patient is outside the normative range (Rolke et al. 2006a, b). It is a very tedious job to establish such normative databases as thresholds vary from location to location and hence reference values from many points are required (Magerl et al. 2010; Neziri et al. 2011).

Not only normative data for many locations are needed but also reactions from different structures (skin, muscle, viscera) are required when a multimodal, multi-tissue QST approach is used.

A recent approach to overcome the localized stimulation using small thermal or mechanical probes has been to activate larger areas and more structures using the same stimulus. A larger volume can be assessed by the computer-controlled cuff-algometry technique. The pain reactions (threshold or rating) related to inflation of a tourniquet applied around an extremity (Polianskis et al. 2002). The pressure applied is distributed throughout the underlying tissues activating a variety of somatic receptors and nociceptors.

Facilitation of provoked central sensitization (Fig. 1): Experimental induction of central sensitization (hyperalgesia and allodynia) can be provoked by intradermal capsaicin. This model has been used to investigate the painful and non-painful legs of patients with unilateral sciatica and compared these with healthy controls. Pain and hyperalgesia responses were enhanced in both legs of patients with unilateral sciatica compared with healthy controls supporting the notion that patients with preexisting neuropathic pain have fundamental differences in central nervous system processing compared with pain-free controls (Aykanat et al. 2012).

The intradermal capsaicin models have also been used to show increased hyperalgesic reactions in patients with rheumatoid arthritis (Morris et al. 1997), fibromyalgia (Morris et al. 1998), and vulvodynia (Foster et al. 2005).

The capsaicin model has been applied to non-painful conditions such as patients with multiple chemical sensitivity and showed facilitated hyperalgesic reactions (Holst et al. 2011).

Transdermal electrical stimulation is another way to induce central sensitization, and larger hyperalgesic areas have been shown in patients with chronic lumbosacral pain including radicular neuropathic features as compared to controls (Lecybyl et al. 2010).

One study has used the capsaicin model in patients for pharmacological profiling. Patients with unilateral sciatica showed heightened responses to intradermal capsaicin compared to pain-free volunteers, and the study by Sumracki et al. (2012) compared the effects of pregabalin (300 mg) and the tetracycline antibiotic and glial attenuator minocycline (400 mg) on capsaicin-induced spontaneous pain, flare, and allodynia and between the affected and unaffected leg. It was concluded that a healthy control group was needed for such studies in order to show differences as a pain patient cannot be used as his/her own control.

Temporal summation and aftersensation (Fig. 2): An important and potent central mechanism in dorsal horn neurons is the temporal integration also termed wind-up-like pain. The initial phase of the wind-up process observed in animals translates into temporal summation in humans (Arendt-Nielsen et al. 1994). If a painful stimulus is repeated 1–3 times per second, the pain will integrate and become more painful with an intensity and stimulus frequency-dependent summation (Arendt-Nielsen et al. 1994). Temporal summation can be elicited using electrical, mechanical, or thermal stimulation modalities and is elicited from the skin, musculoskeletal structures, and viscera (Arendt-Nielsen 1997; Arendt-Nielsen and Yarnitsky 2009).

In many clinical conditions when repeated stimuli are applied in neuropathic, musculoskeletal, and visceral chronic pain patients, temporal summation is

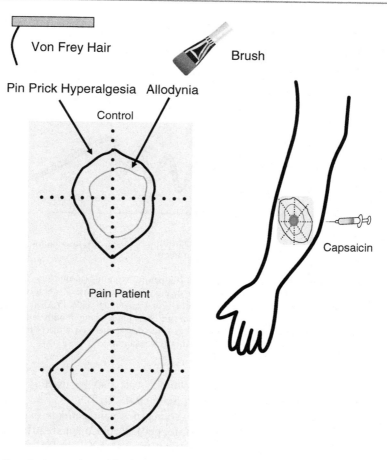

Fig. 1 Provoked central sensitization. Central sensitization can be provoked by intradermal injection of capsaicin. The boundary and hence the area of secondary pinprick hyperalgesia can be assessed by von Frey hair stimulation (change in intensity of the provoked pin prick pain). The boundary and hence the area of dynamic mechanical allodynia can be determined as the change in the perception provoked by brushing the skin from a non-painful sensation to a painful sensation. The areas of hyperalgesia and/or allodynia can be enlarged in chronic pain patients with central sensitization as an indication of increased central gain

substantially facilitated (Arendt-Nielsen et al. 1997a; Graven-Nielsen et al. 2000; Nikolajsen et al. 1996).

In clinical bed site testing, simple devices are used for assessing temporal summation such as tapping the skin with a nylon filament (Nikolajsen et al. 1996). When more standardization is required, however, automated user-independent methods are needed such as thermal (Kong et al. 2013), mechanical (Nie et al. 2009), or electrical stimulation techniques of the skin (Arendt-Nielsen et al. 1994), muscles (Arendt-Nielsen et al. 1997b), or viscera (Drewes et al. 1999).

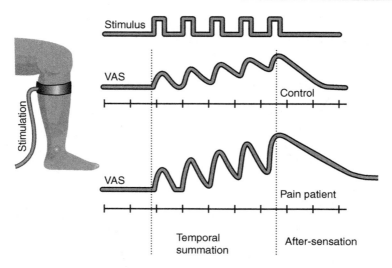

Fig. 2 Temporal summation and aftersensation. If a painful stimulus (in this case a cuff inflated around the leg) of the same intensity is repeated, e.g., five times with 2 s intervals, the pain intensity will gradually increase as assessed on a visual analogue scale (VAS). This is called temporal summation and is a central phenomenon. In a chronic pain patient with central sensitization, the temporal summation is facilitated, and in some cases, the pain will continue for some seconds after the stimulation has been terminated (aftersensation)

The abovementioned cuff-algometry technique can also be used if the cuff is repeatedly inflated (Lemming et al. 2012; Skou et al. 2013).

When repeated stimuli are delivered, sometimes pain patients experience an aftersensation (pain after the stimulus has stopped) (Robinson et al. 2010). This has been observed in patient with neuropathic (Gottrup et al. 2003) and musculoskeletal pain (Staud et al. 2003, 2007). An exclusive facilitation of the aftersensation alone has been proposed to be of diagnostic value (Sato et al. 2012) and supports the basic finding that the summation and the aftersensation are mediated by different underlying central mechanisms (Price et al. 1978; You et al. 2005).

Facilitated temporal summation in chronic pain patients is efficiently inhibited by NMDA receptor antagonists (ketamine and amantadine). This has been found in patients with surgical incisions (Stubhaug et al. 1997), postherpetic neuralgia (Eide et al. 1994), phantom limb pain (Nikolajsen et al. 1996), chronic postsurgical neuropathic pain (Pud et al. 1998), and fibromyalgia (Graven-Nielsen et al. 2000).

Abnormal temporal summation in patients with neuropathic pain does not predict the clinical effect of imipramine or gabapentin (Rasmussen et al. 2004).

Some clinical studies have not shown effect on facilitated temporal summation by lamotrigine (an antiepileptic drug acting on voltage-sensitive sodium channels) in patients with spinal cord injury pain (Finnerup et al. 2002) and by the NMDA-antagonist memantine in patients with phantom limb pain (Nikolajsen et al. 2000) and postherpetic neuralgia (Eisenberg et al. 1998).

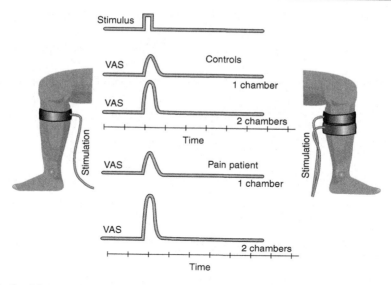

Fig. 3 Spatial summation. The pain intensity provoked by a single stimulus (in this case a cuff inflated around the leg) can be assessed on a VAS. If the area of the stimulation is increased (stimulating with two cuffs instead of one cuff), the pain intensity will increase. This is called spatial summation and is a central phenomenon. In a chronic pain patient with central sensitization, the spatial summation may be facilitated (increased ratio between the pain intensity provoked by the large area and the pain intensity provoked by the small area)

More clinical studies are needed to understand which compounds can modulate temporal summation.

Spatial summation (Fig. 3): Spatial summation is another mechanism, which relies on central networks and the general sensitization status (Bouhassira et al. 1995). In humans, spatial summation can be assessed in different ways where the stimulus is applied to using different stimulation areas, e.g., using thermodes (Nielsen and Arendt-Nielsen 1997), pressure probes (Nie et al. 2009), or cuff algometry (Polianskis et al. 2002). The cuff-algometry technology can utilize one or two cuffs which are automatically inflated, and the volunteer/patient rates the provoked pain intensity (Polianskis et al. 2002). Spatial summation is facilitated in various pain conditions such as fibromyalgia (Staud et al. 2004, 2007), osteoarthritis (Graven-Nielsen et al. 2012), and lateral epicondylitis (Jespersen et al. 2013).

No pharmacological screening studies have used this technique in pain patients.

Reflex receptive fields (Fig. 4): In many animal studies, expansion of receptive fields of dorsal horn neurons has been documented in neuropathic as well as inflammatory (cutaneous, muscle, and viscera) models. Expansion of receptive fields has been a challenging mechanism to assess in humans. In animals, an alternative method for assessment of receptive field expansion of dorsal horn neurons has been developed. This involves quantification of the so-called reflex

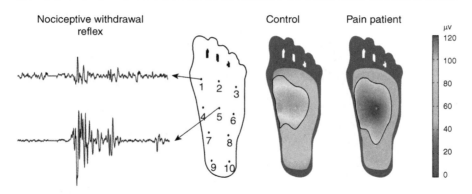

Fig. 4 Reflex receptive fields. Nociceptive withdrawal reflexes evoked by electrical stimulation from ten individual points on the sole of the foot. The reflex recorded from, e.g., the tibialis anterior muscle is large when evoked from, e.g., point 5 and smaller when elicited from, e.g., point 1. Based on the size of the reflex, the so-called reflex receptive field can be constructed and can be found larger in chronic pain patients with central sensitization

receptive field (Schouenborg et al. 1994, 1995) which was found to expand in the presence of central sensitization (Harris and Clarke 2003).

This method based on assessment of the nociceptive withdrawal method has been translated to humans (Andersen et al. 2001), and the reflex receptive field is found to be expanded into spinal cord injured patients (Andersen et al. 2004), in patients with chronic visceral pain (endometriosis) (Neziri et al. 2010), in patients with low back pain (Biurrun Manresa et al. 2013), and in patients with neck pain (Biurrun Manresa et al. 2013).

To date, few pharmacological studies have used this technique for profiling, and the sensitivity is not known. One study showed that the 5-HT-3 receptor antagonist had no effect on the expanded fields in chronic low patients and no effect on the clinical pain (Neziri et al. 2012). This advanced technique may in the future provide new information in clinical studies and for profiling of new compounds.

Descending pain modulation (Fig. 5): There are increasing evidences that the balance between the descending pain inhibition and facilitation may be disturbed in chronic pain and that this phenomenon has a role in maintaining central sensitization (Porreca et al. 2002; Suzuki et al. 2004; You et al. 2010). In humans, status assessment of the descending pathways has recently undergone a revival, and the original DNIC terminology has been renamed to conditioning pain modulation (CPM, Yarnitsky et al. 2010). Less efficient CPM seems to be a general phenomenon in chronic pain conditions (Staud 2012) such as patients with, e.g., myofascial temporomandibular joint pain (Bragdon et al. 2002), chronic low back pain (Peters et al. 1992), fibromyalgia (Kosek and Hansson 1997), whiplash (Daenen et al. 2014), painful OA (Arendt-Nielsen et al. 2010), and chronic tension-type headaches (Sandrini et al. 2006), chronic pancreatitis, and interstitial cystitis (Ness et al. 2014). OA patients with a deficient, descending pain inhibition show a normalization to a pain-free state after surgery (Graven-Nielsen et al. 2012; Kosek

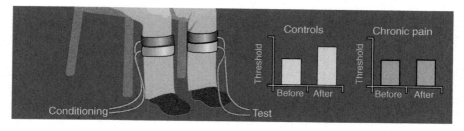

Fig. 5 Descending pain modulation. The descending pain modulation can be assessed by delivering a tonic conditioning pain stimulation to one location (in this case cuff stimulation of the leg), e.g., 2 min. A test stimulus is applied to another location, and the response (e.g., a pain threshold) is assessed before and during the conditioning stimulation. In healthy controls, the threshold increases during the conditioning stimulation (pain inhibition), but in chronic pain patients with central sensitization, this increase is smaller or even absent

and Ordeberg 2000) suggesting that the chronic pain maintained the CPM dysfunction and that the chronic pain saturated the CPM mechanism so the conditioning pain stimulus is less efficient.

The CPM paradigm has in recent years been implemented in pharmacological screening studies, and modulatory effects can be induced by dexmedetomidine (a selective $\alpha(2)$-adrenoceptor agonist) (Baba et al. 2012), buprenorphine (Arendt-Nielsen et al. 2012a, b), fentanyl (Arendt-Nielsen et al. 2012a, b), duloxetine (Yarnitsky et al. 2012), and apomorphine (a non-specific dopamine agonist) (Treister et al. 2013), whereas oxycodone seems not to modulate CPM (Suzan et al. 2013).

The use of CPM for assessing drug efficacy in pain patients has only been used in one study so far where the effect of duloxetine in painful diabetic neuropathy was investigated (Yarnitsky et al. 2012).

Evidently, an alteration in the descending pain modulation could be a promising target for pharmacological intervention (Arendt-Nielsen and Yarnitsky 2009) and should be utilized when new pain drugs are evaluated.

Offset analgesia (Fig. 6): Offset analgesia is provoked by incremental decreases of a nociceptive heat stimulus, and the perception of the provoked pain is less as what is provoked when the same stimulus intensity is given as a single stand-alone stimulus (Grill and Coghill 2002; Yelle et al. 2008). A peripheral component of offset analgesia cannot be excluded but is generally considered as a central inhibitory modulation of pain and as such should be modulated by central sensitization. This is supported by the fact that an association between CPM and offset analgesia has been shown indicating some commonalities of their underlying mechanisms (Honigman et al. 2013).

In healthy volunteers, offset analgesia is found stable in one study (Nilsson et al. 2013) but in another study found to be facilitated when assessed in repeated sessions. Furthermore, the response is reduced in potency with age (Naugle et al. 2013). Induction of thermal hyperalgesia has shown an association with increased magnitude of offset analgesia (Martucci et al. 2012a, b), whereas other

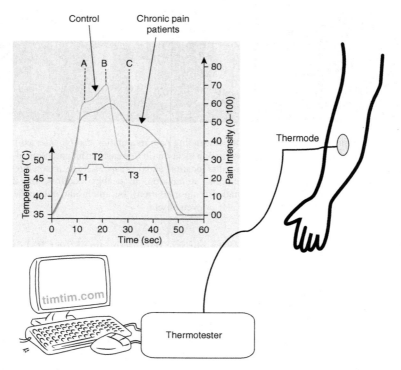

Fig. 6 Offset analgesia. The thermal stimulus paradigm used to evoke offset analgesia. The first phase (T1) consists of a painful heat stimulus applied for 5 s. The second phase (T2) is 1° above T1 also for 5 s. The last phase (T3) is the same temperature as T1 but for 20 s. During the stimulation, the person rates the pain intensity (0–100). The pain rating during T3 is lower than to T1 although the stimulus intensity is the same. This reduction in pain intensity (between *B* and *C*) is called offset analgesia. Chronic pain patients have a less potent offset analgesic response as compared with controls

studies have tried to modulate the potency of offset analgesia by experimental central sensitization without success (Martucci et al. 2012a, b), suggesting that longer-lasting central sensitization is needed.

Only one clinical study has used the offset analgesia paradigm in patients with neuropathic pain and found the response was reduced or absent (Niesters et al. 2011a, b). More clinical studies are needed to confirm the usefulness of this technique.

The offset analgesia is not modulated by intravenous ketamine (Niesters et al. 2011a, b) or morphine (Martucci et al. 2012a, b). More studies are needed to understand the role of this technique in pharmacological screening.

Referred pain (Fig. 7): Referred pain is a central phenomenon predominantly related to pain from deep somatic and visceral structures. Referred pain areas are facilitated as a result of a central sensitization in patients with, e.g., fibromyalgia (Sorensen et al. 1998), whiplash (Koelbaek Johansen et al. 1999), osteoarthritis

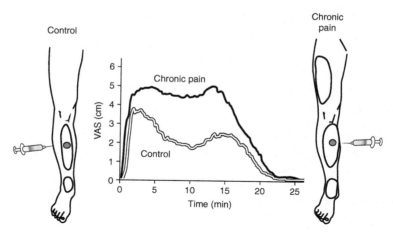

Fig. 7 Referred pain. Local and referred pain can be provoked by intramuscular injection of a bolus of an algogenic substance (e.g., 6 % hypertonic saline). Local pain area develops around the injection site and referred pain develops around a distal location in healthy controls. Referred pain is a central phenomenon. In chronic pain patients with central sensitization, the referred pain area is larger and can spread also to proximal locations. The intensity of the pain provoked is rated stronger by chronic pain patients with central sensitization as compared with controls

(Arendt-Nielsen and Yarnitsky 2009), chronic low back pain (O'Neill et al. 2007), irritable bowel syndrome (Swarbrick et al. 1980), and chronic pancreatitis (Dimcevski et al. 2007). In addition, it has been shown that referred pain areas evoked from muscles are enlarged in visceral pain conditions (Bajaj et al. 2003) as a result of convergence and interaction between sensitization related to deep somatic and visceral structures but also manifested as viscero-visceral sensitization (Giamberardino et al. 2010). Patients with, e.g., urinary calculosis show significant and long-lasting hyperalgesia in the referred pain areas (Giamberardino et al. 1994); patients suffering from a large number of colics show more hyperalgesia than those experiencing a limited number of colics. Similar somatic hyperalgesia is found in referred pain areas in acute appendicitis (Stawowy et al. 2002) and cholecystolithiasis (Stawowy et al. 2005).

Few studies have used the referred pain areas in pain patients for pharmacological profiling. Ketamine (Arendt-Nielsen and Graven-Nielsen 2008), gabapentin (Arendt-Nielsen et al. 2007), and a 5-HT antagonist (Christidis et al. 2008) are found to reduce the area of referred pain. On the contrary, alfentanil and morphine do not reduce referred pain (Schulte et al. 2006).

4 Conclusion and Future Perspectives

Central sensitization can manifest in many different ways, and it is evident that a localized nociceptive focus over time can affect the entire neuroaxis. Development of new, safe, centrally acting compounds targeting central sensitization has not been successful, and new strategies have been taking towards the peripheral target and towards the nonneuronal components.

A number of mechanism-based QST approaches have been developed to assess the potency of the central sensitization mechanisms. Sensory testing can be used to profile chronic pain patients and as a screening tool in a more efficient development of new analgesic compounds.

Profiling chronic pain patients by mechanistic QST in conjunction with pharmacological treatment opportunities may in the future provide the basis for individualized, targeted pain management.

Conflict of Interest No conflict of interest can be reported.

References

Andersen OK, Sonnenborg FA, Arendt-Nielsen L (2001) Reflex receptive fields for human withdrawal reflexes elicited by non-painful and painful electrical stimulation of the foot sole. Clin Neurophysiol 112(4):641–649

Andersen OK, Finnerup NB, Spaich EG, Jensen TS, Arendt-Nielsen L (2004) Expansion of nociceptive withdrawal reflex receptive fields in spinal cord injured humans. Clin Neurophysiol 115(12):2798–2810

Arendt-Nielsen L (1997) Induction and assessment of experimental pain from human skin, muscle and viscera. In: Jensen TS, Turner JA, Wiesenfeld-Hallin Z (eds) Proceedings of the 8th World Congress on pain, Vancouver, Canada, August 17–22: progress in pain research and management, vol 8. IASP Press, Seattle, pp 393–425

Arendt-Nielsen L, Curatolo M (2013) Mechanistic, translational, quantitative pain assessment tools in profiling of pain patients and for development of new analgesic compounds. Scand J Pain 4(4):226–230

Arendt-Nielsen L, Hoeck HC (2011) Optimizing the early phase development of new analgesics by human pain biomarkers. Expert Rev Neurother 11(11):1631–1651

Arendt-Nielsen L, Yarnitsky D (2009) Experimental and clinical applications of quantitative sensory testing applied to skin, muscles and viscera. J Pain 10(6):556–572

Arendt-Nielsen L, Brennum J, Sindrup SH, Bak P (1994) Electrophysiological and psychophysical quantification of temporal summation in the human nociceptive system. Eur J Appl Physiol Occup Physiol 68:266–273

Arendt-Nielsen L, Drewes AM, Hansen JB, Tage-Jensen U (1997a) Gut pain reactions in man: an experimental investigation using short and long duration transmucosal electrical stimulation. Pain 69(3):255–262

Arendt-Nielsen L, Graven-Nielsen T, Svensson P, Jensen TS (1997b) Temporal summation in muscles and referred pain areas: an experimental human study. Muscle Nerve 20(10):1311–1313

Arendt-Nielsen L, Frokjaer JB, Staahl C, Graven-Nielsen T, Huggins JP, Smart TS, Drewes AM (2007) Effects of gabapentin on experimental somatic pain and temporal summation. Reg Anesth Pain Med 2(5):382–388

Arendt-Nielsen L, Graven-Nielsen T (2008) Muscle pain: sensory implications and interaction with motor control. Clin J Pain 24(4):291–298

Arendt-Nielsen L, Nie H, Laursen MB, Laursen BS, Madeleine P, Simonsen OH, Graven-Nielsen T (2010) Sensitization in patients with painful knee osteoarthritis. Pain 149(3):573–581

Arendt-Nielsen L, Andresen T, Malver LP, Oksche A, Mansikka H, Drewes AM (2012a) A double-blind, placebo-controlled study on the effect of buprenorphine and fentanyl on descending pain modulation: a human experimental study. Clin J Pain 28(7):623–627

Arendt-Nielsen L, Graven-Nielsen T, Petrini L (2012b) Experimental human models and assessment of pain in non-pain conditions. In: Giamberardino MA, Jensen TS (eds) Pain comorbidities: understanding and treating the complex patient. IASP Press, Seattle, Chapter 3

Atzeni F, Cazzola M, Benucci M, Di Franco M, Salaffi F, Sarzi-Puttini P (2011) Chronic widespread pain in the spectrum of rheumatological diseases. Best Pract Res Clin Rheumatol 25(2):165–171

Aykanat V, Gentgall M, Briggs N, Williams D, Yap S, Rolan P (2012) Intradermal capsaicin as a neuropathic pain model in patients with unilateral sciatica. Br J Clin Pharmacol 73(1):37–45

Baba Y, Kohase H, Oono Y, Fujii-Abe K, Arendt-Nielsen L (2012) Effects of dexmedetomidine on conditioned pain modulation in humans. Eur J Pain 16(8):1137–1147

Bajaj P, Bajaj P, Graven-Nielsen T, Arendt-Nielsen L (2001a) Osteoarthritis and its association with muscle hyperalgesia: an experimental controlled study. Pain 93(2):107–114

Bajaj P, Graven-Nielsen T, Arendt-Nielsen L (2001b) Trigger points in patients with lower limb osteoarthritis. J Musculoskelet Pain 9:17–33

Bajaj P, Bajaj P, Madsen H, Arendt-Nielsen L (2003) Endometriosis is associated with central sensitization: a psychophysical controlled study. J Pain 4(7):372–380

Balasubramaniam R, de Leeuw R, Zhu H, Nickerson RB, Okeson JP, Carlson CR (2007) Prevalence of temporomandibular disorders in fibromyalgia and failed back syndrome patients: a blinded prospective comparison study. Oral Surg Oral Med Oral Pathol Oral Radiol Endod 104:204–216

Banic B, Petersen-Felix S, Andersen OK, Radanov BP, Villiger PM, Arendt-Nielsen L, Curatolo M (2004) Evidence for spinal cord hypersensitivity in chronic pain after whiplash injury and in fibromyalgia. Pain 107:7–15

Baron R, Saguer M (1994) Axon-reflex reactions in affected and homologous contralateral skin after unilateral peripheral injury of thoracic segmental nerves in humans. Neurosci Lett 165 (1–2):97–100

Basbaum AI, Bautista DM, Scherrer G, Julius D (2009) Cellular and molecular mechanisms of pain. Cell 139:267–284

Benson S, Kattoor J, Wegner A, Hammes F, Reidick D, Grigoleit JS, Engler H, Oberbeck R, Schedlowski M, Elsenbruch S (2012) Acute experimental endotoxemia induces visceral hypersensitivity and altered pain evaluation in healthy humans. Pain 153(4):794–799

Biurrun Manresa JA, Neziri AY, Curatolo M, Arendt-Nielsen L, Andersen OK (2013) Reflex receptive fields are enlarged in patients with musculoskeletal low back and neck pain. Pain 154 (8):1318–1324

Bouhassira D, Gall O, Chitour D, Le Bars D (1995) Dorsal horn convergent neurones: negative feedback triggered by spatial summation of nociceptive afferents. Pain 62(2):195–200

Bragdon EE, Light KC, Costello NL, Sigurdsson A, Bunting S, Bhalang K, Maixner W (2002) Group differences in pain modulation: pain-free women compared to pain-free men and to women with TMD. Pain 96:227–237

Burstein R, Yarnitsky D, Goor-Aryeh I et al (2000) An association between migraine and cutaneous allodynia. Ann Neurol 47:614–624

Burstein R, Jakubowski M, Garcia-Nicas E, Kainz V, Bajwa Z, Hargreaves R, Becerra L, Borsook D (2010) Thalamic sensitization transforms localized pain into widespread allodynia. Ann Neurol 68(1):81–91

Carli G, Suman AL, Biasi G, Marcolongo R (2002) Reactivity to superficial and deep stimuli in patients with chronic musculoskeletal pain. Pain 100:259–269

Cavallone LF, Frey K, Montana MC, Joyal J, Regina KJ, Petersen KL, Gereau RW IV (2013) Reproducibility of the heat/capsaicin skin sensitization model in healthy volunteers. J Pain Res 6:771–784

Christidis N, Ioannidou K, Milosevic M, Segerdahl M, Ernberg M (2008) Changes of hypertonic saline-induced masseter muscle pain characteristics, by an infusion of the serotonin receptor type 3 antagonist granisetron. J Pain 9(10):892–901

Crow M, Denk F, McMahon SB (2013) Genes and epigenetic processes as prospective pain targets. Genome Med 5(2):12

Daenen L, Nijs J, Cras P, Wouters K, Roussel N (2014) Changes in pain modulation occur soon after whiplash trauma but are not related to altered perception of distorted visual feedback. Pain Pract 14(7):588–598

Davis MA, Ettinger WH, Neuhaus JM, Barclay JD, Segal MR (1992) Correlates of knee pain among US adults with and without radiographic knee osteoarthritis. J Rheumatol 19 (12):1943–1949

de Goeij M, van Eijk LT, Vanelderen P, Wilder-Smith OH, Vissers KC, van der Hoeven JG, Kox M, Scheffer GJ, Pickkers P (2013) Systemic inflammation decreases pain threshold in humans in vivo. PLoS One 8(12):e84159

Dimcevski G, Staahl C, Andersen SD, Thorsgaard N, Funch-Jensen P, Arendt-Nielsen L et al (2007) Assessment of experimental pain from skin, muscle, and esophagus in patients with chronic pancreatitis. Pancreas 35(1):22–29

Drewes AM, Petersen P, Qvist P, Nielsen J, Arendt-Nielsen L (1999) An experimental pain model based on electric stimulations of the colon mucosa. Scand J Gastroenterol 34(8):765–771

Eide PK, Jørum E, Stubhaug A, Bremnes J, Breivik H (1994) Relief of post-herpetic neuralgia with the N-methyl-D-aspartic acid receptor antagonist ketamine. A doubleblind, cross-over comparison with morphine and placebo. Pain 58:347–354

Eisenberg E, Kleiser A, Dortort A, Haim T, Yarnitsky D (1998) The NMDA (N-methyl-D-aspartate) receptor antagonist memantine in the treatment of postherpetic neuralgia: a double-blind, placebo-controlled study. Eur J Pain 2(4):321–327

Fernández Carnero J, Fernández-de-las-Peñas C, Sterling M, Souvlis T, Arendt-Nielsen L, Vicenzino B (2009a) Exploration of the extent of somato-sensory impairment in patients with unilateral lateral epicondylalgia. J Pain 10(11):1179–1185

Fernández Carnero J, Fernández-de-las-Peñas C, De La Llave Rincón AI, Ge HY, Arendt-Nielsen L (2009b) Widespread mechanical pain hyper-sensitivity as sign of central sensitization in unilateral lateral epicondylalgia: a blinded, controlled study. Clin J Pain 2(7):555–561

Fernández-de-las-Peñas C, Ge HY, Cuadrado ML, Madeleine P, Pareja JA, Arendt-Nielsen L (2008) Bilateral pressure pain sensitivity mapping of the temporalis muscle in chronic tension type headache. Headache 48(8):1067–1075

Fernández-de-las-Peñas C, Galán del Río F, Fernández Carnero J, Pesquera J, Arendt-Nielsen L, Svensson P (2009) Bilateral widespread mechanical pain sensitivity in myofascial temporomandibular disorder: evidence of impairment in central nociceptive processing. J Pain 10 (11):1170–1178

Fernández-de-las-Peñas C, Galan-del-Rio F, Ortega-Santiago R, Jimenez-Garcia R, Arendt-Nielsen L, Svensson P (2010) Bilateral thermal hyperalgesia in trigeminal and extra-trigeminal regions in patients with myofascial temporomandibular disorders. Exp Brain Res 202:171–179

Fillingim RB, Fillingim LA, Hollins M, Sigurdsson A, Maixner W (1998) Generalized vibrotactile allodynia in a patient with temporomandibular disorder. Pain 78:75–78

Finnerup NB, Sindrup SH, Bach FW, Johannesen IL, Jensen TS (2002) Lamotrigine in spinal cord injury pain: a randomized controlled trial. Pain 96(3):375–383

Foster DC, Dworkin RH, Wood RW (2005) Effects of intradermal foot and forearm capsaicin injections in normal and vulvodynia-afflicted women. Pain 117:128–136

Geber C, Klein T, Azad S, Birklein F, Gierthmühlen J, Huge V, Lauchart M, Nitzsche D, Stengel M, Valet M, Baron R, Maier C, Tölle T, Treede RD (2011) Test-retest and interobserver reliability of quantitative sensory testing according to the protocol of the German Research Network on Neuropathic Pain (DFNS): a multi-centre study. Pain 152(3):548–556

Giamberardino MA (2003) Referred muscle pain/hyperalgesia and central sensitisation. J Rehabil Med 41(Suppl):85–88

Giamberardino MA, de Bigontina P, Martegiani C, Vecchiet L (1994) Effects of extracorporeal shock-wave lithotripsy on referred hyperalgesia from renal/ureteral calculosis. Pain 56(1):77–83

Giamberardino MA, Costantini R, Affaitati G, Fabrizio A, Lapenna D, Tafuri E, Mezzetti A (2010) Viscero-visceral hyperalgesia: characterization in different clinical models. Pain 151(2):307–322

Gottrup H, Kristensen AD, Bach FW, Jensen TS (2003) Aftersensations in experimental and clinical hypersensitivity. Pain 103(1–2):57–64

Graven-Nielsen T, Aspegren Kendall S, Henriksson KG, Bengtsson M, Sörensen J, Johnson A, Gerdle B, Arendt-Nielsen L (2000) Ketamine reduces muscle pain, temporal summation, and referred pain in fibromyalgia patients. Pain 85(3):483–491

Graven-Nielsen T, Wodehouse T, Langford RM, Arendt-Nielsen L, Kidd BL (2012) Normalization of widespread hyperesthesia and facilitated spatial summation of deep-tissue pain in knee osteoarthritis patients after knee replacement. Arthritis Rheum 64(9):2907–2916

Grill JD, Coghill RC (2002) Transient analgesia evoked by noxious stimulus offset. J Neurophysiol 87:2205–2208

Haanpää M (2011) Are neuropathic pain screening tools useful for patients with spinal cord injury? Pain 152:715–716

Hannan MT, Felson DT, Pincus T (2000) Analysis of the discordance between radiographic changes and knee pain in osteoarthritis of the knee. J Rheumatol 27(6):1513–1517

Harris J, Clarke RW (2003) Organisation of sensitisation of hind limb withdrawal reflexes from acute noxious stimuli in the rabbit. J Physiol 546:251–265

Hayes AG, Arendt-Nielsen L, Tate S (2014) Multiple mechanisms have been tested in pain—how can we improve the chances of success? Curr Opin Pharmacol 14:11–17

Herren-Gerber R, Weiss S, Arendt-Nielsen L, Petersen-Felix S, Di Stefano G, Radanov BP, Curatolo M (2004) Modulation of central hypersensitivity by nociceptive input in chronic pain after whiplash injury. Pain Med 5(4):366–376

Holm LW, Carroll LJ, Cassidy JD, Skillgate E, Ahlbom A (2007) Widespread pain following whiplash-associated disorders: incidence, course, and risk factors. J Rheumatol 34:193–200

Holst H, Arendt-Nielsen L, Mosbech H, Elberling J (2011) Increased capsaicin-induced secondary hyperalgesia in patients with multiple chemical sensitivity. Clin J Pain 27(2):156–162

Honigman L, Yarnitsky D, Sprecher E, Weissman-Fogel I (2013) Psychophysical testing of spatial and temporal dimensions of endogenous analgesia: conditioned pain modulation and offset analgesia. Exp Brain Res 228(4):493–501

Hutchinson MR, Buijs M, Tuke J, Kwok YH, Gentgall M, Williams D, Rolan P (2013) Low-dose endotoxin potentiates capsaicin-induced pain in man: evidence for a pain neuroimmune connection. Brain Behav Immun 30:3–11

Jensen TS, Baron R (2002) Translation of symptoms and signs into mechanisms in neuropathic pain: topical review. Pain 102:1–8

Jespersen A, Amris K, Graven-Nielsen T, Arendt-Nielsen L, Bartels EM, Torp-Pedersen S, Bliddal H, Danneskiold-Samsoe B (2013) Assessment of pressure-pain thresholds and central sensitization of pain in lateral epicondylalgia. Pain Med 14(2):297–304

John MT, Miglioretti DL, LeResche L, Von Korff M, Critchlow CW (2003) Widespread pain as a risk factor for dysfunctional temporomandibular disorder pain. Pain 102:257–263

Koelbaek Johansen M, Graven-Nielsen T, Olesen AS, Arendt-Nielsen L (1999) Generalised muscular hyperalgesia in chronic whiplash syndrome. Pain 83(2):229–234

Kong JT, Johnson KA, Balise RR, Mackey S (2013) Test-retest reliability of thermal temporal summation using an individualized protocol. J Pain 14(1):79–88

Konopka KH, Harbers M, Houghton A, Kortekaas R, van Vliet A, Timmerman W, den Boer JA, Struys MM, van Wijhe M (2012) Bilateral sensory abnormalities in patients with unilateral neuropathic pain: a quantitative sensory testing (QST) study. PLoS One 7(5):e37524

Kosek E, Hansson P (1997) Modulatory influence on somatosensory perception from vibration and heterotopic noxious conditioning stimulation (HNCS) in fibromyalgia patients and healthy subjects. Pain 70:41–51

Kosek E, Ordeberg G (2000) Lack of pressure pain modulation by heterotopic noxious conditioning stimulation in patients with painful osteoarthritis before, but not following, surgical pain relief. Pain 88:69–78

Kozin F, McCarty SJ, Sims J, Genant H (1976) The reflex sympathetic dystrophy syndrome. Am J Med 60:320–337

Lapossy E, Maleitzke R, Hrycaj P, Mennet W, Muller W (1995) The frequency of transition of chronic low back pain to fibromyalgia. Scand J Rheumatol 24:29–33

Laursen BS, Bajaj P, Olesen AS, Delmar C, Arendt-Nielsen L (2005) Health related quality of life and quantitative pain measurement in females with chronic non-malignant pain. Eur J Pain 9:267–275

Leblebici B, Pektas ZO, Ortancil O, Hurcan EC, Bagis S, Akman MN (2007) Coexistence of fibromyalgia, temporomandibular disorder, and masticatory myofascial pain syndromes. Rheumatol Int 7:541–544

Lecybyl R, Acosta J, Ghoshdastidar J, Stringfellow K, Hanna M (2010) Validation, reproducibility and safety of trans dermal electrical stimulation in chronic pain patients and healthy volunteers. BMC Neurol 10:5

Lee J, Ellis B, Price C, Baranowski AP (2014) Chronic widespread pain, including fibromyalgia: a pathway for care developed by the British Pain Society. Br J Anaesth 112(1):16–24

Lemming D, Graven-Nielsen T, Sörensen J, Arendt-Nielsen L, Gerdle B (2012) Widespread pain hypersensitivity and facilitated temporal summation of deep tissue pain in whiplash associated disorder: an explorative study of women. J Rehabil Med 44(8):648–657

Macfarlane GJ (1999) Generalized pain, fibromyalgia and regional pain: an epidemiological view. Baillieres Best Pract Res Clin Rheumatol 13:403–414

Magerl W, Krumova EK, Baron R, Tölle T, Treede RD, Maier C (2010) Reference data for quantitative sensory testing (QST): refined stratification for age and a novel method for statistical comparison of group data. Pain 151(3):598–605

Maier C, Baron R, Tölle TR, Binder A, Birbaumer N, Birklein F, Gierthmühlen J, Flor H, Geber C, Huge V, Krumova EK, Landwehrmeyer GB, Magerl W, Maihöfner C, Richter H, Rolke R, Scherens A, Schwarz A, Sommer C, Tronnier V, Uçeyler N, Valet M, Wasner G, Treede RD (2010) Quantitative sensory testing in the German Research Network on Neuropathic Pain (DFNS): somatosensory abnormalities in 1236 patients with different neuropathic pain syndromes. Pain 150(3):439–450

Malan TP, Ossipov MH, Gardell LR, Ibrahim M, Bian D, Lai J, Porreca F (2000) Extraterritorial neuropathic pain correlates with multisegmental elevation of spinal dynorphin in nerve-injured rats. Pain 86(1–2):185–194

Martucci KT, Eisenach JC, Tong C, Coghill RC (2012a) Opioid-independent mechanisms supporting offset analgesia and temporal sharpening of nociceptive information. Pain 153(6):1232–1243

Martucci KT, Yelle MD, Coghill RC (2012b) Differential effects of experimental central sensitization on the time-course and magnitude of offset analgesia. Pain 153(2):463–472

Milligan ED, Watkins LR (2009) Pathological and protective roles of glia in chronic pain. Nat Rev Neurosci 10:23–36

Miranda A, Peles S, Rudolph C, Shaker R, Sengupta JN (2004) Altered visceral sensation in response to somatic pain in the rat. Gastroenterology 126(4):1082–1089

Morris VH, Cruwys SC, Kidd BL (1997) Characterisation of capsaicin-induced mechanical hyperalgesia as a marker for altered nociceptive processing in patients with rheumatoid arthritis. Pain 71:179–186

Morris V, Cruwys S, Kidd B (1998) Increased capsaicin-induced secondary hyperalgesia as a marker of abnormal sensory activity in patients with fibromyalgia. Neurosci Lett 250(3):205–207

Mourão AF, Blyth FM, Branco JC (2010) Generalised musculoskeletal pain syndromes. Best Pract Res Clin Rheumatol 24(6):829–840

Naugle KM, Cruz-Almeida Y, Fillingim RB, Riley JL III (2013) Offset analgesia is reduced in older adults. Pain 154(11):2381–2387

Neogi T, Felson D, Niu J, Nevitt M, Lewis CE, Aliabadi P, Sack B, Torner J, Bradley L, Zhang Y (2009) Association between radiographic features of knee osteoarthritis and pain: results from two cohort studies. BMJ 339:b2844

Ness TJ, Lloyd LK, Fillingim RB (2014) An endogenous pain control system is altered in subjects with interstitial cystitis. J Urol 191(2):364–370

Neziri AY, Haesler S, Petersen-Felix S, Müller M, Arendt-Nielsen L, Manresa JB, Andersen OK, Curatolo M (2010) Generalized expansion of nociceptive reflex receptive fields in chronic pain patients. Pain 151(3):798–805

Neziri AY, Scaramozzino P, Andersen OK, Dickenson AH, Arendt-Nielsen L, Curatolo M (2011) Reference values of mechanical and thermal pain tests in a pain-free population. Eur J Pain 15(4):376–383

Neziri AY, Dickenmann M, Scaramozzino P, Andersen OK, Arendt-Nielsen L, Dickenson AH, Curatolo M (2012) Effect of intravenous tropisetron on modulation of pain and central hypersensitivity in chronic low back pain patients. Pain 153(2):311–318

Nie H, Graven-Nielsen T, Arendt-Nielsen L (2009) Spatial and temporal summation of pain evoked by mechanical pressure stimulation. Eur J Pain 13(6):592–599

Nielsen J, Arendt-Nielsen L (1997) Spatial summation of heat induced pain within and between dermatomes. Somatosens Mot Res 14(2):119–125

Niesters M, Hoitsma E, Sarton E, Aarts L, Dahan A (2011a) Offset analgesia in neuropathic pain patients and effect of treatment with morphine and ketamine. Anesthesiology 115(5):1063–1071

Niesters M, Dahan A, Swartjes M, Noppers I, Fillingim RB, Aarts L, Sarton EY (2011b) Effect of ketamine on endogenous pain modulation in healthy volunteers. Pain 152(3):656–663

Nikolajsen L, Hansen CL, Nielsen J, Keller J, Arendt-Nielsen L, Jensen TS (1996) The effect of ketamine on phantom pain: a central neuropathic disorder maintained by peripheral input. Pain 67(1):69–77

Nikolajsen L, Gottrup H, Kristensen AG, Jensen TS (2000) Memantine (a N-methyl-D-aspartate receptor antagonist) in the treatment of neuropathic pain after amputation or surgery: a randomized, double-blinded, cross-over study. Anesth Analg 91(4):960–966

Nilsson M, Piasco A, Nissen TD, Graversen C, Gazerani P, Lucas MF, Dahan A, Drewes AM, Brock C (2013) Reproducibility of psychophysics and electroencephalography during offset analgesia. Eur J Pain 18(6):824–834

Nishimura I, Thakor D, Lin A, Ruangsri S, Spigelman I (2010) Molecular strategies for therapeutic targeting of primary sensory neurons in chronic pain syndromes. In: Kruger L, Light AR (eds) Translational pain research: from mouse to man. CRC, Boca Raton, Chapter 6

O'Neill S, Manniche C, Graven-Nielsen T, Arendt-Nielsen L (2007) Generalized deep-tissue hyperalgesia in patients with chronic low-back pain. Eur J Pain 11(4):415–420

Oertel BG, Lötsch J (2013) Clinical pharmacology of analgesics assessed with human experimental pain models: bridging basic and clinical research. Br J Pharmacol 168(3):534–553

Peters ML, Schmidt AJ, Van den Hout MA, Koopmans R, Sluijter ME (1992) Chronic back pain, acute postoperative pain and the activation of diffuse noxious inhibitory controls (DNIC). Pain 50:177–187

Petersen KL, Fields HL, Brennum J, Sandroni P, Rowbotham MC (2000) Capsaicin evoked pain and allodynia in post-herpetic neuralgia. Pain 88:125–133

Piché M, Arsenault M, Poitras P, Rainville P, Bouin M (2010) Widespread hypersensitivity is related to altered pain inhibition processes in irritable bowel syndrome. Pain 148(1):49–58

Polianskis R, Graven-Nielsen T, Arendt-Nielsen L (2002) Spatial and temporal aspects of deep tissue pain assessed by cuff algometry. Pain 100:19–26

Porreca F, Ossipov MH, Gebhart GF (2002) Chronic pain and medullary descending facilitation. Trends Neurosci 25:319–325

Price DD, Hayes RL, Ruda M, Dubner R (1978) Neural representation of cutaneous aftersensations by spinothalamic tract neurons. Fed Proc 37(9):2237–2239

Pud D, Eisenberg E, Spitzer A, Adler R, Fried G, Yarnitsky D (1998) The NMDA receptor antagonist amantadine reduces surgical neuropathic pain in cancer patients: a double blind, randomized, placebo controlled trial. Pain 75:349–354

Rasmussen PV, Sindrup SH, Jensen TS, Bach FW (2004) Therapeutic outcome in neuropathic pain: relationship to evidence of nervous system lesion. Eur J Neurol 11(8):545–553

Richards N, McMahon SB (2013) Targeting novel peripheral mediators for the treatment of chronic pain. Br J Anaesth 111(1):46–51

Robinson ME, Bialosky JE, Bishop MD, Price DD, George SZ (2010) Supra-threshold scaling, temporal summation, and after-sensation: relationships to each other and anxiety/fear. J Pain Res 3:25–32

Rolke R, Baron R, Maier C, Tölle TR, Treede RD, Beyer A, Binder A, Birbaumer N, Birklein F, Bötefür IC, Braune S, Flor H, Huge V, Klug R, Landwehrmeyer GB, Magerl W, Maihöfner C, Rolko C, Schaub C, Scherens A, Sprenger T, Valet M, Wasserka B (2006a) Quantitative sensory testing in the German Research Network on Neuropathic Pain (DFNS): standardized protocol and reference values. Pain 123(3):231–243

Rolke R, Magerl W, Campbell KA, Schalber C, Caspari S, Birklein F, Treede RD (2006b) Quantitative sensory testing: a comprehensive protocol for clinical trials. Eur J Pain 10(1):77–88

Sandkühler J (2009) Models and mechanisms of hyperalgesia and allodynia. Physiol Rev 89(2):707–758

Sandrini G, Rossi P, Milanov I, Serrao M, Cecchini AP, Nappi G (2006) Abnormal modulatory influence of diffuse noxious inhibitory controls in migraine and chronic tension-type headache patients. Cephalalgia 26:782–789

Sarzi-Puttini P, Atzeni F, Mease PJ (2011) Chronic widespread pain: from peripheral to central evolution. Best Pract Res Clin Rheumatol 25(2):133–139

Sato H, Saisu H, Muraoka W, Nakagawa T, Svensson P, Wajima K (2012) Lack of temporal summation but distinct aftersensations to thermal stimulation in patients with combined tension-type headache and myofascial temporomandibular disorder. J Orofac Pain 26(4):288–295

Schliessbach J, Siegenthaler A, Streitberger K, Eichenberger U, Nüesch E, Jüni P, Arendt-Nielsen L, Curatolo M (2013) The prevalence of widespread central hypersensitivity in chronic pain patients. Eur J Pain 17(10):1502–1510

Schneider GM, Smith AD, Hooper A, Stratford P, Schneider KJ, Westaway MD, Frizzell B, Olson L (2010) Minimizing the source of nociception and its concurrent effect on sensory hypersensitivity: an exploratory study in chronic whiplash patients. BMC Musculoskelet Disord 11:29

Schomberg D, Olson JK (2012) Immune responses of microglia in the spinal cord: contribution to pain states. Exp Neurol 234:262–270

Schouenborg J, Weng HR, Holmberg H (1994) Modular organization of spinal reflexes: a new hypothesis. NIPS 9:261–265

Schouenborg J, Weng H-R, Kalliomäki J, Holmberg H (1995) A survey of spinal dorsal horn neurones encoding the spinal organization of withdrawal reflexes in the rat. Exp Brain Res 106:19–27

Schulte H, Segerdahl M, Graven-Nielsen T, Grass S (2006) Reduction of human experimental muscle pain by alfentanil and morphine. Eur J Pain 10(8):733–741

Scott D, Jull G, Sterling M (2005) Widespread sensory hypersensitivity is a feature of chronic whiplash-associated disorder but not chronic idiopathic neck pain. Clin J Pain 21:175–181

Shimada A, Cairns BE, Vad N, Ulriksen K, Pedersen AM, Svensson P, Baad-Hansen L (2013) Headache and mechanical sensitization of human pericranial muscles after repeated intake of monosodium glutamate (MSG). J Headache Pain 14(1):2

Skou ST, Graven-Nielsen T, Rasmussen S, Simonsen OH, Laursen MB, Arendt-Nielsen L (2013) Widespread sensitization in patients with chronic pain after revision total knee arthroplasty. Pain 154(9):1588–1594

Smith SB, Maixner DW, Greenspan JD, Dubner R, Fillingim RB, Ohrbach R, Knott C, Slade GD, Bair E, Gibson DG, Zaykin DV, Weir BS, Maixner W, Diatchenko L (2011) Potential genetic risk factors for chronic TMD: genetic associations from the OPPERA case control study. J Pain 12(11 Suppl):T92–T101

Smith SB, Maixner DW, Fillingim RB, Slade G, Gracely RH, Ambrose K, Zaykin DV, Hyde C, John S, Tan K, Maixner W, Diatchenko L (2012) Large candidate gene association study reveals genetic risk factors and therapeutic targets for fibromyalgia. Arthritis Rheum 64(2):584–593

Sorensen J, Graven-Nielsen T, Henriksson KG, Bengtsson M, Arendt-Nielsen L (1998) Hyperexcitability in fibromyalgia. J Rheumatol 25(1):152–155

Staud R (2011) Peripheral pain mechanisms in chronic widespread pain. Best Pract Res Clin Rheumatol 25(2):155–164

Staud R (2012) Abnormal endogenous pain modulation is a shared characteristic of many chronic pain conditions. Expert Rev Neurother 12(5):577–585

Staud R, Cannon RC, Mauderli AP, Robinson ME, Price DD, Vierck CJ Jr (2003) Temporal summation of pain from mechanical stimulation of muscle tissue in normal controls and subjects with fibromyalgia syndrome. Pain 102(1–2):87–95

Staud R, Vierck CJ, Robinson ME, Price DD (2004) Spatial summation of heat pain within and across dermatomes in fibromyalgia patients and pain-free subjects. Pain 111(3):342–350

Staud R, Koo E, Robinson ME, Price DD (2007) Spatial summation of mechanically evoked muscle pain and painful aftersensations in normal subjects and fibromyalgia patients. Pain 130(1–2):177–187

Stawowy M, Rössel P, Bluhme C, Funch-Jensen P, Arendt-Nielsen L, Drewes AM (2002) Somatosensory changes in the referred pain area following acute inflammation of the appendix. Eur J Gastroenterol Hepatol 14(10):1079–1084

Stawowy M, Funch-Jensen P, Arendt-Nielsen L, Drewes AM (2005) Somatosensory changes in the referred pain area in patients with cholecystolithiasis. Eur J Gastroenterol Hepatol 17(8):865–870

Stiasny-Kolster K, Pfau DB, Oertel WH, Treede RD, Magerl W (2013) Hyperalgesia and functional sensory loss in restless legs syndrome. Pain 154(8):1457–1463

Stubhaug A, Breivik H, Eide PK, Kreunen M, Foss A (1997) Mapping of punctuate hyperalgesia around a surgical incision demonstrates that ketamine is a powerful suppressor of central sensitization to pain following surgery. Acta Anaesthesiol Scand 41:1124–1132

Sumracki NM, Hutchinson MR, Gentgall M, Briggs N, Williams DB, Rolan P (2012) The effects of pregabalin and the glial attenuator minocycline on the response to intradermal capsaicin in patients with unilateral sciatica. PLoS One 7(6):e3852

Suzan E, Midbari A, Treister R, Haddad M, Pud D, Eisenberg E (2013) Oxycodone alters temporal summation but not conditioned pain modulation: preclinical findings and possible relations to mechanisms of opioid analgesia. Pain 154(8):1413–1418

Suzuki R, Rygh LJ, Dickenson AH (2004) Bad news from the brain: descending 5-HT pathways that control spinal pain processing. Trends Pharmacol Sci 25(12):613–617

Swarbrick ET, Hegarty JE, Bat L, Williams CB, Dawson AM (1980) Site of pain from the irritable bowel. Lancet 2(8192):443–446

Thompson LR, Boudreau R, Newman AB, Hannon MJ, Chu CR, Nevitt MC, Kent Kwoh C, OAI Investigators (2010) The association of osteoarthritis risk factors with localized, regional and diffuse knee pain. Osteoarthritis Cartilage 18(10):1244–1249

Treister R, Pud D, Eisenberg E (2013) The dopamine agonist apomorphine enhances conditioned pain modulation in healthy humans. Neurosci Lett 548:115–119

van Rijn MA, Marinus J, Putter H, Bosselaar SR, Moseley GL, van Hilten JJ (2011) Spreading of complex regional pain syndrome: not a random process. J Neural Transm 118(9):1301–1309

Veldman PH, Goris RJ (1996) Multiple reflex sympathetic dystrophy. Which patients are at risk for developing a recurrence of reflex sympathetic dystrophy in the same or another limb. Pain 64:463–466

Werner MU, Petersen KL, Rowbotham MC, Dahl JB (2013) Healthy volunteers can be phenotyped using cutaneous sensitization pain models. PLoS One 8(5):e62733

Xu YM, Ge HY, Arendt-Nielsen L (2010) Sustained nociceptive mechanical stimulation of latent myofascial trigger point induces central sensitization in healthy subjects. J Pain 11 (12):1348–1355

Yarnitsky D, Arendt-Nielsen L, Bouhassira D, Edwards RR, Fillingim RB, Granot M, Hansson P, Lautenbacher S, Marchand S, Wilder-Smith O (2010) Recommendations on terminology and practice of psychophysical DNIC testing. Eur J Pain 14(4):339

Yarnitsky D, Granot M, Nahman-Averbuch H, Khamaisi M, Granovsky Y (2012) Conditioned pain modulation predicts duloxetine efficacy in painful diabetic neuropathy. Pain 153 (6):1193–1198

Yelle MD, Rogers JM, Coghill RC (2008) Offset analgesia: a temporal contrast mechanism for nociceptive information. Pain 134:174–186

You HJ, Colpaert FC, Arendt-Nielsen L (2005) The novel analgesic and high-efficacy 5-HT1A receptor agonist F 13640 inhibits nociceptive responses, wind-up, and after-discharges in spinal neurons and withdrawal reflexes. Exp Neurol 191(1):174–183

You HJ, Lei J, Sui MY, Huang L, Tan YX, Tjølsen A, Arendt-Nielsen L (2010) Endogenous descending modulation: spatiotemporal effect of dynamic imbalance between descending facilitation and inhibition of nociception. J Physiol 588(Pt 21):4177–4188

Zakrzewska JM (2013) Multi-dimensionality of chronic pain of the oral cavity and face. J Headache Pain 14:37–46

Nitric Oxide-Mediated Pain Processing in the Spinal Cord

Achim Schmidtko

Contents

1 Expression of NO Synthases in the Spinal Cord and in Dorsal Root Ganglia 104
2 Pro- and Antinociceptive Functions of NO 105
3 Downstream Mechanisms of NO-Mediated Pain Processing 107
    3.1 Activation of NO-GC 107
    3.2 cGMP Signaling 108
    3.3 S-Nitrosylation 110
    3.4 Peroxynitrite Formation 110
4 Conclusion 111
References 111

Abstract

A large body of evidence indicates that nitric oxide (NO) plays an important role in the processing of persistent inflammatory and neuropathic pain in the spinal cord. Several animal studies revealed that inhibition or knockout of NO synthesis ameliorates persistent pain. However, spinal delivery of NO donors caused dual pronociceptive and antinociceptive effects, pointing to multiple downstream signaling mechanisms of NO. This review summarizes the localization and function of NO-dependent signaling mechanisms in the spinal cord, taking account of the recent progress made in this field.

Keywords

Pain • Nociception • Dorsal root ganglia • Spinal cord • Nitric oxide • cGMP

A. Schmidtko (✉)
Institut für Pharmakologie und Toxikologie, Universität Witten/Herdecke, ZBAF, Stockumer Str. 10, 58453 Witten, Germany
e-mail: Achim.Schmidtko@uni-wh.de

© Springer-Verlag Berlin Heidelberg 2015
H.-G. Schaible (ed.), *Pain Control*, Handbook of Experimental Pharmacology 227, DOI 10.1007/978-3-662-46450-2_6

Abbreviations

cGKI	cGMP-dependent protein kinase I (synonym PKG-1, protein kinase G-1)
cGMP	3′, 5′-cyclic guanosine monophosphate
CNG	Cyclic-nucleotide gated
DRG	Dorsal root ganglion
GC-A	Particulate guanylyl cyclase A (synonym NPR-A, natriuretic peptide receptor A)
GC-B	Particulate guanylyl cyclase B (synonym NPR-B, natriuretic peptide receptor B)
HCN	Hyperpolarization activated and cyclic-nucleotide gated
NO	Nitric oxide
NO-GC	NO-sensitive guanylyl cyclase (synonym sGC, soluble guanylyl cyclase)
NOS	NO synthase
PDE	Phosphodiesterase

1 Expression of NO Synthases in the Spinal Cord and in Dorsal Root Ganglia

Nitric oxide (NO) serves as a key biological signal in the regulation of many physiological and pathophysiological functions (Francis et al. 2010). It is a small gaseous molecule with a half-life of several seconds that readily permeates cell membranes. As NO cannot be stored in vesicles and secreted in a controlled fashion, its functions are primarily regulated by the expression and activity of NO synthases (NOSs) that produce NO and L-citrulline from the precursor L-arginine. Three different NOS isoforms have been identified that are encoded by three distinct genes. According to their primary origins or properties, NOS isoforms are referred to as neuronal NOS (nNOS or NOS-1), inducible NOS (iNOS or NOS-2), and endothelial NOS (eNOS or NOS-3). Both nNOS and eNOS are expressed constitutively, exhibit low basal activity, and are stimulated by Ca^{2+} influx and Ca^{2+}/calmodulin binding. iNOS is induced in response to inflammatory stimuli, and its activity does not depend on intracellular Ca^{2+}. The activities of NOS enzymes are regulated by several mechanisms, including phosphorylation, nitrosylation, interaction with other proteins, cofactor/substrate availability, and changes in transcription (Bian et al. 2006; Francis et al. 2010).

A large body of evidence indicates that nNOS is a major source of NO during pain processing in the dorsal horn of the spinal cord. Under basic conditions, nNOS is constitutively expressed in some neurons (5–18 % of total neurons) in laminae I–III (Valtschanoff et al. 1992; Dun et al. 1993; Spike et al. 1993; Zhang et al. 1993; Herdegen et al. 1994; Laing et al. 1994; Saito et al. 1994; Bernardi et al. 1995; Ruscheweyh et al. 2006; Sardella et al. 2011; Gassner et al. 2013). Double-labeling immunostaining experiments detected nNOS in a subpopulation of GABAergic inhibitory neurons which innervate giant projection neurons in lamina I (Puskar et al. 2001) and only sparsely overlap with other subpopulations of inhibitory neurons positive for neuropeptide Y, galanin, and parvalbumin (Laing et al. 1994; Tiong et al. 2011; Polgar et al. 2013). nNOS is also expressed at a relatively low level by excitatory interneurons positive for protein kinase Cγ in laminae II and III of the spinal cord (Hughes et al. 2008; Sardella et al. 2011) and in the somata of a few (<5 %) dorsal root ganglion (DRG) neurons (Aimi et al. 1991; Valtschanoff et al. 1992; Zhang et al. 1993; Henrich et al. 2002; Ruscheweyh et al. 2006).

With regard to the pain-relevant functions of NO, it is important to note that nNOS expression in the dorsal horn and in DRGs is considerably upregulated during the processing of persistent pain. Several animal studies demonstrated that the number of nNOS-immunoreactive dorsal horn neurons and the optical nNOS density in the dorsal horn are increased during inflammatory pain evoked by injection of proinflammatory agents such as formalin, zymosan, or complete Freund's adjuvant into a hindpaw (Herdegen et al. 1994; Yonehara et al. 1997; Maihofner et al. 2000; Chu et al. 2005). In contrast, during neuropathic pain in response to peripheral nerve injury, nNOS expression was primarily upregulated in DRG neurons, leading to an increased number of nNOS-positive DRG neurons and enhanced nNOS immunoreactivity in their central terminals in the dorsal horn of the spinal cord (Zhang et al. 1993; Luo et al. 1999; Guan et al. 2007; Martucci et al. 2008). Hence, nNOS seems to play a particular role in the processing of persistent inflammatory and neuropathic pain in the spinal cord and is expressed in different neuronal populations.

Unlike nNOS, iNOS is, if at all, only weakly expressed in the dorsal horn and in DRGs under basic conditions (Wu et al. 1998; Maihofner et al. 2000; Henrich et al. 2002; Keilhoff et al. 2002; Chu et al. 2005; Ruscheweyh et al. 2006; Tang et al. 2007; Martucci et al. 2008). Data about iNOS induction in response to painful stimuli are not consistent. Whereas some studies reported iNOS induction in the spinal cord during the processing of inflammatory and/or neuropathic pain (Guhring et al. 2000; Tao et al. 2003; Martucci et al. 2008; Hervera et al. 2012), other studies reported that iNOS was not induced by painful stimuli (Keilhoff et al. 2002; Chu et al. 2005; De Alba et al. 2006; Guan et al. 2007). Moreover, the cellular distribution of iNOS in the spinal cord remains unclear. Finally, eNOS is constitutively expressed in vascular structures of the dorsal horn and DRGs (Keilhoff et al. 2002; Chu et al. 2005; Ruscheweyh et al. 2006), and its expression seems not to be regulated during pain processing (Keilhoff et al. 2002; Chu et al. 2005; Guan et al. 2007).

2 Pro- and Antinociceptive Functions of NO

The first evidence for a functional contribution of NO to pain processing was discovered in studies using NOS inhibitors such as L-NAME and L-NMMA, which inhibit all three NOS isoforms in a nonspecific manner. These early studies revealed that intrathecal (i.t.) administration of NOS inhibitors effectively ameliorated the pain behavior in various rodent models of inflammatory and neuropathic pain (for review, see Meller and Gebhart 1993; Luo and Cizkova 2000). Experiments with more selective NOS isoform inhibitors point to an important role of nNOS in the development and maintenance of inflammatory and neuropathic pain (Tao et al. 2004; Chu et al. 2005; Guan et al. 2007; Dableh and Henry 2011) and to a contribution of iNOS to the processing of inflammatory pain (Guhring et al. 2000; Tao et al. 2003).

In addition to NOS inhibitors, NOS isoform-specific knockout mice were used to investigate the pain-relevant functions of NO. Many of these studies revealed that persistent pain behaviors were moderately reduced in nNOS and iNOS but not in eNOS knockout mice. However, the interpretation of the pain behavior in mice lacking a NOS isoform is complicated by the fact that the expression of other NOS isoforms may be compensatory upregulated (Tao et al. 2003, 2004; Boettger et al. 2007; Hervera et al. 2010). Moreover, in the widely used nNOS knockout mouse line with targeted deletion of exon 2, alternatively spliced nNOS variants that are functionally active (such as nNOSβ) are still present in distinct tissues (Eliasson et al. 1997). These obstacles may account for the relatively modest pain phenotypes observed in mice lacking nNOS and/or iNOS (Guhring et al. 2000; Tao et al. 2003, 2004; Chu et al. 2005; Boettger et al. 2007; Guan et al. 2007; Hervera et al. 2010; Kuboyama et al. 2011; Keilhoff et al. 2013).

Because inhibition or knockout of NO synthesis ameliorated persistent pain, NO donors were expected to have mainly pronociceptive effects. Indeed, it has been observed that intrathecally administered NO donors may induce or increase hyperalgesia (Kitto et al. 1992; Meller et al. 1992; Machelska et al. 1998; Ferreira et al. 1999; Lin et al. 1999). However, other studies revealed that NO may also have antinociceptive properties within the spinal cord (Luo and Cizkova 2000). For example, i.t. administration of the NO precursor, L-arginine, reduced the activity of dorsal horn neurons and increased the mechanical threshold for tail withdrawal (Haley et al. 1992; Zhuo et al. 1993). Several studies suggested that the concentration of NO may be an important determinant to explain these dual pro- and antinociceptive effects. For example, neuropathic and postoperative pain behavior of rats was inhibited by administration of low doses of an NO donor, while it was further increased by high doses (Sousa and Prado 2001; Kina et al. 2005). Furthermore, dose-dependent dual NO effects have also been observed in humans: NO administration via a transdermal nitroglycerin patch reduced pain due to shoulder or elbow injury at low NO doses (Berrazueta et al. 1996; Paoloni et al. 2003) and enhanced opioid analgesia (Lauretti et al. 1999a, b). Conversely, high doses of transdermal nitroglycerin patches or ointment induced hyperalgesia (Lauretti et al. 1999a; Cadiou et al. 2007). Altogether, there is considerable evidence that inhibition of NO production in the spinal cord ameliorates persistent inflammatory and neuropathic pain. In contrast, delivery of NO donors may exert both pro- and antinociceptive effects, pointing to different downstream signaling pathways of NO action (Schmidtko et al. 2009).

3 Downstream Mechanisms of NO-Mediated Pain Processing

3.1 Activation of NO-GC

At nanomolar levels, NO binds to a prosthetic heme of NO-sensitive guanylyl cyclase (NO-GC; also referred to as soluble guanylyl cyclase, sGC) and causes the conversion of GTP to cGMP (Francis et al. 2010). NO-GC is a heterodimer consisting of two different subunits termed α and β. Two catalytically active isoforms have been identified ($\alpha_1\beta_1$ and $\alpha_2\beta_1$) in which the β_1 subunit acts as the dimerizing partner for the α_1 or α_2 subunit (Friebe et al. 2007). There is considerable evidence that NO-GC is a major NO target during pain processing. Mice deficient for the β_1 subunit (GC-KO mice), which are completely devoid of NO-GC activity, failed to develop pain sensitization induced by intrathecal administration of NO donors. GC-KO mice also demonstrated considerably reduced pain behaviors in inflammatory and neuropathic pain models, whereas the immediate responses to acute nociceptive stimuli were normal (Schmidtko et al. 2008a). The important role of NO-GC for persistent pain processing in the spinal cord is further supported by antinociceptive effects of the NO-GC inhibitor ODQ after intrathecal injection in models of inflammatory and neuropathic pain (Ferreira et al. 1999; Kawamata and Omote 1999; Tao and Johns 2002; Song et al. 2006). Moreover, similar to NO donors, both pronociceptive and antinociceptive effects were observed after i.t. administration of cGMP analogs (Garry et al. 1994; Iwamoto and Marion 1994; Ferreira et al. 1999; Song et al. 2006), and again the administered dose seems to be a determinant for this dual effect (Tegeder et al. 2002, 2004; Schmidtko et al. 2008b). Dual effects of NO and cGMP were also observed in electrophysiological studies with spinal cord slices, in which superfusion with both NO donors and cGMP analogs inhibited ~50 % but activated ~30 % of dorsal horn neurons (Pehl and Schmid 1997).

The most likely reason for the dual effects of NO donors and cGMP analogs is the presence of different pronociceptive and antinociceptive NO/cGMP downstream signaling mechanisms. Unlike the membrane-permeable gas NO, cGMP mainly acts in intracellular compartments at its site of production. Interestingly, the expression pattern of NO-GC in the spinal cord and in DRGs suggests that NO-mediated cGMP production can modulate pain processing at different sites. In the spinal cord, NO-GC immunoreactivity is enriched in inhibitory interneurons in laminae II and III, i.e., in the area of highest nNOS expression (see above). NO-GC is also expressed in neurokinin 1 (NK_1) receptor-positive projection neurons in lamina I (Ding and Weinberg 2006; Ruscheweyh et al. 2006; Schmidtko et al. 2008a). These cells not only contribute to the ascending conduction of pain but are also essential for NO-dependent long-term potentiation (LTP) at the first synapse in pain pathways (Mantyh and Hunt 2004; Ikeda et al. 2006). In DRGs, however, specific NO-GC immunoreactivity was unexpectedly not detected in neurons. Instead thereof, NO-GC protein seems to be present only in satellite cells and vascular cells (Schmidtko et al. 2008a). This finding is supported by observations that axotomy of the sciatic nerve or incubation of DRG sections with an NO donor initiated cGMP production selectively in non-neuronal DRG cells (Morris et al. 1992; Shi et al. 1998). Considering that peripheral nerve injury leads to nNOS upregulation in somata of DRG neurons (see above) and that satellite cells contain NO-GC, it is likely that NO acts as a paracrine messenger from DRG neurons to satellite cells, thereby possibly contributing to the satellite cell proliferation in response to peripheral nerve injury (Zhuang et al. 2005; Scholz and Woolf 2007; Zhang et al. 2007; Kawasaki et al. 2008). Importantly, the observation that NO-GC is not expressed in primary afferent neurons challenges an earlier hypothesis that NO might act as "retrograde" transmitter which is released by spinal cord neurons and stimulates cGMP production via NO-GC activation in primary afferent neurons (Meller and Gebhart 1993; Luo and Cizkova 2000). Instead thereof, NO seems to be primarily a transmitter that (1) is released from nNOS-positive DRG neurons and dorsal horn interneurons (and possibly from so far unidentified iNOS-positive cells) and (2) induces cGMP production in DRG satellite cells, in lamina I projection neurons, and in laminae II/III inhibitory interneurons (Schmidtko et al. 2009).

3.2 cGMP Signaling

The elucidation of downstream mechanisms of NO/cGMP signaling in the nociceptive system has been complicated by at least two facts: First, cGMP in general signals by various mechanisms including activation of cGMP-dependent protein kinase (cGK; also referred to as protein kinase G, PKG), activation of cyclic-nucleotide-gated (CNG) channels, modulation of hyperpolarization-activated and cyclic-nucleotide-gated (HCN) channels, and modulation of phosphodiesterases (PDEs) (Craven and Zagotta 2006; Feil and Kleppisch 2008). Recent data indicate that all these cGMP targets are present in the nociceptive system. Second, cGMP is produced not only by NO-GC but also in a NO-independent manner by particulate guanylyl cyclases in response to stimulation by natriuretic peptides. Seven particulate guanylyl cyclase isoforms activated by different ligands have been identified in rodents (Garbers et al. 2006), and particulate guanylyl cyclases A and B (GC-A and GC-B; also referred to as natriuretic peptide receptor A [NPR-A] and natriuretic peptide receptor B [NPR-B], respectively) have been detected in DRG neurons (Schmidt et al. 2007; Kishimoto et al. 2008; Schmidtko et al. 2008a; Zhang et al. 2010; Loo et al. 2012).

After the discovery of pain-relevant NO/cGMP signaling in the 1990s, it was initially thought that most effects of NO and cGMP are mediated by cGKI (Qian et al. 1996), corresponding to the functional NO/NO-GC/cGMP/cGKI signaling pathway that exists in many other tissues (Feil and Kleppisch 2008). More recent studies confirmed the important pain-relevant role of cGKI, but cGKI seems to be mainly activated by NO-independent mechanisms during pain processing (see below). Several immunohistochemical studies detected the α-isoform of cGKI in the majority of DRG neurons and their nerve terminals in the spinal cord and in some dorsal horn neurons (Qian et al. 1996; Tao et al. 2000; Sung et al. 2006; Schmidtko et al. 2008b; Luo et al. 2012; Lorenz et al. 2014). After peripheral nerve injury and inflammation, cGKIα is activated in DRG neurons (Sung et al. 2004, 2006; Lorenz et al. 2014), and its expression increases in the spinal cord (Tao et al. 2000; Tegeder et al. 2002; Schmidtko et al. 2003). The essential contribution of cGKIα to persistent pain processing is reflected by the reduced inflammatory and/or neuropathic pain behavior in global or nociceptor-specific cGKI mutants (Tegeder et al. 2004; Luo et al. 2012; Lorenz et al. 2014) and by profound antinociceptive effects of intrathecally administered cGKI inhibitors (Tao et al. 2000; Schmidtko et al. 2003, 2009; Luo et al. 2012; Lorenz et al. 2014). So far identified targets that are phosphorylated by cGKIα in DRG neurons include cysteine-rich protein 4 (CRP4; initially named CRP2, Schmidtko et al. 2008b), vasodilator-stimulated phosphoprotein (VASP), myosin light chains (MLC), inositol 1,4,5-triphosphate receptor 1 (IP_3R1) (Luo et al. 2012), and possibly large-conductance Ca^{2+}-activated K^+ channels (BK_{Ca}) (Zhang et al. 2010; Lu et al. 2014).

However, consistent with the cellular distribution of cGKIα and NO-GC described above, double-immunohistochemical stainings confirmed that cGKI and NO-GC are not colocalized in DRGs and only partially colocalized in the spinal cord (Schmidtko et al. 2008a). This implicates that upstream mechanisms different from NO and NO-GC may activate cGKIα during pain processing. Indeed, several studies demonstrated that the particulate guanylyl cyclases GC-A and GC-B are colocalized with cGKIα in DRG neurons and mediate cGKIα activation after stimulation with natriuretic peptides (Schmidt et al. 2007; Kishimoto et al. 2008; Schmidtko et al. 2008a; Zhang et al. 2010). In addition, an alternate mechanism of cGMP-independent cGKIα activation has been recently discovered in DRG neurons: Oxidants such as hydrogen peroxide (H_2O_2) can cause interprotein disulfide bond formation between two cGKIα cysteine residues, rendering the kinase catalytically active, independently of cGMP (Burgoyne et al. 2007). Interestingly, H_2O_2-induced cGKIα disulfide bond formation was increased in DRGs after peripheral nerve injury, and knock-in mice with impaired H_2O_2 activation but normal cGMP activation of cGKIα demonstrated reduced neuropathic pain behaviors (Lorenz et al. 2014). Hence, both cGMP derived from particulate guanylyl cyclases and H_2O_2 derived from so far unidentified sources activate cGKIα in DRGs during pain processing. In contrast, NO and NO-GC seem to use targets different from cGKIα to mediate their pain-relevant effects in DRGs.

In a recent study, CNG channels were identified as a novel target of NO signaling during pain processing: Using in situ hybridization experiments, the CNG channel subunit CNGA3 was detected in inhibitory neurons of the dorsal horn and in DRG satellite cells. After hindpaw inflammation, CNGA3 expression was upregulated in the dorsal horn and in DRGs, and mice lacking CNGA3 ($CNGA3^{-/-}$) showed increased inflammatory pain behaviors. Moreover, the pain hypersensitivity evoked by i.t. delivery of cGMP analogs and NO donors was increased in $CNGA3^{-/-}$ mice (Heine et al. 2011), indicating that CNGA3-positive CNG channels are a downstream target of NO signaling that contributes in an inhibitory manner to persistent pain processing. Further studies are required to identify additional downstream targets that mediate the pro- and antinociceptive effects of NO-mediated cGMP production.

3.3 S-Nitrosylation

A cGMP-independent mechanism of NO signaling is S-nitrosylation, i.e., the covalent and reversible attachment of NO to a reactive cysteine thiol (Hess et al. 2005). Several recent in vitro and ex vivo studies indicate that S-nitrosylation is a signaling mechanism of NO during pain processing (for review, see Tegeder et al. 2011). For example, whole-cell recordings of rat spinal cord slices revealed that NO may S-nitrosylate voltage-activated Ca^{2+} channels, thereby reducing glutamate release from primary afferent terminals (Jin et al. 2011). Unlike this antinociceptive mechanism, S-nitrosylation of actin was reported to ameliorate inhibitory postsynaptic currents in the spinal dorsal horn (Lu et al. 2011). In DRG neurons, NO was found to activate ATP-sensitive potassium channels by S-nitrosylation of cysteine residues in the SUR1 subunit, and this effect was not blocked by inhibitors of NO-GC or cGKI (Kawano et al. 2009). Furthermore, NO directly activated TRPV1 and TRPA1 channels in isolated inside-out patch recordings (Miyamoto et al. 2009), and it seems likely that this effect is also mediated by S-nitrosylation (Yoshida et al. 2006). In a recent proteomic approach using two-dimensional S-nitrosothiol difference gel electrophoresis and S-nitrosylation-site identification in spinal cord extracts, more than 50 proteins with modified S-nitrosylation in response to peripheral nerve injury were detected. The modified proteins are involved in synaptic signaling, protein folding and transport, mitochondrial function, and redox control (Scheving et al. 2012). The functional contribution of most of these proteins to pain processing is currently unknown; however, it seems very likely that S-nitrosylation essentially contributes to cGMP-independent NO signaling in the nociceptive system.

3.4 Peroxynitrite Formation

Another mechanism of cGMP-independent NO signaling is the reaction of NO with the reactive oxygen species superoxide (O_2^-) to form peroxynitrite ($ONOO^-$) (Beckman et al. 1990). There are numerous potential sources of superoxide within cells, including mitochondria, xanthine oxidase, cyclooxygenases, cytochrome P450 monooxygenases, lipoxygenases, uncoupled endothelial NOS, and nicotinamide adenine dinucleotide phosphate (NADPH) oxidases. The latter comprise a family of enzymes that rely on NADPH for their activity and are increasingly recognized as important sources of reactive oxygen species in the nociceptive system (Ibi et al. 2008; Kim et al. 2010; Kallenborn-Gerhardt et al. 2012, 2013; Lim et al. 2013). Recent studies suggest that peroxynitrite is produced during pain processing and has mainly pronociceptive properties (for review, see Salvemini et al. 2011). Accordingly, peroxynitrite decomposition catalysts attenuated inflammatory and neuropathic pain behaviors in rodents (Ndengele et al. 2008; Chen et al. 2010; Doyle et al. 2012). So far identified targets of peroxynitrite in the spinal cord include cyclooxygenases (Ndengele et al. 2008), cytokines (TNF-α and interleukin 1β, 4, and 10), and glia-derived proteins involved in glutamatergic neurotransmission (glutamate transporters and glutamine synthetase) (Chen et al. 2010; Doyle et al. 2012). Hence, peroxynitrite formation seems to be an additional factor that contributes to the multiple pain-relevant effects of NO in the spinal cord.

4 Conclusion

The processing of persistent inflammatory and neuropathic pain is associated with production of NO in the spinal cord. Over the past decade, our knowledge about the downstream signaling pathways has significantly increased. NO leads to cGMP formation in distinct cells of the nociceptive system and to activation of downstream targets including CNG channels. In addition, NO may signal in a cGMP-independent manner by S-nitrosylation of target proteins and by formation of peroxynitrite. There is strong evidence that inhibition of NO production leads to a profound reduction of inflammatory and neuropathic pain. On the other hand, NO production can also reduce pain under several conditions, because NO activates both pro- and antinociceptive mechanisms. Specific targeting of NO-dependent signaling mechanisms might offer new avenues for the treatment of pain.

Acknowledgments Related work done in the author's laboratory was supported by the Deutsche Forschungsgemeinschaft, Witten/Herdecke University, and Doktor Robert Pfleger-Stiftung.

References

Aimi Y, Fujimura M, Vincent SR, Kimura H (1991) Localization of NADPH-diaphorase-containing neurons in sensory ganglia of the rat. J Comp Neurol 306:382–392

Beckman JS, Beckman TW, Chen J, Marshall PA, Freeman BA (1990) Apparent hydroxyl radical production by peroxynitrite: implications for endothelial injury from nitric oxide and superoxide. Proc Natl Acad Sci U S A 87:1620–1624

Bernardi PS, Valtschanoff JG, Weinberg RJ, Schmidt HH, Rustioni A (1995) Synaptic interactions between primary afferent terminals and GABA and nitric oxide-synthesizing neurons in superficial laminae of the rat spinal cord. J Neurosci 15:1363–1371

Berrazueta JR, Losada A, Poveda J, Ochoteco A, Riestra A, Salas E, Amado JA (1996) Successful treatment of shoulder pain syndrome due to supraspinatus tendinitis with transdermal nitroglycerin. A double blind study. Pain 66:63–67

Bian K, Ke Y, Kamisaki Y, Murad F (2006) Proteomic modification by nitric oxide. J Pharmacol Sci 101:271–279

Boettger MK, Uceyler N, Zelenka M, Schmitt A, Reif A, Chen Y, Sommer C (2007) Differences in inflammatory pain in nNOS-, iNOS- and eNOS-deficient mice. Eur J Pain 11:810–818

Burgoyne JR, Madhani M, Cuello F, Charles RL, Brennan JP, Schroder E, Browning DD, Eaton P (2007) Cysteine redox sensor in PKGIa enables oxidant-induced activation. Science 317:1393–1397

Cadiou H, Studer M, Jones NG, Smith ES, Ballard A, McMahon SB, McNaughton PA (2007) Modulation of acid-sensing ion channel activity by nitric oxide. J Neurosci 27:13251–13260

Chen Z, Muscoli C, Doyle T, Bryant L, Cuzzocrea S, Mollace V, Mastroianni R, Masini E, Salvemini D (2010) NMDA-receptor activation and nitroxidative regulation of the glutamatergic pathway during nociceptive processing. Pain 149:100–106

Chu YC, Guan Y, Skinner J, Raja SN, Johns RA, Tao YX (2005) Effect of genetic knockout or pharmacologic inhibition of neuronal nitric oxide synthase on complete Freund's adjuvant-induced persistent pain. Pain 119:113–123

Craven KB, Zagotta WN (2006) CNG and HCN channels: two peas, one pod. Annu Rev Physiol 68:375–401

Dableh LJ, Henry JL (2011) The selective neuronal nitric oxide synthase inhibitor 7-nitroindazole has acute analgesic but not cumulative effects in a rat model of peripheral neuropathy. J Pain Res 4:85–90

De Alba J, Clayton NM, Collins SD, Colthup P, Chessell I, Knowles RG (2006) GW274150, a novel and highly selective inhibitor of the inducible isoform of nitric oxide synthase (iNOS), shows analgesic effects in rat models of inflammatory and neuropathic pain. Pain 120:170–181

Ding JD, Weinberg RJ (2006) Localization of soluble guanylyl cyclase in the superficial dorsal horn. J Comp Neurol 495:668–678

Doyle T, Chen Z, Muscoli C, Bryant L, Esposito E, Cuzzocrea S, Dagostino C, Ryerse J, Rausaria S, Kamadulski A, Neumann WL, Salvemini D (2012) Targeting the overproduction of peroxynitrite for the prevention and reversal of paclitaxel-induced neuropathic pain. J Neurosci 32:6149–6160

Dun NJ, Dun SL, Wu SY, Forstermann U, Schmidt HH, Tseng LF (1993) Nitric oxide synthase immunoreactivity in the rat, mouse, cat and squirrel monkey spinal cord. Neuroscience 54:845–857

Eliasson MJ, Blackshaw S, Schell MJ, Snyder SH (1997) Neuronal nitric oxide synthase alternatively spliced forms: prominent functional localizations in the brain. Proc Natl Acad Sci U S A 94:3396–3401

Feil R, Kleppisch T (2008) NO/cGMP-dependent modulation of synaptic transmission. Handb Exp Pharmacol (184):529–560

Ferreira J, Santos AR, Calixto JB (1999) The role of systemic, spinal and supraspinal L-arginine-nitric oxide-cGMP pathway in thermal hyperalgesia caused by intrathecal injection of glutamate in mice. Neuropharmacology 38:835–842

Francis SH, Busch JL, Corbin JD, Sibley D (2010) cGMP-dependent protein kinases and cGMP phosphodiesterases in nitric oxide and cGMP action. Pharmacol Rev 62:525–563

Friebe A, Mergia E, Dangel O, Lange A, Koesling D (2007) Fatal gastrointestinal obstruction and hypertension in mice lacking nitric oxide-sensitive guanylyl cyclase. Proc Natl Acad Sci U S A 104:7699–7704

Garbers DL, Chrisman TD, Wiegn P, Katafuchi T, Albanesi JP, Bielinski V, Barylko B, Redfield MM, Burnett JC Jr (2006) Membrane guanylyl cyclase receptors: an update. Trends Endocrinol Metab 17:251–258

Garry MG, Abraham E, Hargreaves KM, Aanonsen LM (1994) Intrathecal injection of cell-permeable analogs of cyclic 3′,5′-guanosine monophosphate produces hyperalgesia in mice. Eur J Pharmacol 260:129–131

Gassner M, Leitner J, Gruber-Schoffnegger D, Forsthuber L, Sandkuhler J (2013) Properties of spinal lamina III GABAergic neurons in naive and in neuropathic mice. Eur J Pain 17:1168–1179

Guan Y, Yaster M, Raja SN, Tao YX (2007) Genetic knockout and pharmacologic inhibition of neuronal nitric oxide synthase attenuate nerve injury-induced mechanical hypersensitivity in mice. Mol Pain 3:29

Guhring H, Gorig M, Ates M, Coste O, Zeilhofer HU, Pahl A, Rehse K, Brune K (2000) Suppressed injury-induced rise in spinal prostaglandin E2 production and reduced early thermal hyperalgesia in iNOS-deficient mice. J Neurosci 20:6714–6720

Haley JE, Dickenson AH, Schachter M (1992) Electrophysiological evidence for a role of nitric oxide in prolonged chemical nociception in the rat. Neuropharmacology 31:251–258

Heine S, Michalakis S, Kallenborn-Gerhardt W, Lu R, Lim HY, Weiland J, Del Turco D, Deller T, Tegeder I, Biel M, Geisslinger G, Schmidtko A (2011) CNGA3: a target of spinal NO/cGMP signaling and modulator of inflammatory pain hypersensitivity. J Neurosci 31:11184–11192

Henrich M, Hoffmann K, Konig P, Gruss M, Fischbach T, Godecke A, Hempelmann G, Kummer W (2002) Sensory neurons respond to hypoxia with NO production associated with mitochondria. Mol Cell Neurosci 20:307–322

Herdegen T, Rudiger S, Mayer B, Bravo R, Zimmermann M (1994) Expression of nitric oxide synthase and colocalisation with Jun, Fos and Krox transcription factors in spinal cord neurons following noxious stimulation of the rat hindpaw. Brain Res Mol Brain Res 22:245–258

Hervera A, Negrete R, Leanez S, Martin-Campos JM, Pol O (2010) The spinal cord expression of neuronal and inducible nitric oxide synthases and their contribution in the maintenance of neuropathic pain in mice. PLoS One 5:e14321

Hervera A, Leanez S, Negrete R, Motterlini R, Pol O (2012) Carbon monoxide reduces neuropathic pain and spinal microglial activation by inhibiting nitric oxide synthesis in mice. PLoS One 7:e43693

Hess DT, Matsumoto A, Kim SO, Marshall HE, Stamler JS (2005) Protein S-nitrosylation: purview and parameters. Nat Rev Mol Cell Biol 6:150–166

Hughes AS, Averill S, King VR, Molander C, Shortland PJ (2008) Neurochemical characterization of neuronal populations expressing protein kinase C gamma isoform in the spinal cord and gracile nucleus of the rat. Neuroscience 153:507–517

Ibi M, Matsuno K, Shiba D, Katsuyama M, Iwata K, Kakehi T, Nakagawa T, Sango K, Shirai Y, Yokoyama T, Kaneko S, Saito N, Yabe-Nishimura C (2008) Reactive oxygen species derived from NOX1/NADPH oxidase enhance inflammatory pain. J Neurosci 28:9486–9494

Ikeda H, Stark J, Fischer H, Wagner M, Drdla R, Jager T, Sandkuhler J (2006) Synaptic amplifier of inflammatory pain in the spinal dorsal horn. Science 312:1659–1662

Iwamoto ET, Marion L (1994) Pharmacologic evidence that spinal muscarinic analgesia is mediated by an L-arginine/nitric oxide/cyclic GMP cascade in rats. J Pharmacol Exp Ther 271:601–608

Jin XG, Chen SR, Cao XH, Li L, Pan HL (2011) Nitric oxide inhibits nociceptive transmission by differentially regulating glutamate and glycine release to spinal dorsal horn neurons. J Biol Chem 286:33190–33202

Kallenborn-Gerhardt W, Schroder K, Del Turco D, Lu R, Kynast K, Kosowski J, Niederberger E, Shah AM, Brandes RP, Geisslinger G, Schmidtko A (2012) NADPH oxidase-4 maintains neuropathic pain after peripheral nerve injury. J Neurosci 32:10136–10145

Kallenborn-Gerhardt W, Schroder K, Geisslinger G, Schmidtko A (2013) NOXious signaling in pain processing. Pharmacol Ther 137:309–317

Kawamata T, Omote K (1999) Activation of spinal N-methyl-D-aspartate receptors stimulates a nitric oxide/cyclic guanosine 3,5-monophosphate/glutamate release cascade in nociceptive signaling. Anesthesiology 91:1415–1424

Kawano T, Zoga V, Kimura M, Liang MY, Wu HE, Gemes G, McCallum JB, Kwok WM, Hogan QH, Sarantopoulos CD (2009) Nitric oxide activates ATP-sensitive potassium channels in mammalian sensory neurons: action by direct S-nitrosylation. Mol Pain 5:12

Kawasaki Y, Xu ZZ, Wang X, Park JY, Zhuang ZY, Tan PH, Gao YJ, Roy K, Corfas G, Lo EH, Ji RR (2008) Distinct roles of matrix metalloproteases in the early- and late-phase development of neuropathic pain. Nat Med 14:331–336

Keilhoff G, Fansa H, Wolf G (2002) Neuronal nitric oxide synthase is the dominant nitric oxide supplier for the survival of dorsal root ganglia after peripheral nerve axotomy. J Chem Neuroanat 24:181–187

Keilhoff G, Schroder H, Peters B, Becker A (2013) Time-course of neuropathic pain in mice deficient in neuronal or inducible nitric oxide synthase. Neurosci Res 77:215–221

Kim D, You B, Jo EK, Han SK, Simon MI, Lee SJ (2010) NADPH oxidase 2-derived reactive oxygen species in spinal cord microglia contribute to peripheral nerve injury-induced neuropathic pain. Proc Natl Acad Sci U S A 107:14851–14856

Kina VA, Villarreal CF, Prado WA (2005) The effects of intraspinal L-NOARG or SIN-1 on the control by descending pathways of incisional pain in rats. Life Sci 76:1939–1951

Kishimoto I, Tokudome T, Horio T, Soeki T, Chusho H, Nakao K, Kangawa K (2008) C-type natriuretic peptide is a Schwann cell-derived factor for development and function of sensory neurons. J Neuroendocrinol 20:1213–1223

Kitto KF, Haley JE, Wilcox GL (1992) Involvement of nitric oxide in spinally mediated hyperalgesia in the mouse. Neurosci Lett 148:1–5

Kuboyama K, Tsuda M, Tsutsui M, Toyohara Y, Tozaki-Saitoh H, Shimokawa H, Yanagihara N, Inoue K (2011) Reduced spinal microglial activation and neuropathic pain after nerve injury in mice lacking all three nitric oxide synthases. Mol Pain 7:50

Laing I, Todd AJ, Heizmann CW, Schmidt HH (1994) Subpopulations of GABAergic neurons in laminae I-III of rat spinal dorsal horn defined by coexistence with classical transmitters, peptides, nitric oxide synthase or parvalbumin. Neuroscience 61:123–132

Lauretti GR, Lima IC, Reis MP, Prado WA, Pereira NL (1999a) Oral ketamine and transdermal nitroglycerin as analgesic adjuvants to oral morphine therapy for cancer pain management. Anesthesiology 90:1528–1533

Lauretti GR, de Oliveira R, Reis MP, Mattos AL, Pereira NL (1999b) Transdermal nitroglycerine enhances spinal sufentanil postoperative analgesia following orthopedic surgery. Anesthesiology 90:734–739

Lim H, Kim D, Lee SJ (2013) Toll-like receptor 2 mediates peripheral nerve injury-induced NADPH oxidase 2 expression in spinal cord microglia. J Biol Chem 288:7572–7579

Lin Q, Palecek J, Paleckova V, Peng YB, Wu J, Cui M, Willis WD (1999) Nitric oxide mediates the central sensitization of primate spinothalamic tract neurons. J Neurophysiol 81:1075–1085

Loo L, Shepherd AJ, Mickle AD, Lorca RA, Shutov LP, Usachev YM, Mohapatra DP (2012) The C-type natriuretic peptide induces thermal hyperalgesia through a noncanonical Gbetagamma-dependent modulation of TRPV1 channel. J Neurosci 32:11942–11955

Lorenz JE, Kallenborn-Gerhardt W, Lu R, Syhr KM, Eaton P, Geisslinger G, Schmidtko A (2014) Oxidant-induced activation of cGMP-dependent protein kinase Iα mediates neuropathic pain after peripheral nerve injury. Antioxid Redox Signal 21(10):1504–1515

Lu J, Katano T, Uta D, Furue H, Ito S (2011) Rapid S-nitrosylation of actin by NO-generating donors and in inflammatory pain model mice. Mol Pain 7:101

Lu R, Lukowski R, Sausbier M, Zhang DD, Sisignano M, Schuh CD, Kuner R, Ruth P, Geisslinger G, Schmidtko A (2014) BKCa channels expressed in sensory neurons modulate inflammatory pain in mice. Pain 155(3):556–565

Luo ZD, Cizkova D (2000) The role of nitric oxide in nociception. Curr Rev Pain 4:459–466

Luo ZD, Chaplan SR, Scott BP, Cizkova D, Calcutt NA, Yaksh TL (1999) Neuronal nitric oxide synthase mRNA upregulation in rat sensory neurons after spinal nerve ligation: lack of a role in allodynia development. J Neurosci 19:9201–9208

Luo C, Gangadharan V, Bali KK, Xie RG, Agarwal N, Kurejova M, Tappe-Theodor A, Tegeder I, Feil S, Lewin G, Polgar E, Todd AJ, Schlossmann J, Hofmann F, Liu DL, Hu SJ, Feil R, Kuner T, Kuner R (2012) Presynaptically localized cyclic GMP-dependent protein kinase 1 is a key determinant of spinal synaptic potentiation and pain hypersensitivity. PLoS Biol 10: e1001283

Machelska H, Przewlocki R, Radomski MW, Przewlocka B (1998) Differential effects of intrathecally and intracerebroventricularly administered nitric oxide donors on noxious mechanical and thermal stimulation. Pol J Pharmacol 50:407–415

Maihofner C, Euchenhofer C, Tegeder I, Beck KF, Pfeilschifter J, Geisslinger G (2000) Regulation and immunohistochemical localization of nitric oxide synthases and soluble guanylyl cyclase in mouse spinal cord following nociceptive stimulation. Neurosci Lett 290:71–75

Mantyh PW, Hunt SP (2004) Setting the tone: superficial dorsal horn projection neurons regulate pain sensitivity. Trends Neurosci 27:582–584

Martucci C, Trovato AE, Costa B, Borsani E, Franchi S, Magnaghi V, Panerai AE, Rodella LF, Valsecchi AE, Sacerdote P, Colleoni M (2008) The purinergic antagonist PPADS reduces pain related behaviours and interleukin-1beta, interleukin-6, iNOS and nNOS overproduction in central and peripheral nervous system after peripheral neuropathy in mice. Pain 137:81–95

Meller ST, Gebhart GF (1993) Nitric oxide (NO) and nociceptive processing in the spinal cord. Pain 52:127–136

Meller ST, Dykstra C, Gebhart GF (1992) Production of endogenous nitric oxide and activation of soluble guanylate cyclase are required for N-methyl-D-aspartate-produced facilitation of the nociceptive tail-flick reflex. Eur J Pharmacol 214:93–96

Miyamoto T, Dubin AE, Petrus MJ, Patapoutian A (2009) TRPV1 and TRPA1 mediate peripheral nitric oxide-induced nociception in mice. PLoS One 4:e7596

Morris R, Southam E, Braid DJ, Garthwaite J (1992) Nitric oxide may act as a messenger between dorsal root ganglion neurones and their satellite cells. Neurosci Lett 137:29–32

Ndengele MM, Cuzzocrea S, Esposito E, Mazzon E, Di Paola R, Matuschak GM, Salvemini D (2008) Cyclooxygenases 1 and 2 contribute to peroxynitrite-mediated inflammatory pain hypersensitivity. FASEB J 22:3154–3164

Paoloni JA, Appleyard RC, Nelson J, Murrell GA (2003) Topical nitric oxide application in the treatment of chronic extensor tendinosis at the elbow: a randomized, double-blinded, placebo-controlled clinical trial. Am J Sports Med 31:915–920

Pehl U, Schmid HA (1997) Electrophysiological responses of neurons in the rat spinal cord to nitric oxide. Neuroscience 77:563–573

Polgar E, Sardella TC, Tiong SY, Locke S, Watanabe M, Todd AJ (2013) Functional differences between neurochemically defined populations of inhibitory interneurons in the rat spinal dorsal horn. Pain 154:2606–2615

Puskar Z, Polgar E, Todd AJ (2001) A population of large lamina I projection neurons with selective inhibitory input in rat spinal cord. Neuroscience 102:167–176

Qian Y, Chao DS, Santillano DR, Cornwell TL, Nairn AC, Greengard P, Lincoln TM, Bredt DS (1996) cGMP-dependent protein kinase in dorsal root ganglion: relationship with nitric oxide synthase and nociceptive neurons. J Neurosci 16:3130–3138

Ruscheweyh R, Goralczyk A, Wunderbaldinger G, Schober A, Sandkuhler J (2006) Possible sources and sites of action of the nitric oxide involved in synaptic plasticity at spinal lamina I projection neurons. Neuroscience 141:977–988

Saito S, Kidd GJ, Trapp BD, Dawson TM, Bredt DS, Wilson DA, Traystman RJ, Snyder SH, Hanley DF (1994) Rat spinal cord neurons contain nitric oxide synthase. Neuroscience 59:447–456

Salvemini D, Little JW, Doyle T, Neumann WL (2011) Roles of reactive oxygen and nitrogen species in pain. Free Radic Biol Med 51:951–966

Sardella TC, Polgar E, Watanabe M, Todd AJ (2011) A quantitative study of neuronal nitric oxide synthase expression in laminae I-III of the rat spinal dorsal horn. Neuroscience 192:708–720

Scheving R, Wittig I, Heide H, Albuquerque B, Steger M, Brandt U, Tegeder I (2012) Protein S-nitrosylation and denitrosylation in the mouse spinal cord upon injury of the sciatic nerve. J Proteomics 75:3987–4004

Schmidt H, Stonkute A, Juttner R, Schaffer S, Buttgereit J, Feil R, Hofmann F, Rathjen FG (2007) The receptor guanylyl cyclase Npr2 is essential for sensory axon bifurcation within the spinal cord. J Cell Biol 179:331–340

Schmidtko A, Ruth P, Geisslinger G, Tegeder I (2003) Inhibition of cyclic guanosine 5′-monophosphate-dependent protein kinase I (PKG-I) in lumbar spinal cord reduces formalin-induced hyperalgesia and PKG upregulation. Nitric Oxide 8:89–94

Schmidtko A, Gao W, Konig P, Heine S, Motterlini R, Ruth P, Schlossmann J, Koesling D, Niederberger E, Tegeder I, Friebe A, Geisslinger G (2008a) cGMP produced by NO-sensitive guanylyl cyclase essentially contributes to inflammatory and neuropathic pain by using targets different from cGMP-dependent protein kinase I. J Neurosci 28:8568–8576

Schmidtko A, Gao W, Sausbier M, Rauhmeier I, Sausbier U, Niederberger E, Scholich K, Huber A, Neuhuber W, Allescher HD, Hofmann F, Tegeder I, Ruth P, Geisslinger G (2008b) Cysteine-rich protein 2, a novel downstream effector of cGMP/cGMP-dependent protein kinase I-mediated persistent inflammatory pain. J Neurosci 28:1320–1330

Schmidtko A, Tegeder I, Geisslinger G (2009) No NO, no pain? The role of nitric oxide and cGMP in spinal pain processing. Trends Neurosci 32:339–346

Scholz J, Woolf CJ (2007) The neuropathic pain triad: neurons, immune cells and glia. Nat Neurosci 10:1361–1368

Shi TJ, Holmberg K, Xu ZQ, Steinbusch H, de Vente J, Hokfelt T (1998) Effect of peripheral nerve injury on cGMP and nitric oxide synthase levels in rat dorsal root ganglia: time course and coexistence. Pain 78:171–180

Song XJ, Wang ZB, Gan Q, Walters ET (2006) cAMP and cGMP contribute to sensory neuron hyperexcitability and hyperalgesia in rats with dorsal root ganglia compression. J Neurophysiol 95:479–492

Sousa AM, Prado WA (2001) The dual effect of a nitric oxide donor in nociception. Brain Res 897:9–19

Spike RC, Todd AJ, Johnston HM (1993) Coexistence of NADPH diaphorase with GABA, glycine, and acetylcholine in rat spinal cord. J Comp Neurol 335:320–333

Sung YJ, Walters ET, Ambron RT (2004) A neuronal isoform of protein kinase G couples mitogen-activated protein kinase nuclear import to axotomy-induced long-term hyperexcitability in Aplysia sensory neurons. J Neurosci 24:7583–7595

Sung YJ, Chiu DT, Ambron RT (2006) Activation and retrograde transport of protein kinase G in rat nociceptive neurons after nerve injury and inflammation. Neuroscience 141:697–709

Tang Q, Svensson CI, Fitzsimmons B, Webb M, Yaksh TL, Hua XY (2007) Inhibition of spinal constitutive NOS-2 by 1400W attenuates tissue injury and inflammation-induced hyperalgesia and spinal p38 activation. Eur J Neurosci 25:2964–2972

Tao YX, Johns RA (2002) Activation and up-regulation of spinal cord nitric oxide receptor, soluble guanylate cyclase, after formalin injection into the rat hind paw. Neuroscience 112:439–446

Tao YX, Hassan A, Haddad E, Johns RA (2000) Expression and action of cyclic GMP-dependent protein kinase Ialpha in inflammatory hyperalgesia in rat spinal cord. Neuroscience 95:525–533

Tao F, Tao YX, Mao P, Zhao C, Li D, Liaw WJ, Raja SN, Johns RA (2003) Intact carrageenan-induced thermal hyperalgesia in mice lacking inducible nitric oxide synthase. Neuroscience 120:847–854

Tao F, Tao YX, Zhao C, Dore S, Liaw WJ, Raja SN, Johns RA (2004) Differential roles of neuronal and endothelial nitric oxide synthases during carrageenan-induced inflammatory hyperalgesia. Neuroscience 128:421–430

Tegeder I, Schmidtko A, Niederberger E, Ruth P, Geisslinger G (2002) Dual effects of spinally delivered 8-bromo-cyclic guanosine mono-phosphate (8-bromo-cGMP) in formalin-induced nociception in rats. Neurosci Lett 332:146–150

Tegeder I, Del Turco D, Schmidtko A, Sausbier M, Feil R, Hofmann F, Deller T, Ruth P, Geisslinger G (2004) Reduced inflammatory hyperalgesia with preservation of acute thermal nociception in mice lacking cGMP-dependent protein kinase I. Proc Natl Acad Sci U S A 101:3253–3257

Tegeder I, Scheving R, Wittig I, Geisslinger G (2011) SNO-ing at the nociceptive synapse? Pharmacol Rev 63:366–389

Tiong SY, Polgar E, van Kralingen JC, Watanabe M, Todd AJ (2011) Galanin-immunoreactivity identifies a distinct population of inhibitory interneurons in laminae I-III of the rat spinal cord. Mol Pain 7:36

Valtschanoff JG, Weinberg RJ, Rustioni A, Schmidt HH (1992) Nitric oxide synthase and GABA colocalize in lamina II of rat spinal cord. Neurosci Lett 148:6–10

Wu J, Lin Q, Lu Y, Willis WD, Westlund KN (1998) Changes in nitric oxide synthase isoforms in the spinal cord of rat following induction of chronic arthritis. Exp Brain Res 118:457–465

Yonehara N, Takemura M, Yoshimura M, Iwase K, Seo HG, Taniguchi N, Shigenaga Y (1997) Nitric oxide in the rat spinal cord in Freund's adjuvant-induced hyperalgesia. Jpn J Pharmacol 75:327–335

Yoshida T, Inoue R, Morii T, Takahashi N, Yamamoto S, Hara Y, Tominaga M, Shimizu S, Sato Y, Mori Y (2006) Nitric oxide activates TRP channels by cysteine S-nitrosylation. Nat Chem Biol 2:596–607

Zhang X, Verge V, Wiesenfeld-Hallin Z, Ju G, Bredt D, Synder SH, Hokfelt T (1993) Nitric oxide synthase-like immunoreactivity in lumbar dorsal root ganglia and spinal cord of rat and monkey and effect of peripheral axotomy. J Comp Neurol 335:563–575

Zhang X, Chen Y, Wang C, Huang LY (2007) Neuronal somatic ATP release triggers neuron-satellite glial cell communication in dorsal root ganglia. Proc Natl Acad Sci U S A 104:9864–9869

Zhang FX, Liu XJ, Gong LQ, Yao JR, Li KC, Li ZY, Lin LB, Lu YJ, Xiao HS, Bao L, Zhang XH, Zhang X (2010) Inhibition of inflammatory pain by activating B-type natriuretic peptide signal pathway in nociceptive sensory neurons. J Neurosci 30:10927–10938

Zhuang ZY, Gerner P, Woolf CJ, Ji RR (2005) ERK is sequentially activated in neurons, microglia, and astrocytes by spinal nerve ligation and contributes to mechanical allodynia in this neuropathic pain model. Pain 114:149–159

Zhuo M, Meller ST, Gebhart GF (1993) Endogenous nitric oxide is required for tonic cholinergic inhibition of spinal mechanical transmission. Pain 54:71–78

The Role of the Endocannabinoid System in Pain

Stephen G. Woodhams, Devi Rani Sagar, James J. Burston, and Victoria Chapman

Contents

1	Cannabinoids and the Endocannabinoid System	121
2	The Cannabinoid Receptors	123
3	Endogenous Ligands: The Endocannabinoids	124
4	Endocannabinoid Synthesis and Degradation	124
5	The Endocannabinoid System and Pain	125
	5.1 The EC System and Peripheral Pain Processing	126
	5.2 The Spinal Endocannabinoid System and Acute Pain Processing	127
	5.3 A Novel Role of Spinal CB_2 Receptors in Chronic Pain States	128
	5.4 CB_2 Receptor Modulation of Spinal Immune Cell Function	130
	5.5 Supraspinal Sites of Action of the Endocannabinoids	130
	5.6 Enhancing EC Signalling: Problems of Plasticity	132
6	Summary	133
References		134

Abstract

Preparations of the *Cannabis sativa* plant have been used to analgesic effect for millenia, but only in recent decades has the endogenous system responsible for these effects been described. The endocannabinoid (EC) system is now known to be one of the key endogenous systems regulating pain sensation, with modulatory actions at all stages of pain processing pathways. The EC system is

---

S.G. Woodhams
Momentum Laboratory of Molecular Neurobiology, Institute of Experimental Medicine, Hungarian Academy of Sciences, Budapest 1083, Hungary

D.R. Sagar • J.J. Burston • V. Chapman (✉)
Arthritis Research UK Pain Centre, University of Nottingham, Nottingham NG7 2UH, UK

School of Life Sciences, Queen's Medical Centre, University of Nottingham, Nottingham NG7 2UH, UK
e-mail: victoria.chapman@nottingham.ac.uk

© Springer-Verlag Berlin Heidelberg 2015
H.-G. Schaible (ed.), *Pain Control*, Handbook of Experimental Pharmacology 227,
DOI 10.1007/978-3-662-46450-2_7

composed of two main cannabinoid receptors (CB_1 and CB_2) and two main classes of endogenous ligands or endocannabinoids (ECs). The receptors have distinct expression profiles, with CB_1 receptors found at presynaptic sites throughout the peripheral and central nervous systems (PNS and CNS, respectively), whilst CB_2 receptor is found principally (but not exclusively) on immune cells. The endocannabinoid ligands are lipid neurotransmitters belonging to either the N-acyl ethanolamine (NAEs) class, e.g. anandamide (AEA), or the monoacylglycerol class, e.g. 2-arachidonoyl glycerol (2-AG). Both classes are short-acting transmitter substances, being synthesised on demand and with signalling rapidly terminated by specific enzymes. ECs acting at CB_1 negatively regulate neurotransmission throughout the nervous system, whilst those acting at CB_2 regulate the activity of CNS immune cells. Signalling through both of these receptor subtypes has a role in normal nociceptive processing and also in the development resolution of acute pain states. In this chapter, we describe the general features of the EC system as related to pain and nociception and discuss the wealth of preclinical and clinical data involving targeting the EC system with focus on two areas of particular promise: modulation of 2-AG signalling via specific enzyme inhibitors and the role of spinal CB_2 in chronic pain states.

Keywords
Pain • Endocannabinoid • Analgesia

Abbreviations

2-AG	2-Arachidonoyl glycerol
Δ^9-THC	Δ^9-Tetrahydrocannabinol
ABHD6	$\alpha\beta$-Hydrolase domain 6
ABHD12	$\alpha\beta$-Hydrolase domain 12
ACC	Anterior cingulate cortex
AEA	Anandamide
AM251	A CB_1 selective receptor inverse agonist/antagonist
CB_1	Cannabinoid type 1 receptor
CB_2	Cannabinoid type 2 receptor
CCI	Chronic constriction injury (neuropathic pain model)
CNS	Central nervous system
DAGLα	Diacylglycerol lipase-α
DRG	Dorsal root ganglia
EC	Endocannabinoid
FAAH	Fatty acid amide hydrolase
fMRI	Functional magnetic resonance imaging

GPCR	G protein-coupled receptor
MAG	Monoacylglycerol
MAGL	Monoacylglycerol lipase
NAE	N-acylethanolamine
NAPE-PLD	N-acyl-phosphatidylethanolamine-hydrolyzing phospholipase D
PAG	Periaqueductal grey matter
PKC	Protein kinase C
PNS	Peripheral nervous system
RVM	Rostral ventromedial medulla
SR144528	A CB_2 selective receptor antagonist
TRPV1	Transient receptor potential vanilloid 1
WDR	Wide dynamic range neuron; a class of spinal neuron with key involvement in the transduction of nociceptive input

1 Cannabinoids and the Endocannabinoid System

The *Cannabis sativa* plant contains over 100 bioactive lipid compounds, known as cannabinoids, which produce a plethora of physiological effects in humans including profound analgesia. However, due to the abuse potential and numerous additional undesirable effects, such as hypomotility (Adams and Martin 1996), deficits in executive function (Pattij et al. 2008), and memory consolidation (Hall and Solowij 1998), the use of cannabis for medicinal purposes has been restricted until recent times. During the 1960s, the major psychoactive component Δ^9-tetrahydrocannabinol (Δ^9-THC) was isolated (Mechoulam and Gaoni 1967), and study of its pharmacology began in earnest. Many years later, a membrane-bound, cannabinoid-sensitive receptor with high expression levels in the nervous system was isolated (Devane et al. 1988; Herkenham et al. 1991). This receptor is now known to form the basis of a key signalling pathway regulating neurotransmission throughout the nervous system, the endocannabinoid (EC) system. The EC system is composed of two G protein-coupled receptors (GPCRs) known as cannabinoid receptors 1 and 2 (CB_1 and CB_2), two families of lipids constituting the EC ligands, and the synthetic and metabolic enzymes which initiate and terminate EC signalling.

In comparison to classical neurotransmitter systems, the EC system possesses several unusual properties from which its key role in control of nociception is derived. Perhaps the most important of these features is the positioning of the EC signalling machinery at neuronal synapses in pain processing pathways (see Fig. 1). ECs are predominantly retrograde neurotransmitters, synthesised in the postsynaptic cell and released into the synapse to travel across and interact with receptors on the presynaptic cell (Katona et al. 1999; Egertova and Elphick 2000), resulting in an inhibition of neurotransmitter release (Freund et al. 2003). Since ECs are not stored in vesicles prior to release, but instead are produced through activity-driven "on-demand" synthesis following strong neuronal activation (Ohno-Shosaku et al. 2001;

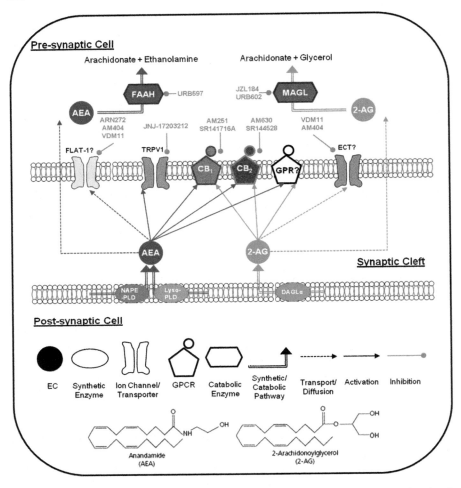

Fig. 1 Endocannabinoid signalling at a notional neuronal synapse. The major synthetic, signalling, and catabolic pathways for AEA and 2-AG are shown. Other GPCRs may be involved in endocannabinoid signalling alongside CB_1 and CB_2. FLAT-1, a truncated form of FAAH, and ECT are the putative EC transporters. The listed compounds are recognised enzyme inhibitors/receptor antagonists which can modulate EC signalling. It should be noted that the synthetic pathway responsible for AEA production in vivo has not been conclusively determined and that some evidence suggests AEA may act as an anterograde transmitter, being synthesised presynaptically and acting at postsynaptic TRPV1 receptors in some CNS regions (Aguiar et al. 2014). *Figure adapted from Burston et al.* (2013)

Wilson and Nicoll 2001) and/or activation of Gq-protein-coupled receptors (Katona and Freund 2012), their signalling is uniquely positioned to act as a brake on neuronal signalling in response to high activity. In pain pathways, this manifests as antinociception or analgesia. Additionally, in contrast to the single ligand-multiple receptor paradigm present in most classical neurotransmitter systems (e.g. glutamate, GABA, 5-HT, etc., Di Marzo and De Petrocellis 2012), the EC system possesses multiple ligands acting at just two major receptors, although the metabolites of these ECs seem to target multiple receptors (Sagar et al. 2012). Given that the tissue concentration of ECs, the size of the metabolic enzyme pool, and the expression levels of EC-sensitive receptors can all vary with changes in physiological and pathophysiological state, this property produces a dynamic lipid network capable of exquisite fine-tuning of cellular signalling (Alexander and Kendall 2007).

2 The Cannabinoid Receptors

The two known cannabinoid receptors have distinct expression patterns, which underlie their separate physiological roles. The CB_1 receptor is found both in the periphery and central nervous system (CNS), although expression in the CNS is far greater. CB_2, in contrast, is typically expressed predominantly by cells of the immune system, including glial cells of the CNS. However, more recent studies have demonstrated functional effects of CB_2 receptor activity on neurones (Racz et al. 2008b; Burston et al. 2013).

The CB_1 receptor is considered to be the most abundant GPCR in the CNS and is highly expressed in the neocortex, cerebellum, and limbic regions (Herkenham et al. 1990, 1991), but is also expressed in the peripheral nervous system (PNS). At peripheral and central terminals of nociceptive sensory nerves, CB_1 gates the transduction of peripheral noxious stimuli into central neuronal pain signals (Guindon and Beaulieu 2006); at the level of the spinal cord, CB_1 receptors can act to reduce or enhance propagation of pain signals to the brain (Sagar et al. 2010a, b; Pernia-Andrade et al. 2009; Woodhams et al. 2012), and those in nociception-associated higher brain regions, such as the periaqueductal grey matter (PAG) and the rostral ventromedial medulla (RVM), can initiate descending inhibition, or block descending facilitation, to the spinal cord nociceptive circuitry (Herkenham et al. 1991; Burns et al. 2007; Rea et al. 2007; Petrosino et al. 2007; Nadal et al. 2013; Martin et al. 1999). Interestingly, and of growing importance in pain research, CB_1 is also highly expressed in the frontal–limbic brain circuits which are central to the affective/emotional aspects of pain in humans (Burns et al. 2007; Lee et al. 2013).

In line with its role in retrograde neurotransmission, CB_1 is generally expressed presynaptically on axon terminals (Katona et al. 1999), where it is enriched in the perisynaptic zone (Nyíri et al. 2005), perfectly positioned to modulate the activity of the N- and P-/Q-type calcium channels which mediate neurotransmitter release. CB_1 is negatively coupled to adenylate cyclase via $G_{i/o}$ proteins, and thus its activation leads to inhibition of calcium channels, activation of potassium channels, and ultimately the reduction of neurotransmitter release (Pertwee 1997). At the circuit level, the net result of CB_1 activity can be excitatory or inhibitory depending on the identity of the presynaptic cell and its location within the neural network. In addition to neuronal CB_1, there is evidence of expression in B cells of the immune system (Graham et al. 2010; Pacher and Mechoulam 2011; Kaplan 2013) and also on glial cells of the CNS (Salio et al. 2002; Stella 2004, 2009; Navarrete and Araque 2008). Indeed, astroglial CB_1 appears to have a pivotal role in both cannabinoid-mediated behaviour (Han et al. 2012) and EC-mediated plasticity (Min and Nevian 2012).

In contrast to the almost ubiquitous expression of CB_1 in the nervous system, CB_2 expression is thought to be confined to immune cells such as macrophages (Han et al. 2009), lymphocytes (Cencioni et al. 2010), and mast cells (Samson et al. 2003) in the periphery and astrocytes and microglia in the CNS (Alkaitis et al. 2010; Salio et al. 2002; Stella 2010). Activation of CB_2 is inhibitory via $G_{i/o}$ proteins, mediating the well-characterised anti-inflammatory effects of the ECs as well as having a role in anti-hyperalgesia in inflammatory pain states (Quartilho et al. 2003; Cabral and Griffin-Thomas 2009; Cabral et al. 2008; Khasabova et al. 2011; Ibrahim et al. 2006). There is a growing debate about the presence of functional CB_2 receptors on some populations of neurons in the CNS (see commentary by Atwood and Mackie 2010), although at present this evidence remains controversial. Despite this restricted expression profile, CB_2 plays an important role in pain signalling and may be of particular importance in the development of chronic pain states. This is discussed in more detail later in this chapter.

3 Endogenous Ligands: The Endocannabinoids

The ECs are lipid signalling molecules formed of two major groups: the N-acylethanolamines (NAEs) and the monoacylglycerols (MAGs). The most widely investigated are anandamide (AEA) and 2-arachidonoyl glycerol (2-AG), respectively (structures shown in Fig. 1). Although AEA was discovered first (Devane et al. 1992), 2-AG is now thought to be the major synaptic EC ligand in nervous tissue, acting as a full agonist at both CB_1 and CB_2 (Sugiura et al. 2000), and present at far higher levels than AEA in brain tissue (Stella et al. 1997). In contrast, AEA is a partial agonist at both cannabinoid receptors, with a slight selectivity for CB_1 over CB_2, but also acts as a full agonist at the ion channel receptor transient receptor potential vanilloid 1 (TRPV1) at higher concentrations (Zygmunt et al. 1999; Ross 2003). As such, AEA can be considered an endocannabinoid/endovanilloid substance and may have a dual role in nociception, being antinociceptive at cannabinoid receptors and pronociceptive at TRPV1 (see Starowicz and Przewlocka 2012 for review).

4 Endocannabinoid Synthesis and Degradation

The major ECs have distinct synthetic and degradative pathways, with the localisation of the enzymatic machinery for each ligand determining its physiological effects. Importantly, both classes of ECs are enzymatically synthesised de novo from membrane phospholipid precursors and are rapidly metabolised by specific enzymes, providing clear targets for the pharmaceutical modulation of EC signalling. A schematic diagram of the actions of ECs at a generic neuronal synapse can be seen in Fig. 1. Synthesis of ECs is initiated either via sustained increases in intracellular calcium following intense stimulation of the cell or through the activation of several classes of GPCRs (Hu et al. 2014). Of particular note is the group 1 metabotropic glutamate receptor mGluR5, which is often located in the perisynaptic zone (Lujan et al. 1996), and has an established role in nociception (Radulovic and Tronson 2012; Hu et al. 2012).

2-AG synthesis proceeds via the production of diacylglycerol (DAG) species from membrane phospholipids by PLCβ; these are then metabolised by the enzymes diacylglycerol lipase-α and β (DAGLα and β) to form 2-AG (Bisogno et al. 2003). Notably, DAG acts as a signalling molecule in its own right, activating protein kinase C (PKC) (Stella et al. 1997)—an enzyme with well-established pronociceptive effects in pain pathways (Velazquez et al. 2007). Synthesis of 2-AG therefore also terminates DAG signalling. Monoacylglycerol lipase (MAGL) is the major metabolic enzyme for 2-AG (Dinh et al. 2002), responsible for 85 % of brain 2-AG hydrolysis, with minor contributions from αβ hydrolase domains 6 and 12 (ABHD6 and ABHD12) (Blankman et al. 2007). ABHD12 is highly expressed in microglia, whilst ABHD6 is postsynaptic and may therefore regulate the release of 2-AG (Savinainen et al. 2011).

Originally, a shared synthetic pathway for AEA and its congener lipids palmitoylethanolamine (PEA) and oleoylethanolamine (OEA) via the enzyme N-acyl-phosphatidylethanolamine-hydrolyzing phospholipase D (NAPE-PLD) was described (Di Marzo et al. 1994; Okamoto et al. 2004). However, NAPE-PLD null mice have normal brain levels of AEA (Leung et al. 2006), and additional synthetic pathways have since been identified (Simon and Cravatt 2006, 2008), calling into question the predominance of this pathway in vivo. At present, the mechanism responsible for the production of AEA in pain pathways has not been conclusively determined. The NAEs PEA and OEA do not directly interact with cannabinoid receptors and are thus not considered to be true ECs. PEA, however, does have antinociceptive and anti-inflammatory properties via interactions with members of the peroxisome proliferator-activated receptor family (D'Agostino et al. 2009; Lo Verme et al. 2005). The metabolism of AEA occurs predominantly via the enzyme fatty acid amide hydrolase (FAAH), although other metabolic pathways have been identified (reviewed in Ueda et al. 2013), and these alterative pathways may have physiological relevance under certain conditions. Of particular relevance is the hydrolysis of both AEA and 2-AG by cyclooxygenase 2 (COX2) (Glaser and Kaczocha 2010), a key enzyme in pain processing, which can produce pronociceptive prostamide EC metabolites (Sang et al. 2006, 2007).

5   The Endocannabinoid System and Pain

Systemic administration of cannabinoid receptor ligands is well known to produce analgesia in animal models of acute and chronic pain (Walker and Huang 2002). Despite the recent increase in use of medicinal marijuana, and the development of licensed cannabinoid drugs such as Sativex for multiple sclerosis (Garcia-Merino 2013), concerns remain over dependence, tolerance, and the cognitive side effects produced by these medications. The undesirable effects of cannabinoids are caused by the global activation of CNS CB_1 receptors, and as a result research has focussed on presumed site-specific modulation of endogenous ligand activity or on effects at the non-psychotropic CB_2 receptor. In the next sections, we will describe the differing contributions of ECs to acute nociception and the development of pathophysiological pain states in each physiological compartment: at the periphery, in the spinal cord, and in supraspinal regions. Particular emphasis will be placed on two promising areas of EC research, namely, the modulation of 2-AG signalling and the novel role of spinal CB_2 receptors in chronic pain states.

5.1 The EC System and Peripheral Pain Processing

In the periphery, CB_1 receptors localised on sensory afferent terminals gate the transduction of pain signals from noxious stimuli (Stander et al. 2005). These receptors have a significant contribution to cannabinoid-mediated analgesia, since their selective deletion in mice greatly reduces the efficacy of both locally and systemically administered cannabinoids in models of acute and chronic pain (Agarwal et al. 2007). Numerous studies have demonstrated antinociceptive efficacy of local administrations of both AEA (Guindon et al. 2006; Guindon and Beaulieu 2006) and 2-AG (Desroches et al. 2008; Guindon et al. 2007), and elevated peripheral tissue levels of both ECs have been detected in numerous preclinical models of inflammatory pain (Beaulieu et al. 2000; Maione et al. 2007), suggesting endogenous activity of the EC system. These effects are not solely mediated via neuronal CB_1, since blocking CB_1 and/or CB_2 receptors prior to formalin administration increases nociceptive responses (Guindon et al. 2007). In fact, the antinociceptive effect of peripherally administered 2-AG on inflammatory pain has been described as mainly CB_2 mediated (Guindon et al. 2007). The mechanism is likely multifaceted, involving inhibition of the production and release of proinflammatory and pronociceptive mediators such as reactive oxygen species (Hao et al. 2010) and cytokines (Cencioni et al. 2010) by peripheral immune cells and also the peripheral release of endogenous opioids (Ibrahim et al. 2005; Desroches et al. 2014a).

Since ECs are rapidly degraded in vivo, their effects are short-lived, and they are thus unsuitable for use as analgesics. However, the recent development of specific inhibitors of the major catabolic enzymes (MAGL and FAAH) has enabled researchers to prolong the effect of endogenously generated ECs. This is a very promising analgesic strategy, since ECs are specifically generated at sites of nociceptive activity, and such an approach may avoid the unwanted effects of global CB_1 receptor agonism. Systemic administration of either MAGL inhibitors (Long et al. 2009a; Ignatowska-Jankowska et al. 2014; Kinsey et al. 2009, 2013; Bisogno et al. 2009) or FAAH inhibitors (Jayamanne et al. 2006; Chang et al. 2006; Lichtman et al. 2004; Kathuria et al. 2003; Fegley et al. 2005; Russo et al. 2007) has been shown to be antinociceptive in models of acute and chronic pain. Local effects of MAGL and FAAH inhibition are best studied in rodent models of inflammatory pain, in which peripheral administration of an irritant produces oedema and pain behaviour. The MAGL inhibitor JZL184 (Long et al. 2009a) effectively blocks the pain behaviour and thermal hypersensitivity induced by intra-plantar capsaicin administration (Spradley et al. 2010). Intra-plantar JZL184 administration also inhibits local MAGL activity, increases 2-AG tissue levels, and blocks pain behaviour in the formalin model through mechanisms involving both CB_1 and CB_2 (Guindon et al. 2011). Peripheral FAAH inhibition via intra-plantar URB597 reduced carrageenan-induced hyperalgesia (Jhaveri et al. 2008) and also ameliorates an electrophysiological marker of central sensitisation in rat spinal cord (Sagar et al. 2008). Unlike the actions of 2-AG, peripheral effects of AEA may not involve CB_2. URB937, a peripherally restricted FAAH inhibitor which cannot cross the blood–brain barrier, blocks the hypersensitivity induced by both inflammation and peripheral nerve injury via a CB_1 receptor-mediated mechanism (Clapper et al. 2010). Enhancing peripheral EC signalling also has potential in the relief of chronic pain states, since inhibition of MAGL or FAAH is antinociceptive via CB_1 and CB_2 in a rodent model of neuropathic pain (Desroches et al. 2014b).

5.2 The Spinal Endocannabinoid System and Acute Pain Processing

The dorsal horn of the spinal cord is a key region in the pain processing pathway, receiving and encoding sensory input from the periphery and integrating the descending modulatory signals from higher brain regions. Intrathecal administration of cannabinoids is antinociceptive (Welch and Stevens 1992), whilst blocking spinal CB_1 receptors produces hyperalgesia in mice (Richardson et al. 1997) and enhances nociception-evoked firing of wide dynamic range (WDR) neurones (Chapman 1999). These data suggest a powerful regulatory role of EC signalling at this level of the pain pathway.

Although the nociceptive circuitry of the dorsal horn has yet to be fully described (Todd 2010), the expression pattern of the EC system within some of its key components has been mapped. CB_1 receptors are highly expressed presynaptically on the central terminals of nociceptive primary afferents and on populations of excitatory interneurons within the dorsal horn (Hegyi et al. 2009). At these synapses, the 2-AG signalling machinery has a complementary expression pattern, with DAGLα present postsynaptically (Nyilas et al. 2009) and MAGL localised within presynaptic terminals (Horvath et al. 2014). In contrast, the localisation of AEA-related machinery in the nociceptive circuitry remains unclear (Hegyi et al. 2012).

The simplest picture arising from these data is that spinal 2-AG signalling initiated by excessive nociceptive activity negatively modulates nociceptive signalling via inhibiting the release of pronociceptive neurotransmitters from primary afferent terminals. Indeed, under naïve conditions, spinal administration of the

MAGL inhibitor JZL184 selectively inhibits acute mechanically evoked nociceptive neurotransmission in a CB_1-receptor-mediated manner in rats (Woodhams et al. 2012), whilst FAAH inhibition is ineffective (unpublished data from our group). However, CB_1 is also expressed on inhibitory interneurons and glial cells (Hegyi et al. 2009), and spinal levels of both AEA and 2-AG are elevated in animal models of acute and chronic pain (Sagar et al. 2010a, b, 2012). Spinal administration of AEA (Welch et al. 1995) or elevation of spinal AEA via the FAAH inhibitor URB597 (Jhaveri et al. 2006) is also antinociceptive, suggesting AEA signalling is also important. Furthermore, pronociceptive actions of CB_1 agonism on inhibitory interneurons have been described following severe noxious stimulation (Pernia-Andrade et al. 2009), suggesting that the effects of EC signalling vary depending upon the physiological context.

During sustained noxious stimulation, spinal 2-AG and AEA signalling is temporally segregated. Assessing spinal levels of ECs in the surgical incision model of a resolving pain state in rats has revealed a time course of changes (Alkaitis et al. 2010). In the hours following a peripheral surgical incision, no changes in spinal 2-AG are observed, but there was a marked decrease in AEA levels corresponding to maximal mechanical hypersensitivity, which returns to baseline as nociceptive behaviour subsides. In contrast, 2-AG levels increase at later time points, corresponding with glial cell activation, upregulation of CB_2 receptors, and resolution of the pain state. These data suggest that reduced AEA signalling may be involved in the onset of a pain state, whilst enhanced 2-AG signalling may be involved in pain resolution. Targeting AEA signals during the onset of pain behaviour may therefore not be an effective analgesic strategy, but it may instead be possible to block the onset of hypersensitivity by prolonging the immediate effects of 2-AG. A preliminary report indicates that spinal administration of JZL184 after carrageenan administration is able to block WDR receptive field expansion, a correlate of central sensitisation, in rats (Woodhams et al. 2012).

5.3 A Novel Role of Spinal CB_2 Receptors in Chronic Pain States

In persistent pain states, such as neuropathic pain resulting from peripheral nerve damage, central sensitisation leads to a reorganisation of spinal nociceptive circuitry and the development of hypersensitivity. This manifests as both hyperalgesia (excessive pain following a nociceptive stimulus) and allodynia (perception of a normally innocuous stimulus as painful), for which current analgesics are largely ineffective. It is in such chronic pain states that EC research has the greatest potential. Spinal EC signalling is significantly enhanced in models of neuropathic pain, with elevations of both cannabinoid receptors (Lim et al. 2003; Zhang et al. 2003) and ECs reported (Sagar et al. 2012). Given the role of spinal 2-AG signalling at CB_2 receptors in resolution of pain states highlighted above (Alkaitis et al. 2010), targeting CB_2 in chronic pain states is a novel area of research with great promise.

A number of studies have reported upregulation of CB_2 receptor mRNA and protein in the spinal cord in models of neuropathic pain (see references in Sagar et al. 2012), and spinal administration of CB_2 agonists attenuates both neuronal (Sagar et al. 2005) and behavioural (Romero-Sandoval et al. 2008a; Yamamoto et al. 2008) nociceptive responses. Interestingly, and in contrast to both mixed CB_1 and CB_2 agonists and selective CB_1 agonists, activation of spinal CB_2 receptors was not effective in control rats. Similarly, CB_2 agonism alters spinal nociceptive neuronal activity in a model of osteoarthritic pain, but has no effect in control animals (Burston et al. 2013). These data suggest a pain state-specific role for this receptor. Specificity of CB_2 agonists has been demonstrated by their lack of effect in CB_2 knockout mice in these models (Yamamoto et al. 2008). However, there is some conflicting evidence in the literature, with a recent study reporting no effect of the CB_2 agonists GW405833 and JWH-133 on mechanical allodynia in the chronic constriction injury (CCI) model of neuropathy (Brownjohn and Ashton 2012). This study also reported no elevation of CB_2 at either the protein or mRNA level, in contrast to numerous other studies.

The majority of preclinical studies of neuropathic pain utilise ligation or severance of peripheral nerves to induce a chronic pain state, but painful neuropathy is also associated with disease states such as diabetes, or as a result of chemotherapy treatments (Windebank and Grisold 2008). The few studies modelling these more clinically relevant conditions have, however, revealed evidence of a role for spinal CB_2 receptors in attenuating pain behaviour. In a model of chemotherapy-induced neuropathy, spinal administration of the pan cannabinoid receptor agonist WIN55, 212-2 reduced mechanical allodynia via CB_1 and CB_2 receptors (Rahn et al. 2007). Likewise, in the rat STZ model of diabetes and diabetic neuropathy, CB_1 receptor is downregulated in nociceptive primary afferent fibres (Zhang et al. 2007), but systemic administration of WIN55, 212-2 is still able to produce significant antinociceptive effects (Doğrul et al. 2004; Ulugol et al. 2004).

Targeting CB_2 may also be effective in other chronic pain states, such as osteoarthritis. Mice over-expressing the CB_2 receptor develop less severe pain behaviour following intra-articular injection of monosodium iodoacetate, whilst mice lacking CB_2 receptors show enhanced pain responses on the contralateral side, indicating a role of CB_2 receptors in this pain state (La Porta et al. 2013). Additionally, we have recently demonstrated that spinal CB_2 expression is elevated in a rat model of osteoarthritic pain and that chronic systemic administration of the CB_2 agonist JWH-133 can block pain behaviour, whilst acute spinal administration attenuates mechanically evoked nociceptive neurotransmission (Burston et al. 2013). Moreover, we found that spinal CB_2 mRNA expression was negatively correlated with (macroscopic) knee chondropathy in end-stage OA patients, suggesting that disease progression may lead to a decrease in CB_2 receptor expression and a facilitation of spinal hyperexcitability in human OA.

5.4 CB$_2$ Receptor Modulation of Spinal Immune Cell Function

CB$_2$ is primarily expressed on glial cells of the CNS, and these cells are therefore most likely to mediate the effects of CB$_2$ agonists in models of persistent pain. Studies of cultured microglial cells have revealed that 2-AG signalling via CB$_2$ receptors inhibits microglial migration (Walter et al. 2003) and that CB$_2$ agonists reduce both the activation and migration of cultured primary microglia challenged with bacterial lipopolysaccharide (Romero-Sandoval et al. 2009). In a unilateral spinal nerve transsection model of neuropathic pain, CB$_2$ upregulation on spinal microglia and perivascular cells has been reported (Romero-Sandoval et al. 2008b), and spinal administration of the CB$_2$ receptor agonist JWH015 reduced both mechanical hypersensitivity and markers of microglia activation in this model. CB$_2$ receptors are upregulated on both microglia and astrocytes following spared nerve injury in mice, and chronic systemic administration of the CB$_2$ agonist NESS400 reduces pain behaviour, astrogliosis, microglial activation, and levels of proinflammatory cytokines, whilst promoting levels of anti-inflammatory cytokines (Luongo et al. 2010).

Genetic manipulations in mice have revealed the likely function of upregulated CB$_2$ to be in attenuating the development of central sensitisation. Mice lacking CB$_2$ receptors have enhanced pain responses to sciatic nerve ligation compared to their wild-type littermates, including the development of hypersensitivity in the contralateral side (Racz et al. 2008b). These behavioural effects are accompanied by large increases in spinal glial cell activation, which can be significantly attenuated by over-expression of CB$_2$ receptors. Activation of CB$_2$ receptors can reduce the release of proinflammatory cytokines from glial cells, and this appears to be important in the analgesic mechanism. Neuropathy induces a large IFN-γ response in CB$_2$ knockout mice, whilst deletion of both IFN-γ and CB$_2$ blocks the development of neuropathy-induced pain behaviour (Racz et al. 2008a).

It should be noted that the role of CB$_2$ in persistent pain states may not be restricted to immune cells, as there is some evidence for neuronal CB$_2$ expression in the spinal cords of neuropathic (Racz et al. 2008b) and osteoarthritic animals (Burston et al. 2013). However, at the present time, a lack of specific tools to map the expression of this receptor has hampered efforts to conclusively demonstrate neuronal expression.

5.5 Supraspinal Sites of Action of the Endocannabinoids

At the supraspinal level, the EC system can influence ascending pain signals in the thalamus, descending modulatory signals in the brainstem, and the affective/emotional aspects of pain sensation through actions in frontal–limbic circuits. In the rat, direct microinjection of cannabinoid agonists into the thalamus, PAG, dorsal raphe nucleus, and RVM all produce antinociceptive effects in acute pain tests (Martin et al. 1995, 1996, 1999), which can be blocked via antagonism of CB$_1$ (Lichtman et al. 1996). The involvement of ECs at these sites has been revealed by the release of AEA following electrical stimulation of the PAG or after peripheral inflammatory insult (Walker et al. 1999). PAG levels of AEA and 2-AG are also elevated in animal models of neuropathic pain (Petrosino et al. 2007), highlighting this area as a key region in the pain matrix. Prolonging AEA action in the PAG via inhibition of FAAH is antinociceptive in acute pain tests, although a biphasic effect suggests higher AEA concentrations can result in pronociceptive via actions at TRPV1 (Maione et al. 2006). The mechanism of action of ECs in the RVM has been revealed through electrophysiological studies. In vivo recordings from the RVM in rats demonstrated that cannabinoids inhibit the firing of ON cells, whilst promoting the firing of OFF cells (Meng and Johansen 2004). As their names suggest, these two cell types have opposing effects, with ON cells facilitating and OFF cells inhibiting nociceptive activity.

The PAG has excitatory projections to the RVM, and the action of AEA in this region is thought to be via disinhibition of PAG neurons, leading to activation of OFF cells in the RVM, and induction of descending inhibitory GABAergic signalling to the spinal cord (Vaughan et al. 1999, 2000). EC signalling in the PAG also mediates stress-induced analgesia (SIA), the well-characterised reduction in nociceptive responses observed following the presentation of an external environmental stressor. In animal studies, mice or rats exposed to mild electrical foot shock show reduced pain behaviour in a subsequent test of acute nociception (Hohmann et al. 2005). This effect is mediated by CB_1 and accompanied by the release of both AEA and 2-AG in the PAG. The effect size of SIA can be augmented through the use of either FAAH or MAGL inhibitors (Suplita et al. 2005, 2006), though it should be noted the MAGL inhibitor utilised in this study has low efficacy and selectivity (Vandevoorde et al. 2007). More recent evidence suggests that the involvement of 2-AG in the dorsolateral PAG is critical and suggests a mechanism involving the activation of descending inhibitory pathways (Gregg et al. 2012). The link between stress and pain in humans is far more complex than in animal models, but recent clinical studies suggest it is no less important (Vachon-Presseau et al. 2013). Since the EC system is also strongly linked to anxiety/anxiolysis (reviewed in Mechoulam and Parker 2013), it seems highly likely that EC signalling will be involved in this interaction. Indeed, a recent report has demonstrated that impaired endocannabinoid signalling in the RVM underpins hyperresponsivity to noxious stimuli in the Wistar Kyoto rat, a strain with a heightened stress/anxiety profile (Rea et al. 2014).

One of the major unresolved issues in pain research is how to match the objectively measurable neuronal and behavioural nociceptive output of nociceptive responses, with the subjective, integrative experience of pain. Since non-human animals cannot report their internal states, and invasive electrophysiological recordings cannot be performed in humans, it has long seemed that this was a gap which could not be bridged. However, the advent of imaging techniques such as functional magnetic resonance imaging (fMRI) has allowed researchers to begin to address this. Cannabinoids alter affective or emotional states in humans via effects on the frontal–limbic circuitry, and this likely contributes to their analgesic effects. Dissociating the affective component from the somatosensory is complex, but a recent neuroimaging study has attempted to do so. Subjects were asked to rate the intensity of mechanically evoked allodynic responses following cutaneous capsaicin treatment, in the presence and absence of systemically administered Δ^9-THC, whilst brain activity was monitored via fMRI (Lee et al. 2013). The cannabinoid treatment had no significant effect on scores of pain intensity, but reportedly reduced the unpleasantness of the pain sensation. This effect was accompanied by reduced activity in the anterior cingulate cortex (ACC) and enhanced activity in the amygdala. The ability of EC-directed perturbations to mimic this affective component of cannabinoid analgesia has not yet been tested, but this study indicates that this will be an important question to answer. Some recent preclinical data mirror these fMRI findings and further highlight the ACC as an important area in the aversive component of pain (Lu et al. 2011; Chen et al. 2012). Astrocyte activity in the ACC correlates with the maintenance of pain aversion behaviour, and ablation of these astrocytes reverses the behavioural phenotype. Since astrocyte activity is regulated by cannabinoid receptors, it seems logical to assume that altering EC tone within the ACC could prove a useful strategy for treating the affective component of chronic pain states.

5.6 Enhancing EC Signalling: Problems of Plasticity

The antinociceptive efficacy of 2-AG elevation via MAGL inhibition in preclinical models is well established, presenting a very promising area for therapeutic research. Systemic administration of the MAGL inhibitor JZL184 (Long et al. 2009a) elevates levels of 2-AG in brain and peripheral tissues (Long et al. 2009b) and increases acute pain thresholds in mice. Robust antinociceptive effects have also been demonstrated in models of peripheral inflammatory pain (Ghosh et al. 2013), gastrointestinal pain (Busquets-Garcia et al. 2011; Kinsey et al. 2011), neuropathic pain (Kinsey et al. 2009, 2010), chemotherapy-induced neuropathy (Guindon et al. 2013), and bone cancer pain (Khasabova et al. 2011). Many of these conditions are refractory to standard analgesics, and thus targeting MAGL/2-AG has great potential. However, some recent evidence suggests that there may be a problem with using MAGL inhibitors. High doses of JZL184 induce some cannabinoid-like behaviours (Long et al. 2009a), suggesting that this approach may not entirely avoid the unwanted side effects of cannabinoids. Worse still, sustained global elevation of 2-AG via genetic deletion of MAGL or persistent blockade of MAGL activity with JZL184 produces functional antagonism of the brain EC system, resulting in profound downregulation and desensitisation of CB_1 receptors in nociception-associated regions and a loss of analgesic phenotype (Chanda et al. 2010; Schlosburg et al. 2010). Chronic JZL184 administration may also result in physical dependence since repeated high dose JZL184 treatment produces behavioural symptoms of withdrawal when precipitated with a CB_1 antagonist (Schlosburg et al. 2009).

However, these may not be insurmountable problems. Chronic partial inhibition of MAGL produces sustained analgesia in the absence of cannabinoid side effects in mice (Busquets-Garcia et al. 2011; Kinsey et al. 2011). Furthermore, a novel inhibitor, KML129, with a more attractive therapeutic profile has been developed (Ignatowska-Jankowska et al. 2014). This compound has greater selectivity for MAGL over FAAH than JZL184 and produces antinociceptive, but not cannabimimetic, effects in mouse models of acute and chronic pain. It remains to be seen whether this compound will also avoid functional antagonism of the EC system upon chronic administration.

FAAH inhibitors are also robustly antinociceptive in models of acute and chronic pain (Jayamanne et al. 2006; Chang et al. 2006; Lichtman et al. 2004; Kathuria et al. 2003; Fegley et al. 2005; Russo et al. 2007) and appear to lack the cannabimimetic effects associated with MAGL inhibition (Schlosburg et al. 2010; Busquets-Garcia et al. 2011). Although there is a wealth of preclinical data to support the use of EC-targeted compounds in clinical pain trials, the only major attempt to date was a failure. Pfizer investigated the ability of a highly selective FAAH inhibitor, PF-04457845, to produce analgesia in an osteoarthritic patient population (Huggins et al. 2012). The drug treatment was well tolerated and produced significant elevations in circulating AEA, but no analgesic effect was observed. Some preclinical evidence suggests a loss of antinociceptive activity with chronic administration of a FAAH inhibitor (Okine et al. 2012), but it is also possible that the major analgesic effect of cannabinoids in humans is dissociative, rather than sensory (Lee et al. 2013), an effect which has not been demonstrated with FAAH or MAGL inhibitors. This perhaps highlights a limitation of current clinical trials of analgesics, which focus on the sensory analgesic profiles of novel compounds without assessing effects on the affective/emotional aspects of pain. Based on this rather disappointing failure of PF-04457845, it remains to be seen whether the highly promising findings from animal studies can be translated into the clinic.

6 Summary

The EC system is a critical regulator of nociceptive function, active at all levels of the pain processing pathways. It is also a highly plastic system, with altered expression and function occurring with switches in physiological state. A wealth of preclinical data suggests that pharmacological modulation of EC function via the use of specific enzyme inhibitors is an efficacious analgesic approach and may be of particular importance in chronic, refractory pain states. However, clinical research in this area is still at an early stage, and some initial setbacks introduce a note of caution.

Acknowledgements D.S. and J.J.B. are funded by Arthritis Research UK Pain Centre funding (grant no. 18769).

References

Adams IB, Martin BR (1996) Cannabis: pharmacology and toxicology in animals and humans. Addiction 91:1585–1614

Agarwal N, Pacher P, Tegeder I, Amaya F, Constantin CE, Brenner GJ, Rubino T, Michalski CW, Marsicano G, Monory K, Mackie K, Marian C, Batkai S, Parolaro D, Fischer MJ, Reeh P, Kunos G, Kress M, Lutz B, Woolf CJ, Kuner R (2007) Cannabinoids mediate analgesia largely via peripheral type 1 cannabinoid receptors in nociceptors. Nat Neurosci 10:870–879

Aguiar DC, Moreira FA, Terzian AL, Fogaca MV, Lisboa SF, Wotjak CT, Guimaraes FS (2014) Modulation of defensive behavior by Transient Receptor Potential Vanilloid Type-1 (TRPV1) channels. Neurosci Biobehav Rev 46(Pt 3):418–428

Alexander SPH, Kendall DA (2007) The complications of promiscuity: endocannabinoid action and metabolism. Br J Pharmacol 152:602–623

Alkaitis MS, Solorzano C, Landry RP, Piomelli D, Deleo JA, Romero-Sandoval EA (2010) Evidence for a role of endocannabinoids, astrocytes and p38 phosphorylation in the resolution of postoperative pain. PLoS One 5:e10891

Atwood BK, Mackie K (2010) CB2: a cannabinoid receptor with an identity crisis. Br J Pharmacol 160:467–479

Beaulieu P, Bisogno T, Punwar S, Farquhar-Smith WP, Ambrosino G, Di Marzo V, Rice AS (2000) Role of the endogenous cannabinoid system in the formalin test of persistent pain in the rat. Eur J Pharmacol 396:85–92

Bisogno T, Howell F, Williams G, Minassi A, Cascio MG, Ligresti A, Matias I, Schiano-Moriello-A, Paul P, Williams EJ (2003) Cloning of the first sn1-DAG lipases points to the spatial and temporal regulation of endocannabinoid signalling in the brain. J Cell Biol 163:463–468

Bisogno T, Burston JJ, Rai R, Allarà M, Saha B, Mahadevan A, Razdan RK, Wiley JL, Di Marzo V (2009) Synthesis and pharmacological activity of a potent inhibitor of the biosynthesis of the endocannabinoid 2-arachidonoylglycerol. ChemMedChem 4:946–950

Blankman JL, Simon GM, Cravatt BF (2007) A comprehensive profile of brain enzymes that hydrolyze the endocannabinoid 2-arachidonoylglycerol. Chem Biol 14:1347–1356

Brownjohn PW, Ashton JC (2012) Spinal cannabinoid CB2 receptors as a target for neuropathic pain: an investigation using chronic constriction injury. Neuroscience 203:180–193

Burns HD, Van Laere K, Sanabria-Bohórquez S, Hamill TG, Bormans G, Eng W-S, Gibson R, Ryan C, Connolly B, Patel S, Krause S, Vanko A, Van Hecken A, Dupont P, De Lepeleire I, Rothenberg P, Stoch SA, Cote J, Hagmann WK, Stoch SA, Cote J, Hagmann WK, Jewell JP, Lin LS, Liu P, Goulet MT, Gottesdiener K, Wagner JA, de Hoon J, Mortelmans L, Fong TM, Hargreaves RJ (2007) [18F]MK-9470, a positron emission tomography (PET) tracer for in vivo human PET brain imaging of the cannabinoid-1 receptor. Proc Natl Acad Sci U S A 104:9800–9805

Burston JJ, Sagar DR, Shao P, Bai M, King E, Brailsford L, Turner JM, Hathway GJ, Bennett AJ, Walsh DA, Kendall DA, Lichtman A, Chapman V (2013) Cannabinoid CB2 receptors regulate central sensitization and pain responses associated with osteoarthritis of the knee joint. PLoS One 8:e80440

Busquets-Garcia A, Puighermanal E, Pastor A, de la Torre R, Maldonado R, Ozaita A (2011) Differential role of anandamide and 2-arachidonoylglycerol in memory and anxiety-like responses. Biol Psychiatry 70:479–486

Cabral GA, Griffin-Thomas L (2009) Emerging role of the cannabinoid receptor CB2 in immune regulation: therapeutic prospects for neuroinflammation. Expert Rev Mol Med 11:e3

Cabral GA, Raborn ES, Griffin L, Dennis J, Marciano-Cabral F (2008) CB2 receptors in the brain: role in central immune function. Br J Pharmacol 153:240–251

Cencioni MT, Chiurchiu V, Catanzaro G, Borsellino G, Bernardi G, Battistini L, Maccarrone M (2010) Anandamide suppresses proliferation and cytokine release from primary human T-lymphocytes mainly via CB2 receptors. PLoS One 5:e8688

Chanda PK, Gao Y, Mark L, Btesh J, Strassle BW, Lu P, Piesla MJ, Zhang M-Y, Bingham B, Uveges A, Kowal D, Garbe D, Kouranova EV, Ring RH, Bates B, Pangalos MN, Kennedy JD, Whiteside GT, Samad TA (2010) Monoacylglycerol lipase activity is a critical modulator of the tone and integrity of the endocannabinoid system. Mol Pharmacol 78:996–1003

Chang L, Luo L, Palmer JA, Sutton S, Wilson SJ, Barbier AJ, Breitenbucher JG, Chaplan SR, Webb M (2006) Inhibition of fatty acid amide hydrolase produces analgesia by multiple mechanisms. Br J Pharmacol 148:102–113

Chapman V (1999) The cannabinoid CB1 receptor antagonist, SR141716A, selectively facilitates nociceptive responses of dorsal horn neurones in the rat. Br J Pharmacol 127:1765–1767

Chen FL, Dong YL, Zhang ZJ, Cao DL, Xu J, Hui J, Zhu L, Gao YJ (2012) Activation of astrocytes in the anterior cingulate cortex contributes to the affective component of pain in an inflammatory pain model. Brain Res Bull 87:60–66

Clapper JR, Moreno-Sanz G, Russo R, Guijarro A, Vacondio F, Duranti A, Tontini A, Sanchini S, Sciolino NR, Spradley JM, Hohmann AG, Calignano A, Mor M, Tarzia G, Piomelli D (2010) Anandamide suppresses pain initiation through a peripheral endocannabinoid mechanism. Nat Neurosci 13:6

D'Agostino G, La Rana G, Russo R, Sasso O, Iacono A, Esposito E, Mattace Raso G, Cuzzocrea S, Loverme J, Piomelli D (2009) Central administration of palmitoylethanolamide reduces hyperalgesia in mice via inhibition of NF-kappaB nuclear signalling in dorsal root ganglia. Eur J Pharmacol 613:54–59

Desroches J, Guindon J, Lambert C, Beaulieu P (2008) Modulation of the anti-nociceptive effects of 2-arachidonoyl glycerol by peripherally administered FAAH and MGL inhibitors in a neuropathic pain model. Br J Pharmacol 155:913–924

Desroches J, Bouchard JF, Gendron L, Beaulieu P (2014a) Involvement of cannabinoid receptors in peripheral and spinal morphine analgesia. Neuroscience 261:23–42

Desroches J, Charron S, Bouchard JF, Beaulieu P (2014b) Endocannabinoids decrease neuropathic pain-related behavior in mice through the activation of one or both peripheral CB(1) and CB (2) receptors. Neuropharmacology 77:441–452

Devane WA, Dysarz FA, Johnson MR, Melvin LS, Howlett AC (1988) Determination and characterization of a cannabinoid receptor in rat-brain. Mol Pharmacol 34:605–613

Devane WA, Hanus L, Breuer A, Pertwee RG, Stevenson LA, Griffin G, Gibson D, Mandelbaum A, Etinger A, Mechoulam R (1992) Isolation and structure of a brain constituent that binds to the cannabinoid receptor. Science 258:1946–1949

Di Marzo V, De Petrocellis L (2012) Why do cannabinoid receptors have more than one endogenous ligand? Philos Trans R Soc Lond B Biol Sci 367:3216–3228

Di Marzo V, Fontana A, Cadas H, Schinelli S, Cimino G, Schwartz J-C, Piomelli D (1994) Formation and inactivation of endogenous cannabinoid anandamide in central neurons. Nature 372:686–691

Dinh TP, Carpenter D, Leslie FM, Freund TF, Katona I, Sensi SL, Kathuria S, Piomelli D (2002) Brain monoglyceride lipase participating in endocannabinoid inactivation. Proc Natl Acad Sci U S A 99:10819–10824

Doğrul A, Gül H, Yildiz O, Bilgin F, Güzeldemir ME (2004) Cannabinoids blocks tactile allodynia in diabetic mice without attenuation of its antinociceptive effect. Neurosci Lett 368:82–86

Egertova M, Elphick MR (2000) Localisation of cannabinoid receptors in the rat brain using antibodies to the intracellular C-terminal tail of CB1. J Comp Neurol 422:159–171

Fegley D, Gaetani S, Duranti A, Tontini A, Mor M, Tarzia G, Piomelli D (2005) Characterization of the fatty acid amide hydrolase inhibitor cyclohexyl carbamic acid 3'-carbamoyl-biphenyl-3-yl ester (URB597): effects on anandamide and oleoylethanolamide deactivation. J Pharmacol Exp Ther 313:352–358

Freund TF, Katona I, Piomelli D (2003) Role of endogenous cannabinoids in synaptic signalling. Physiol Rev 83(3):1017–1066

Garcia-Merino A (2013) Endocannabinoid system modulator use in everyday clinical practice in the UK and Spain. Expert Rev Neurother 13:9–13

Ghosh S, Wise LE, Chen Y, Gujjar R, Mahadevan A, Cravatt BF, Lichtman AH (2013) The monoacylglycerol lipase inhibitor JZL184 suppresses inflammatory pain in the mouse carrageenan model. Life Sci 92:498–505

Glaser ST, Kaczocha M (2010) Cyclooxygenase-2 mediates anandamide metabolism in the mouse brain. J Pharmacol Exp Ther 335:380–388

Graham ES, Angel CE, Schwarcz LE, Dunbar PR, Glass M (2010) Detailed characterisation of CB2 receptor protein expression in peripheral blood immune cells from healthy human volunteers using flow cytometry. Int J Immunopathol Pharmacol 23:25–34

Gregg LC, Jung K-M, Spradley JM, Nyilas R, Suplita RL, Zimmer A, Watanabe M, Mackie K, Katona IN, Piomelli D, Hohmann AG (2012) Activation of type 5 metabotropic glutamate receptors and diacylglycerol lipase-alpha initiates 2-arachidonoylglycerol formation and endocannabinoid-mediated analgesia. J Neurosci 32:9457–9468

Guindon J, Beaulieu P (2006) Antihyperalgesic effects of local injections of anandamide, ibuprofen, rofecoxib and their combinations in a model of neuropathic pain. Neuropharmacology 50:814–823

Guindon J, Loverme J, De Lean A, Piomelli D, Beaulieu P (2006) Synergistic antinociceptive effects of anandamide, an endocannabinoid, and nonsteroidal anti-inflammatory drugs in peripheral tissue: a role for endogenous fatty-acid ethanolamides? Eur J Pharmacol 550:68–77

Guindon J, Desroches J, Beaulieu P (2007) The antinociceptive effects of intraplantar injections of 2-arachidonoyl glycerol are mediated by cannabinoid CB2 receptors. Br J Pharmacol 150:693–701

Guindon J, Guijarro A, Piomelli D, Hohmann AG (2011) Peripheral antinociceptive effects of inhibitors of monoacylglycerol lipase in a rat model of inflammatory pain. Br J Pharmacol 163:1464–1478

Guindon J, Lai Y, Takacs SM, Bradshaw HB, Hohmann AG (2013) Alterations in endocannabinoid tone following chemotherapy-induced peripheral neuropathy: effects of endocannabinoid deactivation inhibitors targeting fatty-acid amide hydrolase and monoacylglycerol lipase in comparison to reference analgesics following cisplatin treatment. Pharmacol Res 67:94–109

Hall W, Solowij N (1998) Adverse effects of cannabis. Lancet 352:1611–1616

Han KH, Lim S, Ryu J, Lee CW, Kim Y, Kang JH, Kang SS, Ahn YK, Park CS, Kim JJ (2009) CB1 and CB2 cannabinoid receptors differentially regulate the production of reactive oxygen species by macrophages. Cardiovasc Res 84:378–386

Han J, Kesner P, Metna-Laurent M, Duan T, Xu L, Georges F, Koehl M, Abrous DN, Mendizabal-Zubiaga J, Grandes P, Liu Q, Bai G, Wang W, Xiong L, Ren W, Marsicano G, Zhang X (2012) Acute cannabinoids impair working memory through astroglial CB1 receptor modulation of hippocampal LTD. Cell 148:1039–1050

Hao M-X, Jiang L-S, Fang N-Y, Pu J, Hu L-H, Shen L-H, Song W, He B (2010) The cannabinoid WIN55,212-2 protects against oxidized LDL-induced inflammatory response in murine macrophages. J Lipid Res 51:2181–2190

Hegyi Z, Kis G, Holló K, Ledent C, Antal M (2009) Neuronal and glial localization of the cannabinoid-1 receptor in the superficial spinal dorsal horn of the rodent spinal cord. Eur J Neurosci 30:251–262

Hegyi Z, Holló K, Kis G, Mackie K, Antal M (2012) Differential distribution of diacylglycerol lipase-alpha and N-acylphosphatidylethanolamine-specific phospholipase d immunoreactivity in the superficial spinal dorsal horn of rats. Glia 60(9):1316–1329

Herkenham M, Lynn AB, Little MD, Johnson MR, Melvin LS, Decosta BR, Rice KC (1990) Cannabinoid receptor localization in brain. Proc Natl Acad Sci U S A 87:1932–1936

Herkenham M, Lynn AB, Johnson MR, Melvin LS, Decosta BR, Rice KC (1991) Characterization and localization of cannabinoid receptors in rat-brain—a quantitative in vitro autoradiographic study. J Neurosci 11:563–583

Hohmann AG, Suplita RL, Bolton NM, Neely MH, Fegley D, Mangieri R, Krey JF, Walker JM, Holmes PV, Crystal JD, Duranti A, Tontini A, Mor M, Tarzia G, Piomelli D (2005) An endocannabinoid mechanism for stress-induced analgesia. Nature 435:1108–1112

Horvath E, Woodhams SG, Nyilas R, Henstridge CM, Kano M, Sakimura K, Watanabe M, Katona I (2014) Heterogeneous presynaptic distribution of monoacylglycerol lipase, a multipotent regulator of nociceptive circuits in the mouse spinal cord. Eur J Neurosci 39:419–434

Hu JH, Yang L, Kammermeier PJ, Moore CG, Brakeman PR, Tu J, Yu S, Petralia RS, Li Z, Zhang PW, Park JM, Dong X, Xiao B, Worley PF (2012) Preso1 dynamically regulates group I metabotropic glutamate receptors. Nat Neurosci 15:836–844

Hu SS-J, Ho Y-C, Chiou L-C (2014) No more pain upon Gq-protein-coupled receptor activation: role of endocannabinoids. Eur J Neurosci 39:467–484

Huggins JP, Smart TS, Langman S, Taylor L, Young T (2012) An efficient randomised, placebo-controlled clinical trial with the irreversible fatty acid amide hydrolase-1 inhibitor PF-04457845, which modulates endocannabinoids but fails to induce effective analgesia in patients with pain due to osteoarthritis of the knee. Pain 153:1837–1846

Ibrahim MM, Porreca F, Lai J, Albrecht PJ, Rice FL, Khodorova A, Davar G, Makriyannis A, Vanderah TW, Mata HP, Malan TP (2005) CB2 cannabinoid receptor activation produces antinociception by stimulating peripheral release of endogenous opioids. Proc Natl Acad Sci U S A 102:3093–3098

Ibrahim MM, Rude ML, Stagg NJ, Mata HP, Lai J, Vanderah TW, Porreca F, Buckley NE, Makriyannis A, Malan TP (2006) CB2 cannabinoid receptor mediation of antinociception. Pain 122:36–42

Ignatowska-Jankowska BM, Ghosh S, Crowe MS, Kinsey SG, Niphakis MJ, Abdullah RA, Tao Q, O'Neal ST, Walentiny DM, Wiley JL, Cravatt BF, Lichtman AH (2014) In vivo characterization of the highly selective monoacylglycerol lipase inhibitor KML29: antinociceptive activity without cannabimimetic side effects. Br J Pharmacol 171(6):1392–1407

Jayamanne A, Greenwood R, Mitchell VA, Aslan S, Piomelli D, Vaughan CW (2006) Actions of the FAAH inhibitor URB597 in neuropathic and inflammatory chronic pain models. Br J Pharmacol 147:281–288

Jhaveri MD, Richardson D, Kendall DA, Barrett DA, Chapman V (2006) Analgesic effects of fatty acid amide hydrolase inhibition in a rat model of neuropathic pain. J Neurosci 26:13318–13327

Jhaveri MD, Richardson D, Robinson I, Garle MJ, Patel A, Sun Y, Sagar DR, Bennett AJ, Alexander SPH, Kendall DA, Barrett DA, Chapman V (2008) Inhibition of fatty acid amide hydrolase and cyclooxygenase-2 increases levels of endocannabinoid related molecules and produces analgesia via peroxisome proliferator-activated receptor-alpha in a model of inflammatory pain. Neuropharmacology 55:85–93

Kaplan BLF (2013) The role of CB1 in immune modulation by cannabinoids. Pharmacol Ther 137:365–374

Kathuria S, Gaetani S, Fegley D, Valino F, Duranti A, Tontini A, Mor M, Tarzia G, La Rana G, Calignano A, Giustino A, Tattoli M, Palmery M, Cuomo V, Piomelli D (2003) Modulation of anxiety through blockade of anandamide hydrolysis. Nat Med 9:76–81

Katona I, Freund T (2012) Multiple functions of endocannabinoid signalling in the brain. Annu Rev Neurosci 35:529–558

Katona I, Sperlagh B, Sik A, Kafalvi A, Vizi ES, Mackie K, Freund TF (1999) Presynaptically located CB1 cannabinoid receptors regulate GABA release from axon terminals of specific hippocampal interneurons. J Neurosci 19:4544–4558

Khasabova IA, Chandiramani A, Harding-Rose C, Simone DA, Seybold VS (2011) Increasing 2-arachidonoyl glycerol signalling in the periphery attenuates mechanical hyperalgesia in a model of bone cancer pain. Pharmacol Res 64:60–67

Kinsey SG, Long JZ, O'Neal ST, Abdullah RA, Poklis JL, Boger DL, Cravatt BF, Lichtman AH (2009) Blockade of endocannabinoid-degrading enzymes attenuates neuropathic pain. J Pharmacol Exp Ther 330:902–910

Kinsey SG, Long JZ, Cravatt BF, Lichtman AH (2010) Fatty acid amide hydrolase and monoacylglycerol lipase inhibitors produce anti-allodynic effects in mice through distinct cannabinoid receptor mechanisms. J Pain 11:1420–1428

Kinsey SG, Nomura DK, O'Neal ST, Long JZ, Mahadevan A, Cravatt BF, Grider JR, Lichtman AH (2011) Inhibition of monoacylglycerol lipase attenuates nonsteroidal anti-inflammatory drug-induced gastric hemorrhages in mice. J Pharmacol Exp Ther 338:795–802

Kinsey SG, Wise LE, Ramesh D, Abdullah R, Selley DE, Cravatt BF, Lichtman AH (2013) Repeated low dose administration of the monoacylglycerol lipase inhibitor JZL184 retains CB1 receptor mediated antinociceptive and gastroprotective effects. J Pharmacol Exp Ther 345(3):492–501

La Porta C, Bura SA, Aracil-Fernandez A, Manzanares J, Maldonado R (2013) Role of CB1 and CB2 cannabinoid receptors in the development of joint pain induced by monosodium iodoacetate. Pain 154:160–174

Lee MC, Ploner M, Wiech K, Bingel U, Wanigasekera V, Brooks J, Menon DK, Tracey I (2013) Amygdala activity contributes to the dissociative effect of cannabis on pain perception. Pain 154:124–134

Leung D, Saghatelian A, Simon GM, Cravatt BF (2006) Inactivation of N-acyl phosphatidylethanolamine phospholipase D reveals multiple mechanisms for the biosynthesis of endocannabinoids. Biochemistry 45:4720–4726

Lichtman AH, Cook SA, Martin BR (1996) Investigation of brain sites mediating cannabinoid-induced antinociception in rats: evidence supporting periaqueductal gray involvement. J Pharmacol Exp Ther 276:585–593

Lichtman A, Leung D, Shelton C, Saghatelian A, Hardouin C, Boger D, Cravatt B (2004) Reversible inhibitors of fatty acid amide hydrolase that promote analgesia: evidence for an unprecedented combination of potency and selectivity. J Pharmacol Exp Ther 311:441–448

Lim G, Sung B, Ji RR, Mao J (2003) Upregulation of spinal cannabinoid-1-receptors following nerve injury enhances the effects of Win 55,212-2 on neuropathic pain behaviors in rats. Pain 105:275–283

Lo Verme J, Fu J, Astarita G, La Rana G, Russo R, Calignano A, Piomelli D (2005) The nuclear receptor peroxisome proliferator-activated receptor-alpha mediates the anti-inflammatory actions of palmitoylethanolamide. Mol Pharmacol 67:15–19

Long JZ, Li WW, Booker L, Burston JJ, Kinsey SG, Schlosburg JE, Pavon FJ, Serrano AM, Selley DE, Parsons LH, Lichtman AH, Cravatt BF (2009a) Selective blockade of 2-arachidonoylglycerol hydrolysis produces cannabinoid behavioral effects. Nat Chem Biol 5:37–44

Long JZ, Nomura DK, Cravatt BF (2009b) Characterization of monoacylglycerol lipase inhibition reveals differences in central and peripheral endocannabinoid metabolism. Chem Biol 16:744–753

Lu Y, Zhu L, Gao YJ (2011) Pain-related aversion induces astrocytic reaction and proinflammatory cytokine expression in the anterior cingulate cortex in rats. Brain Res Bull 84:178–182

Lujan R, Nusser Z, Roberts JD, Shigemoto R, Somogyi P (1996) Perisynaptic location of metabotropic glutamate receptors mGluR1 and mGluR5 on dendrites and dendritic spines in the rat hippocampus. Eur J Neurosci 8:1488–1500

Luongo L, Palazzo E, Tambaro S, Giordano C, Gatta L, Scafuro MA, Rossi FS, Lazzari P, Pani L, De Novellis V, Malcangio M, Maione S (2010) 1-(2′,4′-dichlorophenyl)-6-methyl-N-cyclohexylamine-1,4-dihydroindeno[1,2-c]pyraz ole-3-carboxamide, a novel CB2 agonist, alleviates neuropathic pain through functional microglial changes in mice. Neurobiol Dis 37:177–185

Maione S, Bisogno T, De Novellis V, Palazzo E, Cristino L, Valenti M, Petrosino S, Guglielmotti V, Rossi F, Di Marzo V (2006) Elevation of endocannabinoid levels in the ventrolateral periaqueductal grey through inhibition of fatty acid amide hydrolase affects descending nociceptive pathways via both cannabinoid receptor type 1 and transient receptor potential vanilloid type-1 receptors. J Pharmacol Exp Ther 316:969–982

Maione S, De Petrocellis L, De Novellis V, Moriello AS, Petrosino S, Palazzo E, Rossi FS, Woodward DF, Di Marzo V (2007) Analgesic actions of N-arachidonoyl-serotonin, a fatty acid amide hydrolase inhibitor with antagonistic activity at vanilloid TRPV1 receptors. Br J Pharmacol 150:766–781

Martin WJ, Patrick SL, Coffin PO, Tsou K, Walker JM (1995) An examination of the central sites of action of cannabinoid-induced antinociception in the rat. Life Sci 56:2103–2109

Martin WJ, Hohmann AG, Walker JM (1996) Suppression of noxious stimulus-evoked activity in the ventral posterolateral nucleus of the thalamus by a cannabinoid agonist: correlation between electrophysiological and antinociceptive effects. J Neurosci 16:6601–6611

Martin WJ, Coffin PO, Attias E, Balinsky M, Tsou K, Walker JM (1999) Anatomical basis for cannabinoid-induced antinociception as revealed by intracerebral microinjections. Brain Res 822:237–242

Mechoulam R, Gaoni Y (1967) Absolute configuration of delta1-tetrahydrocannabinol major active constituent of hashish. Tetrahedron Lett 12:1109–1111

Mechoulam R, Parker LA (2013) The endocannabinoid system and the brain. Annu Rev Psychol 64:21–47

Meng ID, Johansen JP (2004) Antinociception and modulation of rostral ventromedial medulla neuronal activity by local microinfusion of a cannabinoid receptor agonist. Neuroscience 124:685–693

Min R, Nevian T (2012) Astrocyte signalling controls spike timing-dependent depression at neocortical synapses. Nat Neurosci 15:746–753

Nadal X, La Porta C, Andreea Bura S, Maldonadois R (2013) Involvement of the opioid and cannabinoid systems in pain control: new insights from knockout studies. Eur J Pharmacol 716 (1–3):142

Navarrete M, Araque A (2008) Endocannabinoids mediate neuron-astrocyte communication. Neuron 57:883–893

Nyilas R, Gregg LC, Mackie K, Watanabe M, Zimmer A, Hohmann AG, Katona I (2009) Molecular architecture of endocannabinoid signalling at nociceptive synapses mediating analgesia. Eur J Neurosci 29:1964–1978

Nyíri G, Cserép C, Szabadits E, Mackie K, Freund TF (2005) CB1 cannabinoid receptors are enriched in the perisynaptic annulus and on preterminal segments of hippocampal GABAergic axons. Neuroscience 136:811–822

Ohno-Shosaku T, Maejima T, Kano M (2001) Endogenous cannabinoids mediate retrograde signals from depolarized postsynaptic neurons to presynaptic terminals. Neuron 29:729–738

Okamoto Y, Morishita J, Tsuboi K, Tonai T, Ueda N (2004) Molecular characterization of a phospholipase D generating anandamide and its congeners. J Biol Chem 279:5298–5305

Okine BN, Norris LM, Woodhams S, Burston J, Patel A, Alexander SP, Barrett DA, Kendall DA, Bennett AJ, Chapman V (2012) Lack of effect of chronic pre-treatment with the FAAH inhibitor URB597 on inflammatory pain behaviour: evidence for plastic changes in the endocannabinoid system. Br J Pharmacol 167:627–640

Pacher P, Mechoulam R (2011) Is lipid signalling through cannabinoid 2 receptors part of a protective system? Prog Lipid Res 50:193–211

Pattij T, Wiskerke J, Schoffelmeer ANM (2008) Cannabinoid modulation of executive functions. Eur J Pharmacol 585:458–463

Pernia-Andrade AJ, Kato A, Witschi R, Nyilas R, Katona I, Freund TF, Watanabe M, Filitz J, Koppert W, Schuttler J, Ji G, Neugebauer V, Marsicano G, Lutz B, Vanegas H, Zeilhofer HU (2009) Spinal endocannabinoids and CB1 receptors mediate C-fiber-induced heterosynaptic pain sensitization. Science 325:760–764

Pertwee RG (1997) Pharmacology of cannabinoid CB1 and CB2 receptors. Pharmacol Ther 74:129–180

Petrosino S, Palazzo E, De Novellis V, Bisogno T, Rossi F, Maione S, Di Marzo V (2007) Changes in spinal and supraspinal endocannabinoid levels in neuropathic rats. Neuropharmacology 52:415–422

Quartilho A, Mata HP, Ibrahim MM, Vanderah TW, Porreca F, Makriyannis A, Malan TP (2003) Inhibition of inflammatory hyperalgesia by activation of peripheral CB2 cannabinoid receptors. Anesthesiology 99:955–960

Racz I, Nadal X, Alferink J, Banos JE, Rehnelt J, Martin M, Pintado B, Gutierrez-Adan A, Sanguino E, Bellora N (2008a) Interferon-gamma is a critical modulator of CB(2) cannabinoid receptor signalling during neuropathic pain. J Neurosci 28:12136–12145

Racz I, Nadal X, Alferink J, Banos JE, Rehnelt J, Martin M, Pintado B, Gutierrez-Adan A, Sanguino E, Manzanares J, Zimmer A, Maldonado R (2008b) Crucial role of CB2 cannabinoid receptor in the regulation of central immune responses during neuropathic pain. J Neurosci 28:12125–12135

Radulovic J, Tronson NC (2012) Preso1, mGluR5 and the machinery of pain. Nat Neurosci 15:805–807

Rahn EJ, Makriyannis A, Hohmann AG (2007) Activation of cannabinoid CB1 and CB2 receptors suppresses neuropathic nociception evoked by the chemotherapeutic agent vincristine in rats. Br J Pharmacol 152:765–777

Rea K, Roche M, Finn DP (2007) Supraspinal modulation of pain by cannabinoids: the role of GABA and glutamate. Br J Pharmacol 152:633–648

Rea K, Olango WM, Okine BN, Madasu MK, Mcguire IC, Coyle K, Harhen B, Roche M, Finn DP (2014) Impaired endocannabinoid signalling in the rostral ventromedial medulla underpins genotype-dependent hyper-responsivity to noxious stimuli. Pain 155:69–79

Richardson JD, Aanonsen L, Hargreaves KM (1997) SR 141716A, a cannabinoid receptor antagonist, produces hyperalgesia in untreated mice. Eur J Pharmacol 319:R3–R4

Romero-Sandoval A, Chai N, Nutile-Mcmenemy N, Deleo JA (2008a) A comparison of spinal Iba1 and GFAP expression in rodent models of acute and chronic pain. Brain Res 1219:116–126

Romero-Sandoval A, Nutile-Mcmenemy N, Deleo JA (2008b) Spinal microglial and perivascular cell cannabinoid receptor type 2 activation reduces behavioral hypersensitivity without tolerance after peripheral nerve injury. Anesthesiology 108:722–734

Romero-Sandoval EA, Horvath R, Landry RP, Deleo JA (2009) Cannabinoid receptor type 2 activation induces a microglial anti-inflammatory phenotype and reduces migration via MKP induction and ERK dephosphorylation. Mol Pain 5:25

Ross RA (2003) Anandamide and vanilloid TRPV1 receptors. Br J Pharmacol 140:790–801

Russo R, Loverme J, La Rana G, Compton TR, Parrott JA, Duranti A, Tontini A, Mor M, Tarzia G, Calignano A, Piomelli D (2007) The fatty acid amide hydrolase inhibitor URB597 (cyclohexylcarbamic acid 3'-carbamoylbiphenyl-3-yl ester) reduces neuropathic pain after oral administration in mice. J Pharmacol Exp Ther 322:236–242

Sagar DR, Kelly S, Millns PJ, O'Shaughnessey CT, Kendall DA, Chapman V (2005) Inhibitory effects of CB1 and CB2 receptor agonists on responses of DRG neurons and dorsal horn neurons in neuropathic rats. Eur J Neurosci 22:371–379

Sagar DR, Kendall DA, Chapman V (2008) Inhibition of fatty acid amide hydrolase produces PPAR-alpha-mediated analgesia in a rat model of inflammatory pain. Br J Pharmacol 155 (8):1297–1306

Sagar DR, Jhaveri M, Richardson D, Gray RA, De Lago E, Fernandez-Ruiz J, Barrett D, Kendall D, Chapman V (2010a) Endocannabinoid regulation of spinal nociceptive processing in a model of neuropathic pain. Eur J Neurosci 31:8

Sagar DR, Staniaszek LE, Okine BN, Woodhams S, Norris LM, Pearson RG, Garle MJ, Alexander SPH, Bennett AJ, Barrett DA, Kendall DA, Scammell BE, Chapman V (2010b) Tonic modulation of spinal hyperexcitability by the endocannabinoid receptor system in a rat model of osteoarthritis pain. Arthritis Rheum 62:3666–3676

Sagar DR, Burston JJ, Woodhams SG, Chapman V (2012) Dynamic changes to the endocannabinoid system in models of chronic pain. Philos Trans R Soc Lond B Biol Sci 367:3300–3311

Salio C, Fischer J, Franzoni MF, Conrath M (2002) Pre- and postsynaptic localizations of the CB1 cannabinoid receptor in the dorsal horn of the rat spinal cord. Neuroscience 110:755–764

Samson MT, Small-Howard A, Shimoda LMN, Koblan-Huberson M, Stokes AJ, Turner H (2003) Differential roles of CB1 and CB2 cannabinoid receptors in mast cells. J Immunol 170:4953–4962

Sang N, Zhang J, Chen C (2006) PGE2 glycerol ester, a COX-2 oxidative metabolite of 2-arachidonoyl glycerol, modulates inhibitory synaptic transmission in mouse hippocampal neurons. J Physiol 572:735–745

Sang N, Zhang J, Chen C (2007) COX-2 oxidative metabolite of endocannabinoid 2-AG enhances excitatory glutamatergic synaptic transmission and induces neurotoxicity. J Neurochem 102:1966–1977

Savinainen JR, Saario SM, Laitinen JT (2011) The serine hydrolases MAGL, ABHD6 and ABHD12 as guardians of 2-arachidonoylglycerol signalling through cannabinoid receptors. Acta Physiol 204:267–276

Schlosburg J, Carlson B, Ramesh D, Abdullah R, Long J, Cravatt B, Lichtman A (2009) Inhibitors of endocannabinoid-metabolizing enzymes reduce precipitated withdrawal responses in THC-dependent mice. AAPS J 11:342–352

Schlosburg JE, Blankman JL, Long JZ, Nomura DK, Pan B, Kinsey SG, Nguyen PT, Ramesh D, Booker L, Burston JJ, Thomas EA, Selley DE, Sim-Selley LJ, Liu Q-S, Lichtman AH, Cravatt BF (2010) Chronic monoacylglycerol lipase blockade causes functional antagonism of the endocannabinoid system. Nat Neurosci 13:1113–1119

Simon GM, Cravatt BF (2006) Endocannabinoid biosynthesis proceeding through glycerophospho-N-acyl ethanolamine and a role for α/β-hydrolase 4 in this pathway. J Biol Chem 281:26465–26472

Simon GM, Cravatt BF (2008) Anandamide biosynthesis catalyzed by the phosphodiesterase GDE1 and detection of glycerophospho-N-acyl ethanolamine precursors in mouse brain. J Biol Chem 283:9341–9349

Spradley JM, Guindon J, Hohmann AG (2010) Inhibitors of monoacylglycerol lipase, fatty-acid amide hydrolase and endocannabinoid transport differentially suppress capsaicin-induced behavioral sensitization through peripheral endocannabinoid mechanisms. Pharmacol Res 62:249–258

Stander S, Schmelz M, Metze D, Luger T, Rukwied R (2005) Distribution of cannabinoid receptor 1 (CB1) and 2 (CB2) on sensory nerve fibers and adnexal structures in human skin. J Dermatol Sci 38:177–188

Starowicz K, Przewlocka B (2012) Modulation of neuropathic-pain-related behaviour by the spinal endocannabinoid/endovanilloid system. Philos Trans R Soc Lond B Biol Sci 367:3286–3299

Stella N (2004) Cannabinoid signalling in glial cells. Glia 48:267–277

Stella N (2009) Endocannabinoid signalling in microglial cells. Neuropharmacology 56(Suppl 1):244–253

Stella N (2010) Cannabinoid and cannabinoid-like receptors in microglia, astrocytes, and astrocytomas. Glia 58:1017–1030

Stella N, Schweitzer P, Piomelli D (1997) A second endogenous cannabinoid that modulates long-term potentiation. Nature 388:773–778

Sugiura T, Kondo S, Kishimoto S, Miyashita T, Nakane S, Kodaka T, Suhara Y, Takayama H, Waku K (2000) Evidence that 2-arachidonoylglycerol but not N-palmitoylethanolamine or anandamide is the physiological ligand for the cannabinoid CB2 receptor: comparison of the agonistic activities of various cannabinoid receptor ligands in HL-60 cells. J Biol Chem 275:605–612

Suplita RL, Farthing JN, Gutierrez T, Hohmann AG (2005) Inhibition of fatty-acid amide hydrolase enhances cannabinoid stress-induced analgesia: sites of action in the dorsolateral periaqueductal gray and rostral ventromedial medulla. Neuropharmacology 49:1201–1209

Suplita RL, Gutierrez T, Fegley D, Piomelli D, Hohmann AG (2006) Endocannabinoids at the spinal level regulate, but do not mediate, nonopioid stress-induced analgesia. Neuropharmacology 50:372–379

Todd AJ (2010) Neuronal circuitry for pain processing in the dorsal horn. Nat Rev Neurosci 11:823–836

Ueda N, Tsuboi K, Uyama T (2013) Metabolism of endocannabinoids and related N-acylethanolamines: canonical and alternative pathways. FEBS J 280(9):1874–1894

Ulugol A, Karadag HC, Ipci Y, Tamer M, Dokmeci I (2004) The effect of WIN 55,212-2, a cannabinoid agonist, on tactile allodynia in diabetic rats. Neurosci Lett 371:167–170

Vachon-Presseau E, Martel M-O, Roy M, Caron E, Albouy GV, Marin M-F, Plante I, Sullivan MJ, Lupien SJ, Rainville P (2013) Acute stress contributes to individual differences in pain and pain-related brain activity in healthy and chronic pain patients. J Neurosci 33:6826–6833

Vandevoorde S, Jonsson KO, Labar G, Persson E, Lambert DM, Fowler CJ (2007) Lack of selectivity of URB602 for 2-oleoylglycerol compared to anandamide hydrolysis in vitro. Br J Pharmacol 150:186–191

Vaughan CW, Mcgregor IS, Christie MJ (1999) Cannabinoid receptor activation inhibits GABAergic neurotransmission in rostral ventromedial medulla neurons in vitro. Br J Pharmacol 127:935–940

Vaughan CW, Connor M, Bagley EE, Christie MJ (2000) Actions of cannabinoids on membrane properties and synaptic transmission in rat periaqueductal gray neurons in vitro. Mol Pharmacol 57:288–295

Velazquez KT, Mohammad H, Sweitzer SM (2007) Protein kinase C in pain: involvement of multiple isoforms. Pharmacol Res 55:578–589

Walker M, Huang S (2002) Cannabinoid analgesia. Pharmacol Ther 95:127–135

Walker JM, Huang SM, Strangman NM, Tsou K, Sanudo-Pena MC (1999) Pain modulation by release of the endogenous cannabinoid anandamide. Proc Natl Acad Sci U S A 96:12198–12203

Walter L, Franklin A, Witting A, Wade C, Xie YH, Kunos G, Mackie K, Stella N (2003) Nonpsychotropic cannabinoid receptors regulate microglial cell migration. J Neurosci 23:1398–1405

Welch SP, Stevens DL (1992) Antinociceptive activity of intrathecally administered cannabinoids alone, and in combination with morphine, in mice. J Pharmacol Exp Ther 262:10–18

Welch SP, Dunlow LD, Patrick GS, Razdan RK (1995) Characterization of anandamide- and fluoroanandamide-induced antinociception and cross-tolerance to delta 9-THC after intrathecal administration to mice: blockade of delta 9-THC-induced antinociception. J Pharmacol Exp Ther 273:1235–1244

Wilson RI, Nicoll RA (2001) Endogenous cannabinoids mediate retrograde signalling at hippocampal synapses. Nature 410:588–592

Windebank AJ, Grisold W (2008) Chemotherapy-induced neuropathy. J Peripher Nerv Syst 13:27–46

Woodhams SG, Wong A, Barrett DA, Bennett AJ, Chapman V, Alexander SPH (2012) Spinal administration of the monoacylglycerol lipase inhibitor JZL184 produces robust inhibitory effects on nociceptive processing and the development of central sensitization in the rat. Br J Pharmacol 167:1609–1619

Yamamoto W, Mikami T, Iwamura H (2008) Involvement of central cannabinoid CB2 receptor in reducing mechanical allodynia in a mouse model of neuropathic pain. Eur J Pharmacol 583:56–61

Zhang J, Hoffert C, Vu HK, Groblewski T, Ahmad S, O'Donnell D (2003) Induction of CB2 receptor expression in the rat spinal cord of neuropathic but not inflammatory chronic pain models. Eur J Neurosci 17:2750–2754

Zhang F, Hong S, Stone V, Smith PJW (2007) Expression of cannabinoid CB1 receptors in models of diabetic neuropathy. J Pharmacol Exp Ther 323:508–515

Zygmunt PM, Petersson J, Andersson DA, Chuang HH, Sorgard M, Di Marzo V, Julius D, Hogestatt ED (1999) Vanilloid receptors on sensory nerves mediate the vasodilator action of anandamide. Nature 400:452–457

The Role of Glia in the Spinal Cord in Neuropathic and Inflammatory Pain

Elizabeth Amy Old, Anna K. Clark, and Marzia Malcangio

Contents

1 Origin and Function of Glia 146
2 Acute and Chronic Pain 148
   2.1 Inflammatory and Neuropathic Pain 149
3 Spinal Glia Changes in Models of Neuropathic Pain 150
   3.1 Microglial Responses to Injury or Insult 150
   3.2 Astrocytic Responses to Injury or Insult 152
   3.3 CX3CL1, CX3CR1 and Cathepsin S 153
   3.4 TNF and TNFR 154
   3.5 IL-1β and IL-1R 156
4 Spinal Glia During Inflammatory Pain 157
5 Spinal Glia During Rheumatoid Arthritis Pain 158
6 Concluding Remarks 160
References 161

Abstract

Chronic pain, both inflammatory and neuropathic, is a debilitating condition in which the pain experience persists after the painful stimulus has resolved. The efficacy of current treatment strategies using opioids, NSAIDS and anticonvulsants is limited by the extensive side effects observed in patients, underlining the necessity for novel therapeutic targets. Preclinical models of chronic pain have recently provided evidence for a critical role played by glial cells in the mechanisms underlying the chronicity of pain, both at the site of damage in the periphery and in the dorsal horn of the spinal cord. Here microglia and astrocytes respond to the increased input from the periphery and change morphology, increase in number and release pro-nociceptive mediators such as

E.A. Old • A.K. Clark • M. Malcangio (✉)
Wolfson Centre for Age Related Diseases, King's College London, London, UK
e-mail: marzia.malcangio@kcl.ac.uk

ATP, cytokines and chemokines. These gliotransmitters can sensitise neurons by activation of their cognate receptors thereby contributing to central sensitization which is fundamental for the generation of allodynia, hyperalgesia and spontaneous pain.

Keywords

Glia • Microglia • Astrocytes • Neuropathic pain • Inflammatory pain • Spinal cord • CX3CL1/R1 • IL-1β • TNF • Rheumatoid arthritis

1 Origin and Function of Glia

The term neuroglia was coined to describe the interstitial substance surrounding neurons of the CNS (central nervous system) and was later determined to consist of distinct neuroglial cells. The name astrocyte was introduced to describe 'star-shaped' neuroglial cells which were observed to form the supportive system of the CNS. A third cellular element of the CNS was acknowledged in 1913 when improved staining techniques for neuroglial cells led to the recognition of non-neuronal cells that were distinct from astrocytes. During the 1920s del Rio-Hortega employed silver carbon staining and light microscopy to visualise a population of cells that appeared different from the macroglia (oligodendrocytes and astrocytes) that had previously been described in the CNS. These cells, termed microglia, were thought to arise either from CNS invasion of blood mononuclear (monocytic) cells or mesodermal pial elements (Del Rio-Hortega 1932, 2012a, b; Kettenmann et al. 2011; Prinz and Mildner 2011). Whilst alternatives for the origin of microglia have been hypothesised, evidence to support the hypothesis that these cells are of monocytic lineage was later strengthened using autoradiography. Leblond and colleagues demonstrated that microglia possess specific monocytic characteristics, specifically the ability to transform from an amoeboid to a ramified cell (Imamoto and Leblond 1978; Ling et al. 1980). The monocytic/myeloid origins of microglia has now been confirmed conclusively by the absence of microglia from the CNS of a genetically altered mouse strain that lack PU.1, a key transcription factor in the control of myeloid cell differentiation (McKercher et al. 1996; Beers et al. 2006).

Microglia appear at an early stage during embryogenesis where they originate from macrophages in the foetal yolk sac that migrate into the CNS (Saijo and Glass 2011); evidence from postnatal fluorescently labelled cells demonstrates that in this phase of development, microglia are capable of differentiating from monocytes entering the CNS from the circulation (Perry et al. 1985). Conversely, in adult rodents, circulating monocytes only enter the CNS under conditions where there is disruption to the blood-brain barrier (BBB) (King et al. 2009); however, as demonstrated by the use of immune-irradiated mice, the CNS colonisation by these cells is transient and does not contribute to the resident microglial cell pool

(Ajami et al. 2011). Rather the microglial population is maintained locally and independently of circulating monocytes via the proliferation of existing cells (Lawson et al. 1992; Ginhoux et al. 2010).

Microglia are specialised phagocytes of the CNS and constitute 5–20 % of the total glial cell population (Saijo and Glass 2011). Under physiological conditions microglia exist in a 'resting' or 'quiescent' state; these cells can be distinguished morphologically by their small soma and ramified process that perform immune surveillance of the surrounding area. Additionally they express receptors for complement components (Fcγ receptor of IgG) and exhibit low expression of cell surface antigens (Nimmerjahn et al. 2005). Several mechanisms by which these cells maintain a quiescent state have been proposed, including interaction of microglial CX3CR1 receptor with its neuronal ligand the chemokine CX3CL1 and inhibitory signalling through the microglial cell surface proteins CD172, CD200R and CD45 with their neuronal ligands CD47, CD200 and CD22, respectively (Ransohoff and Perry 2009; Ransohoff and Cardona 2010; Cardona et al. 2006; Saijo and Glass 2011).

Within the healthy CNS, microglia have a number of key roles in addition to their function as immune surveyors. As well as monitoring extrasynaptic regions, microglial processes transiently contact synapses, including presynaptic terminals and perisynaptic clefts (Tremblay et al. 2010). Additionally these cells possess a number of neurotransmitter receptors; combined these attributes allow microglia to monitor synaptic function (Salter and Beggs 2014). Microglial cells are present in the CNS from the early stages of development and due to their phagocytic capacity are able to contribute to the elimination of excess neurons that form as part of normal development (Marin-Teva et al. 2011). However, rather than simply removing waste, microglia can initiate apoptosis of cells via the release of several factors including superoxide ions and TNF (Marin-Teva et al. 2004; Sedel et al. 2004; Salter and Beggs 2014). Additionally, a number of inappropriate synaptic connections are made between neurons during development, and microglia play an important role in the regulation of these contacts via the process of synaptic pruning—the phagocytosis of both pre- and postsynaptic elements in a complement cascade-dependent manner (Salter and Beggs 2014; Stevens et al. 2007; Schafer et al. 2012). Microglia contribute to the maturation of established synapses; the functional properties of synapses develop abnormally in mice deficient in several microglial proteins, including CX3CR1, for example, the absence of which results in increased excitatory neurotransmission (Paolicelli et al. 2011; Salter and Beggs 2014). The involvement of microglia to the maintenance of synaptic plasticity has also been investigated in the adult CNS, where they contribute to both homeostatic and activity-triggered plasticity by releasing a number of mediators such as TNF and TGFβ (Butovsky et al. 2014; Koeglsperger et al. 2013).

Of all of the glial cell populations within the CNS, astrocytes are by far the most abundant. In mammals astrocytogenesis begins in late embryogenesis and continues into the postnatal phase. The origin of astrocytes is likely diverse, varying throughout the stages of development. For example, within the cerebral cortex, astrocytes develop from two distinct sources: from radial glia in the ventricular zone during embryogenesis and from subventricular zone (SVZ) progenitors during the postnatal period (Wang and Bordey 2008). Additionally, studies using in vivo retroviral gene transfer have demonstrated that SVZ progenitors can generate both grey and white matter astrocytes as well as oligodendrocytes (Levison and Goldman 1993, 1997; Levison et al. 1993).

The typical image of an astrocyte is that of a stellate cell; however, astrocytes have a complex and heterogeneous morphology (Wang and Bordey 2008). Whilst the nomenclature is considered outdated by some, astrocytes are typically classified into one of two subtypes, protoplasmic and fibrous, where the former are found throughout the grey matter and possess branches that give rise to uniformly distributed processes and the latter are distributed within the white matter and exhibit long fibrelike processes (Sofroniew and Vinters 2010). Furthermore, the branching processes of protoplasmic astrocytes envelop synapses and have endfeet that encase blood vessels, whilst the endfeet of the unbranched processes of fibrous astrocytes envelop nodes of Ranvier (Wang and Bordey 2008). Immunohistochemically astrocytes are commonly identified by the presence of glial fibrils, the major component of which is glial fibrillary acidic protein (GFAP). Antibodies against S100B, a member of the S100 family of EF-band calcium binding proteins, are also used to identify astrocytes; however, this antigen is only expressed by a subset of mature astrocytes. Thus, its expression is not representative of the astrocyte population as a whole (Baudier et al. 1986; Deloulme et al. 2004; Hachem et al. 2005).

Due to the absence of several biophysical properties such as the ability to generate action potentials under physiological conditions, astrocytes were once described as passive (Steinhauser et al. 1992); however, this term is now rarely used. Whilst the initial function of astrocytes was understood to be as little more than an inert scaffold to neurons, it is now widely accepted that they play a critical role in a vast number of processes within the CNS (Volterra and Meldolesi 2005). The close association of astrocytes with many neuronal elements allows astrocytes to regulate the extraneuronal environment around synapses and provide support and nourishment to these cells (Gao and Ji 2010), for example, by buffering extracellular potassium and glutamate. Additionally astrocytes have been demonstrated to significantly contribute to neuronal survival and maturation (via the synthesis and release of growth and trophic factors such as NGF), synapse formation and the regulation of angiogenesis and are a major source of adhesion and extracellular matrix proteins within the CNS that can both promote and inhibit neurite growth (Wang and Bordey 2008).

2 Acute and Chronic Pain

Pain is a subjective sensory experience associated with actual or potential tissue damage (www.iasp.org). It is always unpleasant and therefore an emotional experience. Acute pain is mediated by multi-synaptic pathways beginning with specialised primary afferent fibres (nociceptors) in the periphery, whose cell bodies lie in the dorsal root ganglia (DRG), via the dorsal horn of the spinal cord to supraspinal sites such as the parabrachial area (PBA), periaqueductal grey (PAG), thalamus and cortex. The nociceptors convert the energy of noxious stimuli into electrical impulses which are transmitted to the dorsal horn of the spinal cord. Within the spinal cord, nociceptive transmission is mediated predominately by glutamate acting upon postsynaptic ionotropic receptors; however, the co-release of substance P and calcitonin gene-related peptide (CGRP) and the subsequent activation of their respective NK1 and CGRP receptors postsynaptically modulates glutamatergic transmission (Go and Yaksh 1987; Cheunsuang and Morris 2000; Malcangio and Bowery 1994; Lever et al. 2003; Oku et al. 1987; Seybold et al. 2003). Modulatory control of acute pain is provided by descending inhibitory and facilitatory pathways from the brain and from inhibitory and excitatory interneurons within the spinal cord (Todd 2010; D'Mello and Dickenson 2008).

Acute pain is a vital protective mechanism that persists only for the duration of tissue damage or the presence of a noxious stimulus. Evidence for the importance of acute pain is provided by individuals expressing the rare phenotype of insensitivity to pain; these patients experience regular inadvertent self-mutilation through injury and have a significantly lower than average life expectancy (Cox et al. 2006; Verpoorten et al. 2006).

Commonly pain outlives its usefulness and becomes chronic pain, which is profoundly different from acute pain, as it is the result of plastic changes within the pain pathways. Chronic pain is a debilitating condition lasting longer than 3 months from the noxious stimuli and causes pain to be perceived as out of proportion to the initial inciting injury. Chronic pain can be classified into different types depending on its cause: inflammatory pain is commonly due to tissue inflammation as observed in arthritis, and neuropathic pain arises upon injury to the nervous system.

2.1 Inflammatory and Neuropathic Pain

In both cases a combination of mechanisms results in augmented nociceptive transmission due to peripheral and central sensitisation. Neuropathic pain is that arising from a lesion to the PNS (peripheral nervous system) or CNS as a consequence of physical trauma or disease pathogenesis. Chronic inflammatory pain is typically associated with inflammatory diseases such as arthritis, where the presence of algogenic mediators results in nociceptor sensitisation in the affected tissue, such as the joints in the case of arthritis. Chronic pain is characterised by the presence of a range of symptoms including hyperalgesia (increased sensitivity to noxious stimuli), allodnyia (a painful response to a previously innocuous stimuli) and spontaneous pain (Woolf and Mannion 1999). The treatment of chronic pain is a substantial problem within the clinic as many patients do not respond adequately to available analgesics. The difficulty in treating chronic pain conditions is thought to be a consequence of the heterogeneity of the molecular mechanisms underlying the development and maintenance of chronic pain, many of which are not well understood.

The mechanisms by which these maladaptive pain states arise can be categorised into peripheral sensitisation (changes in peripheral nerves) and central sensitisation (including immune responses in the spinal cord). Peripheral sensitisation due to injury to a localised area leads to primary hyperalgesia and results in increased nociceptive transmission from the periphery which causes changes in dorsal horn neuron activation. Dorsal horn neurons exhibit reduced thresholds to noxious stimuli applied to the periphery, de novo excitation by previously innocuous stimuli, expansion of their receptive fields and increased spontaneous activity. These changes increase excitability of CNS neurons and result in central sensitisation, which is fundamental for the generation of allodynia, secondary hyperalgesia and spontaneous pain (Kuner 2010; Latremoliere and Woolf 2009; Sandkuhler 2009). The resulting increase in glutamate released from central terminals of sensitised neurons trigger phosphorylation of NMDA receptors and post-translational modifications of neuronal proteins. Glial cells within the spinal cord respond to the enhanced nociceptive input by proliferating and switching to a responsive state whereby they release mediators which contribute to dorsal horn mechanisms of chronic pain (Clark and Malcangio 2012; Old and Malcangio 2012). Enhanced activity within the spinal cord triggers cortical and subcortical structures to facilitate excitation and signalling and constitutes higher centre modulation (D'Mello and Dickenson 2008).

Here we will describe and define some of the key mechanisms by which microglia and astrocyte responses in the dorsal horn contribute to the development and maintenance of chronic neuropathic and inflammatory pain states, with a specific focus on chemokine/cytokine signalling and second messenger activation.

3 Spinal Glia Changes in Models of Neuropathic Pain

3.1 Microglial Responses to Injury or Insult

Microglia are the resident immune cells of the CNS and as such respond to pathological insult or tissue injury and subsequent release of mediators from damaged cells. These cells respond to a number of mediators that are increased in the dorsal horn following peripheral injury. They express receptors for the neurotransmitters released from the central terminals of primary afferents: glutamate NMDA and AMPA receptors, the SP receptor NK1 and the CGRP receptor, TrkB receptors for brain derived neurotrophic factor (BDNF) and purinergic receptors including P2X7 and P2X4, many of which are upregulated following injury to the periphery (Rasley et al. 2002; Pezet et al. 2002; McMahon and Malcangio 2009; Ransohoff and Perry 2009).

Microglia respond quickly to an insult by proliferating to expand their population; this process is often termed microgliosis. An insult need not be central in nature, for example it could be the result of a CNS infection or spinal cord injury. In rodent peripheral nerve injury models of neuropathic pain, microglia within the dorsal horn of the spinal cord respond swiftly to augmented primary afferent fibre input. This phenotypic shift is often referred to as the 'activation' of microglia. Input from the periphery is vital for this process; blockade of peripheral nerve conduction by the application of a local anaesthetic prevents nerve injury-induced microglial response (Wen et al. 2007; Hathway et al. 2009; Suter et al. 2009). In the region of the dorsal horn in which the injured primary afferents terminate, microglia increase in their numbers and appear more amoeboid, with a hypertrophied soma and thick retracted processes. Additionally they alter their expression of cell surface antigens that play a critical role in immune responses; for example, MHC class II is increased allowing the presentation of antigens by microglia (McMahon and Malcangio 2009). Activated microglia also synthesise and release a variety of pro-nociceptive mediators into the extracellular environment, including cytokines (e.g. TNF and IL-1β), chemokines (e.g. CCL2), reactive oxygen species (ROS) and nitric oxide (NO). Prominent signalling pathways in the development of neuropathic pain are the CX3CR1/L1 loop (discussed below) and the purinergic pathway of microglial activation and subsequent BDNF release. ATP is known to stimulate microglia both in vivo and in vitro via purinergic receptors on the surface membrane of the cells (Honda et al. 2001; Tsuda et al. 2003, 2012; Davalos et al. 2005). Microglial P2X4 is of particular importance for the development of neuropathic pain; it is upregulated in microglia as early as 24 hours after peripheral nerve injury, and the pharmacological inhibition of this receptor transiently attenuates mechanical allodynia in rodents. Furthermore, the intrathecal administration of ATP-treated microglia in naive mice results in the development of mechanical hypersensitivity (Tsuda et al. 2003; Coull et al. 2005). Current evidence indicated that downstream effects of the activation of microglial purinergic receptors are mediated by the release of BDNF via the phosphorylation of p38 MAPK; the administration of an inhibitor of the BDNF receptor, TrkB, prevents the behavioural changes observed following the administration of activated microglia (Coull et al. 2005), and a p38 MAPK inhibitor is able to prevent release of BDNF from microglia (Trang et al. 2009).

The temporal profile of microglial activation within the spinal cord is well documented in models of neuropathic pain. Real-time PCR of spinal cord tissue extracts demonstrates alterations in the levels of microglia-specific mRNA transcripts (e.g. TLR4 and CD14) within hours of injury (Tanga et al. 2004), and the phosphorylation of the microglial MAPK p38 is evident within minutes from peripheral nerve injury (Svensson et al. 2005b; Clark et al. 2007b). Morphological changes (the presence of amoeboid cells and increased immunoreactivity for the microglial marker Ox42) have been observed immunohistochemically as early as 3 days after nerve injury and are maintained for at least 50 days post-surgery (Clark et al. 2007a). Furthermore, the temporal profile of microglial activation is concomitant to the development of behavioural hypersensitivity (Clark et al. 2007b; Zhang et al. 2007; Peters et al. 2007) and the intrathecal administration of compounds that inhibit glial activation to neuropathic rodents is able to attenuate pain behaviours (Clark et al. 2007a; Tawfik et al. 2007), demonstrating that the activation of these cells does indeed contribute to aberrant pain signalling.

3.2 Astrocytic Responses to Injury or Insult

Due to their many functions in the maintenance of CNS homeostasis, under physiological conditions astrocytes are often referred to as 'active'. Following a change in their environment, such as that occurring in the spinal cord in models of chronic pain, these cells shift to a 'reactive' phenotype; astrocytes undergo hypertrophy and increase their expression of a number of cellular proteins including GFAP, S100β and vimentin, which are consequently often used as markers of astrocyte reactivity (Ridet et al. 1997; Pekny and Nilsson 2005). This process is referred to as astrogliosis and has been demonstrated in a number of surgical models of neuropathic pain (Garrison et al. 1991; Colburn et al. 1999; Sweitzer et al. 1999). Astrocyte reactivity can occur through the activation of several pathways; these cells express the receptors for the neurotransmitters NMDA, SP and CGRP (Porter and McCarthy 1997) and thus respond to increased transmitter release in the dorsal horn of the spinal cord subsequent to peripheral nerve injury. In addition, astrocytes possess a variety of cytokine and chemokine receptors that can be activated by their ligands released from microglia; key examples include IL-18/IL-18R, TNF/TNFR and CCL2/CCR2 (Miyoshi et al. 2008; Gao et al. 2009, 2010b). One of the astrocytic changes most likely to be responsible for the contribution of these cells to the maintenance of pain is a decrease in the glutamate transporters GLT1 and GLAST. These transporters function to regulate extracellular glutamate concentrations at non-toxic levels. A decrease in their expression or function results in an elevation in the concentration of spinal extracellular glutamate which can elicit nociceptive hypersensitivity via the activation of NMDA and AMPA receptors (Liaw et al. 2005; Weng et al. 2006). Recent evidence indicates that astrocytes, like microglia, play a critical role in aberrant pain signalling, as the administration of glial inhibitors, such as fluorocitrate, attenuates pain behaviours in rodents (Milligan et al. 2003; Watkins et al. 1997; Okada-Ogawa et al. 2009). Interestingly, the intrathecal administration of reactive astrocytes, briefly incubated with TNF, induces the development of mechanical allodynia in uninjured mice (Gao et al. 2010b). As with microglia, the temporal profile of astrocyte activation has been assessed; astrogliosis becomes apparent after the microglial response, several days after peripheral nerve injury, and lasts for over 4 weeks post-injury (Colburn et al. 1999; Tanga et al. 2004; Romero-Sandoval et al. 2008). Additionally genetically altered mice that are deficient in GFAP develop a shorter lasting mechanical allodynia than their wild-type counterparts. Finally the administration of a GFAP antisense mRNA that prevents translation of GFAP reverses established mechanical allodynia when administered to rats 6 weeks following a peripheral nerve injury (Kim et al. 2009). Together these data demonstrate a role for astrocytes in the maintenance of neuropathic pain.

3.3 CX3CL1, CX3CR1 and Cathepsin S

CX3CL1, also known as fractalkine, is the only member of the CX3C family of cytokines. The cx3cl1 gene was originally described as abundant in the brain and heart but present in all tissues assessed, except peripheral blood leukocytes (Pan et al. 1997). Recent evidence obtained from a series of in situ hybridisation and immunohistochemical studies has determined that neurons are the principle source of CX3CL1 in the CNS (Hughes et al. 2002; Nishiyori et al. 1998; Tarozzo et al. 2003). Additionally, the development of a transgenic mouse expressing a red fluorescent reporter under the control of the CX3CL1 promoter has demonstrated that within the spinal cord CX3CL1 is expressed in dorsal horn neurons (Kim et al. 2011). Both the CX3CL1 protein and mRNA are expressed here, but the protein is not upregulated following peripheral nerve injury (Verge et al. 2004; Clark et al. 2009; Lindia et al. 2005). Furthermore, it has been demonstrated that cytoplasmic glutamate-containing vesicles are present within CX3CL1-positive neurons, indicating that CX3CL1 is expressed within excitatory neurons (Tong et al. 2000).

First described as a potent chemoattractant of T-cells and monocytes in the late 1990s (Bazan et al. 1997; Pan et al. 1997), CX3CL1 exists in two forms. As a membrane-tethered protein it plays a critical role in the firm adhesion of leukocytes to the endothelium to facilitate transmigration (Imai et al. 1997; Fong et al. 1998; Corcione et al. 2009). Soluble CX3CL1 is produced by metalloproteases (ADAM10/17) or protease (Cathepsin S) -mediated cleavage of membrane-bound CX3CL1. Whilst ADAM 10 is responsible for constitutive shedding of CX3CL1, ADAM 17 facilitates inducible shedding of this protein (Bazan et al. 1997; Hundhausen et al. 2003, 2007; Clark et al. 2011). Cathepsin S, on the other hand, is expressed by antigen-presenting cells such as microglia where its release is dependent on the activation of the purinergic receptor P2X7 (Clark et al. 2010) and is enhanced by pro-inflammatory mediators (Liuzzo et al. 1999a, b). CX3CL1 exerts its biological effects by binding to CX3CR1 for which it is the only ligand. Expression analysis of CX3CR1 has demonstrated its mRNA to be abundant in the spleen and peripheral blood leukocytes as well as the brain. Furthermore, the expression of CX3CR1 mRNA is tenfold higher in cultured microglial than whole brain samples, suggesting the receptor is of microglial origin. FACS analysis and calcium imaging of microglia have demonstrated that these cells possess the functional CX3CR1 protein as well as the mRNA (Harrison et al. 1998). This expression profile of CX3CR1 has been confirmed conclusively by the development of a transgenic mouse in which the CX3CR1 contains a green fluorescent protein reporter allowing visualisation of the transcribed protein; here, in the mouse, CX3CR1 protein was observed in microglia throughout the CNS but was not present in astrocytes or oligodendrocytes (Jung et al. 2000).

Within the CNS the CX3CL1–CX3CR1 interaction contributes the maintenance of a quiescent phenotype in microglia and suppresses the release of pro-inflammatory mediators (Mizuno et al. 2003; Lyons et al. 2009). As such, under these conditions, this protein–protein relationship is thought to be neuroprotective. In the context of chronic pain, several observations support a pro-nociceptive role of CX3CL1. The administration of the soluble chemokine domain of CX3CL1 into the intrathecal space at the lumbar level is pro-nociceptive and causes otherwise naive animals to exhibit nocifensive behaviours (Zhuang et al. 2007; Clark et al. 2007b; Milligan et al. 2004, 2005a). Consistently, the intrathecal administration of anti-CX3CL1 antibodies to neuropathic rodents attenuates pain-related behaviour (Clark et al. 2007b). Interestingly, the CSF levels of CX3CL1 increase in neuropathic animals compared to sham controls (Clark et al. 2007b, 2009). CX3CR1 exhibits similar pro-nociceptive attributes under aberrant pain conditions. Enhanced expression of the protein within the dorsal horn of the spinal cord is associated with microgliosis following peripheral nerve injury (Zhuang et al. 2007; Staniland et al. 2010). One mechanism described is IL-6 dependent; IL-6 mRNA and protein expression is induced in neurons following the peripheral nerve injury (Arruda et al. 1998; Lee et al. 2009), and prophylactic treatment with an IL-6 neutralising antibody prevents increased CX3CR1 expression, whilst, conversely, the administration of recombinant IL-6 significantly augments CX3CR1 expression (Lee et al. 2010). Supporting a pro-nociceptive role of CX3CR1 and a critical role for CX3CL1–CX3CR1 interaction in the development of pathological pain responses, CX3CR1-deficient mice do not develop hyperalgesia and/or allodynia in models of nerve injury and exhibit reduced microgliosis when compared to their wild-type littermate controls (Staniland et al. 2010). Similarly, the intrathecal administration of an anti-CX3CR1 antibody attenuates both the behavioural and microglial responses to injury (Zhuang et al. 2007; Milligan et al. 2004, 2005a). Within spinal cord microglia activation of CX3CR1 by CX3CL1 results in increased intracellular calcium concentrations (Harrison et al. 1998), the phosphorylation of p38 MAPK and subsequent release of pro-nociceptive molecules such as IL-6, NO and IL-1β (Zhuang et al. 2007; Clark et al. 2007b).

3.4 TNF and TNFR

TNF, previously known as TNFα, is a small pro-inflammatory cytokine first described in activated macrophages as a molecule with tumour-regression activity (Carswell et al. 1975). TNF belongs to a superfamily of ligand/receptor proteins that share a structural motif — the TNF homology domain. The TNF receptors are the other members of this family of proteins; two have been identified and are either constitutively expressed (TNFR1/p55-R) or inducible (TNFR2/p75-R) (Bodmer et al. 2002; Leung and Cahill 2010). Under physiological conditions TNF is expressed at very low levels in the spinal cord; however, it is rapidly upregulated in both microglia and astrocytes (glia) and neurons following peripheral injury (DeLeo et al. 1997; Ohtori et al. 2004; Hao et al. 2007; Youn et al. 2008). Similarly both TNFR1 and TNFR2 are expressed within glia and neurons in the spinal cord (Gruber-Schoffnegger et al. 2013; Ohtori et al. 2004; Hao et al. 2007). The use of receptor-specific protein ligands has demonstrated that it is TNFR1 activation in the spinal cord that is responsible for the pro-nociceptive effects of TNF under steady-state conditions; TNFR2, on the other hand, contributes to the pro-nociceptive effects of TNF under chronic pain states following its upregulation after nerve injury (Liu et al. 2007; Clark et al. 2013).

In the context of peripheral nerve injury models of neuropathic pain, TNF is detectable in the periphery at the site of injury where it is upregulated and exerts pro-nociceptive functions (George et al. 1999; Shubayev and Myers 2000; Sommer and Schafers 1998; Leung and Cahill 2010). Within the spinal cord TNF exerts its pro-nociceptive effects via actions on both neurons and glia. The spinal administration of TNF is associated with the development of mechanical allodynia and thermal hypersensitivity (Youn et al. 2008; Zhang et al. 2011). Additionally the application of exogenous TNF to spinal cord slices induces rapid modulation of Aδ- and C-fibre (nociceptive-fibre)-mediated neurotransmission (Youn et al. 2008), enhances dorsal horn neuronal responses to C-fibre stimulation (Reeve et al. 2000) and augments excitatory neurotransmission in lamina II neurons in a predominantly TNFR1-dependent manner (Zhang et al. 2011). Additionally long-term potentiation (LTP) induced by stimulation of the sciatic nerve is attenuated in TNFR knock-out mice (Park et al. 2011). These effects may not be mediated directly by the activation of neuronal TNF receptors, but rather indirectly by the activation of receptors located on glial cells within the dorsal horn (Gruber-Schoffnegger et al. 2013). Indeed, TNF is also able to modulate glial cell activation within the spinal cord as blockade of TNF signalling is associated with a reduction in glial cell activity in models of peripheral nerve injury (Svensson et al. 2005b; Nadeau et al. 2011). Via P2X7 signalling ATP is able to induce TNF production in microglia by a P38 MAPK-dependent pathway (Suzuki et al. 2004; Lister et al. 2007); in rats with a peripheral nerve injury, concomitant increases of TNF and p-P38 are observed in the spinal cord, and blockade of TNF suppresses P38 phosphorylation and microgliosis (Schafers et al. 2003; Marchand et al. 2009). Evidence from cultured astrocytes demonstrates that TNF transiently activates JNK (Gao et al. 2009), a mitogen-activated kinase (MAPK) associated with the synthesis and release of pro-nociceptive mediators (Ji et al. 2009). In particular this pathway is associated with the upregulation of astrocytic CCL2; CCL2 upregulation by TNF is dose dependently inhibited by a JNK inhibitor (Gao et al. 2009). Similarly, the spinal injection of TNF produces JNK-dependent behavioural hypersensitivity, whilst the administration of a JNK inhibitor to nerve-injured animals suppresses pain-associated behaviour (Gao et al. 2009). Peripheral nerve injury induces a slow but persistent activation of JNK MAPK particularly within astrocytes in the spinal cord. Whilst the administration of a specific JNK inhibitor has no effect on acute pain responses, it potently prevents and reverses nerve injury-induced mechanical allodynia in rodents (Zhuang et al. 2006).

3.5 IL-1β and IL-1R

Interleukin-1β (IL-1β) is a small pro-inflammatory cytokine first described as a pyrogenic factor released from leukocytes (Dinarello 2007), and was one of the first cytokines to be implicated in the mechanisms underlying enhanced nociception in rodents following peripheral nerve injury (Clark et al. 2013). IL-1β, a 17.5 kDa protein, is part of the well-characterised IL-1 family of proteins that also includes IL-1α, the IL-1 receptor (IL-1R1), a decoy receptor (IL-1R2) and the endogenous receptor antagonist (IL-1ra). IL-1ra is structurally similar to IL-1β and binds to IL-1R1 with similar affinity; however, it lacks the ability to activate the receptor and stimulate downstream signalling. Similarly, IL-1R2 is able to bind IL-1β with comparable affinity to IL-1R1 but lacks the appropriate cytosolic region to activate downstream signalling proteins (Weber et al. 2010; Dunn et al. 2001). Within the CNS IL-1β is expressed at low levels by both neurons and glial cells (Ren and Torres 2009; Clark et al. 2006; Copray et al. 2001; Guo et al. 2007; Sommer and Kress 2004), and the expression of IL-1R1 is equally widespread within the spinal cord (Zhang et al. 2008; Sweitzer et al. 1999; Gruber-Schoffnegger et al. 2013).

Under physiological conditions IL-1β has a number of homeostatic functions such as regulation of sleep and temperature (Dinarello 1996). More importantly, IL-1β is a key regulator of pro-inflammatory responses and the control of innate immune responses to pathogen-associated danger signals [pathogen-associated molecular patterns (PAMPS) and danger-associated molecular patterns (DAMPS)], such as lipopolysaccharide (LPS) and bacterial DNA. The mechanism by which IL-1β is secreted from cells differs from the classical ER–Golgi route of protein secretion due to a lack of a leader sequence (Rubartelli et al. 1990). Rather, IL-1β is synthesised as a larger precursor protein and is cleaved into a mature form; the mechanism behind this has now been elucidated and involves caspase-1, which cleaves IL-1β, and a series of accessory proteins that form a complex known as the inflammasome that provides a platform for the activation of caspase (Martinon et al. 2002; Schroder and Tschopp 2010).

Similarly to TNF, IL-1β is pro-nociceptive when administered centrally; the intrathecal administration of exogenous IL-1β results in thermal and mechanical hypersensitivity (Gruber-Schoffnegger et al. 2013; Reeve et al. 2000; Sung et al. 2004; Kawasaki et al. 2008). Further evidence for a pro-nociceptive effect of IL-1β comes from rodents lacking this protein which exhibit attenuated behavioural responses to peripheral nerve injury (Wolf et al. 2006). Furthermore it has been observed that human patients with a range of painful neuropathies exhibit enhanced CSF levels of IL-1β (Alexander et al. 2005; Backonja et al. 2008). Following peripheral nerve injury, IL-1β is upregulated in the spinal cord where it is expressed primarily by glial cells but also by neurons (Clark et al. 2013). Inhibition of normal IL-1β signalling is anti-nociceptive in animal models of neuropathic pain; intrathecal administration of IL-1ra can both prevent and reverse (Milligan et al. 2005b, 2006) nocifensive behaviours in these animals, as can inhibition of caspase-1 which prevents the proteolytic activation of IL-1β, resulting in decreased secretion of the mature form from microglia (Clark et al. 2006).

IL-1β released from glial cells is able to modulate neuronal activity. This cytokine can facilitate glutamatergic transmission via the phosphorylation of the NMDA receptor (Gruber-Schoffnegger et al. 2013) which enhances behavioural hypersensitivity; treatment with IL-1ra attenuates both the behavioural phenotype and receptor phosphorylation. The mechanism underlying this modulation is complex and calcium dependent (Viviani et al. 2003), and involves several intracellular signalling proteins including PKC, IP3, PLC, PLA2 and src kinases. Inhibitors of these proteins are able to block IL-1β-induced phosphorylation of either the NR1 subunit or NR2A/B subunit of the NMDA receptor in vitro similarly to IL-1ra (Viviani et al. 2003). The application of exogenous IL-1β to spinal cord slices is able to increase AMPA-mediated currents in lamina II neurons (Kawasaki et al. 2008) and induce LTP within lamina I neurons (Gruber-Schoffnegger et al. 2013). Patch clamp studies of IL-1β-treated spinal cord slices demonstrate that IL-1β enhances excitatory neurotransmission whilst attenuating inhibitory neurotransmission (Kawasaki et al. 2008). Overall these data indicate that IL-1β contributes to a potentiation of excitatory transmission and a suppression of inhibitory transmission, which together correlate with the facilitation of behavioural hypersensitivity.

As with many other cytokines, as well as a neuronal effect, IL-1β is thought to contribute to a pro-nociceptive state via the activation of glia. Additionally glia may contribute to the effects of IL-1β on neuronal excitability as these are absent following the inhibition of glial cell function (Gruber-Schoffnegger et al. 2013; Liu et al. 2013).

4   Spinal Glia During Inflammatory Pain

Inflammation is a critical protective mechanism occurring in response to injury, infection or irritation. It is characterised by five defining components: redness, heat, swelling, loss of function and pain. Under physiological conditions inflammation allows removal or repair of damaged tissue following an injury to the organism. Under these circumstances the role of inflammatory pain is protective, limiting the use of the affected area and preventing further damage during the healing process. However, in patients with chronic inflammatory conditions, such as arthritis (see section below), chronic pain hypersensitivity is a common complaint (Walsh and McWilliams 2014). The continued presence of algogenic mediators results in the sensitisation of the peripheral tissues (Schaible et al. 2009). Commonly models which involve the direct administration of an exogenous algogenic substance into the hind-paw of the rodent have been utilised in order to investigate how spinal glial mechanisms contribute to inflammatory pain. However, increasing spinal glial mechanisms are being studied in more clinically relevant models of inflammatory pain, such as arthritis.

An extensive body of evidence supports a role for spinal microglia and astrocytes (glia) in inflammatory pain mechanisms. Peripheral inflammation results in sensitization of spinal neurons (Vazquez et al. 2012; Konig et al. 2014) and glial cell reactivity within the dorsal horn of the spinal cord. Spinal astrogliosis and/or microgliosis have been reported in many inflammatory pain models (Watkins et al. 1997; Sweitzer et al. 1999; Clark et al. 2007a; Raghavendra et al. 2004). Critically, inhibition of glial cell activity during peripheral inflammation reduces pain behaviours substantially (Meller et al. 1994; Watkins et al. 1997; Clark et al. 2007a), suggesting that the activity of these cells in the spinal cord is vital for the full development of inflammatory pain.

A number of the spinal glia mechanisms that have to identified to contribute to neuropathic pain may also play a role in pain that occurs as result of peripheral inflammation. The cytokine/chemokine glial mechanisms described in relation to neuropathic pain above also play a role in inflammatory pain. One such astrocytic mechanism is inflammation-induced phosphorylation of JNK in spinal astrocytes (Gao et al. 2010a). Intraplantar injection of CFA (Gao et al. 2010a) or carrageenan (Bas et al. 2015) induces JNK phosphorylation, and JNK inhibition attenuates inflammatory pain behaviours (Gao et al. 2010a; Bas et al. 2015). This inflammation-induced JNK phosphorylation is regulated by spinal TNF secretion (Bas et al. 2015) as is the case during neuropathic pain (Gao et al. 2010a). One spinal microglial mechanism that regulates inflammatory pain is the phosphorylation of p38 MAPK. Peripheral inflammation induces extensive p38 phosphorylation which is restricted to microglial cells (Svensson et al. 2003a, 2005a). Inhibition of p38 in inflammatory pain models results in attenuation of pain behaviours (Svensson et al. 2003b, 2005a). The activation of many microglial receptors results in intracellular p38 phosphorylation, suggesting that this is a key intracellular signalling pathway during both inflammatory and neuropathic pain. Indeed, it has previously been demonstrated that the microglial CX3CR1 receptor leads to p38 phosphorylation (Clark et al. 2007b) and is critical for the full expression of inflammatory pain (Staniland et al. 2010), as well as neuropathic pain (see above section).

5 Spinal Glia During Rheumatoid Arthritis Pain

Rheumatoid arthritis (RA) is a chronic autoimmune disease characterised by synovial inflammation and joint destruction. The disease aetiology remains unclear but is thought to comprise a complex interplay between environmental and genetic factors. The clinical signs of RA are accompanied by chronic pain, representing a major unmet clinical need. Despite the availability of disease-modifying agents that reduce the clinical signs of RA, the treatment of chronic pain remains inadequate at present (Kidd et al. 2007; Sokka et al. 2007; Dray 2008; Walsh and McWilliams 2014). The mechanisms of RA pathology have been studied extensively using rodent models. Although pain is common in RA patients, it is only relatively recently that pain has been studied in rodent models that replicate the complex mechanisms of this inflammatory disorder. Whilst inflammatory pain states associated with models of mono-arthritis have been extensively characterised, pain that occurs as a result of more clinically relevant poly-arthritic models has only recently been examined.

A number of poly-arthritic rodent models, all of which mimic some aspect of human RA (Williams 1998; Vincent et al. 2012), have been studied in the context of pain. Critically the activation of glial cells within the spinal cord represents a commonality between these models. Collagen-induced arthritis (CIA) represents the rodent model of RA that is most widely used for pathogenesis studies. In CIA immunisation with type II collagen results in an immune response directed against the joints, which closely resembles many aspects of human RA (Trentham 1982; Williams 2004). In addition, passive transfer of collagen antibody cocktails (collagen antibody-induced arthritis; CAIA) or of serum from the spontaneous arthritic KxB/N mouse is also commonly used to model RA.

The models detailed above have all been identified as inducing pain behaviours in the rodent hind-paw. In the CIA model, DBA/1 mice develop mechanical and thermal hypersensitivity from the onset of arthritis (Inglis et al. 2007). In the rat, CIA-induced mechanical hypersensitivity is present before the onset of inflammation (Clark et al. 2012), mirroring the sensory changes reported in models of other autoimmune disorders such as multiple sclerosis, in which hypersensitivity is present before the onset of clinical scores (Olechowski et al. 2009; Clark and Malcangio 2012). Furthermore, pain associated with the CAIA and K/BxN serum transfer models of experimental arthritis in mice has been recently characterised (Christianson et al. 2010, 2011; Bas et al. 2012). In both models joint inflammation is transient, whereas mechanical hypersensitivity is long-lasting, persisting long after joint inflammation has subsided. Importantly, arthritis-induced hypersensitivity is attenuated by analgesic agents used clinically for the treatment of RA, such as anti-TNF agents and NSAIDs (Inglis et al. 2007; Christianson et al. 2010; Bas et al. 2012; Boettger et al. 2010), suggesting good correlation between the inflammatory pain in these models and pain in RA patients. Interestingly in both the CAIA and K/BxN models, mechanical hypersensitivity after the resolution of joint inflammation is NSAID insensitive, suggesting that the early inflammatory stage of these models is followed by a late phase which is non-inflammatory (Christianson et al. 2010; Bas et al. 2012).

The contribution of CNS changes in these RA models is not fully understood, however, spinal glial cell reactivity has been reported. Astrogliosis is observed in the lumbar dorsal horn of CIA (Inglis et al. 2007; Clark et al. 2012), CAIA (Bas et al. 2012; Agalave et al. 2014) and K/BxN serum transfer treated animals (Christianson et al. 2010, 2011). It is evident that glial cells contribute to arthritis-induced hypersensitivity, as intrathecal administration of the glial inhibitor pentoxifylline is able to reverse late-stage CAIA-induced hypersensitivity (Bas et al. 2012). However, it is currently unclear whether this analgesic effect is due to inhibition of astrocytes and/or microglia cell activity. Intrathecal administration of a JNK inhibitor is also able to reverse late-phase hypersensitivity in the CAIA model (Bas et al. 2012). JNK phosphorylation occurs in spinal astrocytes following peripheral inflammation (Gao et al. 2010a) and peripheral nerve injury (Zhuang et al. 2006).

CAIA also enhances spinal JNK phosphorylation, however, the cellular localisation of p-JNK in arthritis is yet to be determined (Bas et al. 2012). Thus, it remains to be fully established whether changes in astrocyte activity contribute to arthritis-induced hypersensitivity.

A role for spinal microglia in arthritis-induced hypersensitivity seems increasingly likely. In the rat CIA model, increased microglial cell activity during mechanical hypersensitivity is observed as phosphorylation of p38 MAPK (p-p38) (Clark et al. 2012). It has recently been demonstrated that spinal inhibition of the microglia protease CatS or the chemokine FKN, which constitutes a key neuron-microglial signalling system during neuropathic pain (see above section), is able to reverse established mechanical hypersensitivity in CIA and reduce microglial p-p38 (Clark et al. 2012). This analgesic effect of CatS or FKN inhibition is independent of alteration in disease progression, with inflammation and clinical score unaffected by treatment (Clark et al. 2012). Increased microglia p-p38 has also been observed in adjuvant-induced arthritis (Boyle et al. 2006). Whilst Boyle and colleagues did not examine pain hypersensitivity in detail in this study, spinal administration of a p38 inhibitor was significantly more effective at reducing joint inflammation than systemic administration of the same compound (Boyle et al. 2006). A similar reduction in inflammation was also evident following intrathecal treatment with the anti-TNF agent etanercept, which attenuated arthritis-induced p-p38 (Boyle et al. 2006), suggesting that spinal microglial mechanisms may also regulate peripheral inflammation under some conditions. Indeed, in RA patients, TNF neutralisation inhibits pain prior to reducing inflammation in the joint (Hess et al. 2011), raising the possibility of a spinal microglia contribution. Spinal TLR-4 also appears to regulate microglial cell activity during both K/BxN serum transfer arthritis and CAIA. In TLR-4 knock-out mice, the late phase of K/BxN mechanical hypersensitivity, and spinal gliosis, is attenuated compared to wild-type mice (Christianson et al. 2011). Interestingly, during CAIA, the damage-associated molecular pattern (DAMP) molecule, extracellular high-mobility group box-1 protein (HMGB1) acting as a TLR-4 ligand, is critical for arthritis-induced hypersensitivity (Agalave et al. 2014). Thus, a number of microglial targets may contribute to hypersensitivity in models of RA and may represent promising therapeutic targets for the treatment of established pain in RA patients.

6   Concluding Remarks

The concept that glial-mediated mechanisms modulate neuronal processing in the spinal cord and contribute to the facilitation of pain signalling has developed considerably in recent years. That a substantial number of pre-clinical studies have flourished in this field reinforces the premise that glial targets may be exploited as novel approaches to the treatment of chronic pain.

References

Agalave NM, Larsson M, Abdelmoaty S, Su J, Baharpoor A, Lundback P, Palmblad K, Andersson U, Harris H, Svensson CI (2014) Spinal HMGB1 induces TLR4-mediated long-lasting hypersensitivity and glial activation and regulates pain-like behavior in experimental arthritis. Pain 155(9):1802–1813

Ajami B, Bennett JL, Krieger C, McNagny KM, Rossi FM (2011) Infiltrating monocytes trigger EAE progression, but do not contribute to the resident microglia pool. Nat Neurosci 14:1142–1149

Alexander GM, van Rijn MA, van Hilten JJ, Perreault MJ, Schwartzman RJ (2005) Changes in cerebrospinal fluid levels of pro-inflammatory cytokines in CRPS. Pain 116:213–219

Arruda JL, Colburn RW, Rickman AJ, Rutkowski MD, DeLeo JA (1998) Increase of interleukin-6 mRNA in the spinal cord following peripheral nerve injury in the rat: potential role of IL-6 in neuropathic pain. Brain Res Mol Brain Res 62:228–235

Backonja MM, Coe CL, Muller DA, Schell K (2008) Altered cytokine levels in the blood and cerebrospinal fluid of chronic pain patients. J Neuroimmunol 195:157–163

Bas DB, Su J, Sandor K, Agalave NM, Lundberg J, Codeluppi S, Baharpoor A, Nandakumar KS, Holmdahl R, Svensson CI (2012) Collagen antibody-induced arthritis evokes persistent pain with spinal glial involvement and transient prostaglandin dependency. Arthritis Rheum 64:3886–3896

Bas DB, Abdelmoaty S, Sandor K, Codeluppi S, Fitzsimmons B, Steinauer J, Hua XY, Yaksh TL, Svensson CI (2015) Spinal release of tumour necrosis factor activates c-Jun N-terminal kinase and mediates inflammation-induced hypersensitivity. Eur J Pain 19(2):260–270

Baudier J, Glasser N, Gerard D (1986) Ions binding to S100 proteins. I. Calcium- and zinc-binding properties of bovine brain S100 alpha alpha, S100a (alpha beta), and S100b (beta beta) protein: Zn2+ regulates Ca2+ binding on S100b protein. J Biol Chem 261:8192–8203

Bazan JF, Bacon KB, Hardiman G, Wang W, Soo K, Rossi D, Greaves DR, Zlotnik A, Schall TJ (1997) A new class of membrane-bound chemokine with a CX3C motif. Nature 385:640–644

Beers DR, Henkel JS, Xiao Q, Zhao W, Wang J, Yen AA, Siklos L, McKercher SR, Appel SH (2006) Wild-type microglia extend survival in PU.1 knockout mice with familial amyotrophic lateral sclerosis. Proc Natl Acad Sci U S A 103:16021–16026

Bodmer JL, Schneider P, Tschopp J (2002) The molecular architecture of the TNF superfamily. Trends Biochem Sci 27:19–26

Boettger MK, Weber K, Grossmann D, Gajda M, Bauer R, Bar KJ, Schulz S, Voss A, Geis C, Brauer R, Schaible HG (2010) Spinal tumor necrosis factor alpha neutralization reduces peripheral inflammation and hyperalgesia and suppresses autonomic responses in experimental arthritis: a role for spinal tumor necrosis factor alpha during induction and maintenance of peripheral inflammation. Arthritis Rheum 62:1308–1318

Boyle DL, Jones TL, Hammaker D, Svensson CI, Rosengren S, Albani S, Sorkin L, Firestein GS (2006) Regulation of peripheral inflammation by spinal p38 MAP kinase in rats. PLoS Med 3:e338

Butovsky O, Jedrychowski MP, Moore CS, Cialic R, Lanser AJ, Gabriely G, Koeglsperger T, Dake B, Wu PM, Doykan CE, Fanek Z, Liu L, Chen Z, Rothstein JD, Ransohoff RM, Gygi SP, Antel JP, Weiner HL (2014) Identification of a unique TGF-beta-dependent molecular and functional signature in microglia. Nat Neurosci 17:131–143

Cardona AE, Pioro EP, Sasse ME, Kostenko V, Cardona SM, Dijkstra IM, Huang D, Kidd G, Dombrowski S, Dutta R, Lee JC, Cook DN, Jung S, Lira SA, Littman DR, Ransohoff RM (2006) Control of microglial neurotoxicity by the fractalkine receptor. Nat Neurosci 9:917–924

Carswell EA, Old LJ, Kassel RL, Green S, Fiore N, Williamson B (1975) An endotoxin-induced serum factor that causes necrosis of tumors. Proc Natl Acad Sci U S A 72:3666–3670

Cheunsuang O, Morris R (2000) Spinal lamina I neurons that express neurokinin 1 receptors: morphological analysis. Neuroscience 97:335–345

Christianson CA, Corr M, Firestein GS, Mobargha A, Yaksh TL, Svensson CI (2010) Characterization of the acute and persistent pain state present in K/BxN serum transfer arthritis. Pain 151:394–403

Christianson CA, Dumlao DS, Stokes JA, Dennis EA, Svensson CI, Corr M, Yaksh TL (2011) Spinal TLR4 mediates the transition to a persistent mechanical hypersensitivity after the resolution of inflammation in serum-transferred arthritis. Pain 152:2881–2891

Clark AK, Malcangio M (2012) Microglial signalling mechanisms: Cathepsin S and Fractalkine. Exp Neurol 234:283–292

Clark AK, D'Aquisto F, Gentry C, Marchand F, McMahon SB, Malcangio M (2006) Rapid co-release of interleukin 1beta and caspase 1 in spinal cord inflammation. J Neurochem 99:868–880

Clark AK, Gentry C, Bradbury EJ, McMahon SB, Malcangio M (2007a) Role of spinal microglia in rat models of peripheral nerve injury and inflammation. Eur J Pain 11:223–230

Clark AK, Yip PK, Grist J, Gentry C, Staniland AA, Marchand F, Dehvari M, Wotherspoon G, Winter J, Ullah J, Bevan S, Malcangio M (2007b) Inhibition of spinal microglial cathepsin S for the reversal of neuropathic pain. Proc Natl Acad Sci U S A 104:10655–10660

Clark AK, Yip PK, Malcangio M (2009) The liberation of fractalkine in the dorsal horn requires microglial cathepsin S. J Neurosci 29:6945–6954

Clark AK, Staniland AA, Marchand F, Kaan TK, McMahon SB, Malcangio M (2010) P2X7-dependent release of interleukin-1beta and nociception in the spinal cord following lipopolysaccharide. J Neurosci 30:573–582

Clark AK, Staniland AA, Malcangio M (2011) Fractalkine/CX3CR1 signalling in chronic pain and inflammation. Curr Pharm Biotechnol 12:1707–1714

Clark AK, Grist J, Al-Kashi A, Perretti M, Malcangio M (2012) Spinal cathepsin S and fractalkine contribute to chronic pain in the collagen-induced arthritis model. Arthritis Rheum 64:2038–2047

Clark AK, Old EA, Malcangio M (2013) Neuropathic pain and cytokines: current perspectives. J Pain Res 6:803–814

Colburn RW, Rickman AJ, DeLeo JA (1999) The effect of site and type of nerve injury on spinal glial activation and neuropathic pain behavior. Exp Neurol 157:289–304

Copray JC, Mantingh I, Brouwer N, Biber K, Kust BM, Liem RS, Huitinga I, Tilders FJ, Van Dam AM, Boddeke HW (2001) Expression of interleukin-1 beta in rat dorsal root ganglia. J Neuroimmunol 118:203–211

Corcione A, Ferretti E, Bertolotto M, Fais F, Raffaghello L, Gregorio A, Tenca C, Ottonello L, Gambini C, Furtado G, Lira S, Pistoia V (2009) CX3CR1 is expressed by human B lymphocytes and mediates [corrected] CX3CL1 driven chemotaxis of tonsil centrocytes. PLoS One 4:e8485

Coull JA, Beggs S, Boudreau D, Boivin D, Tsuda M, Inoue K, Gravel C, Salter MW, De Koninck Y (2005) BDNF from microglia causes the shift in neuronal anion gradient underlying neuropathic pain. Nature 438:1017–1021

Cox JJ, Reimann F, Nicholas AK, Thornton G, Roberts E, Springell K, Karbani G, Jafri H, Mannan J, Raashid Y, Al-Gazali L, Hamamy H, Valente EM, Gorman S, Williams R, McHale DP, Wood JN, Gribble FM, Woods CG (2006) An SCN9A channelopathy causes congenital inability to experience pain. Nature 444:894–898

D'Mello R, Dickenson AH (2008) Spinal cord mechanisms of pain. Br J Anaesth 101:8–16

Davalos D, Grutzendler J, Yang G, Kim JV, Zuo Y, Jung S, Littman DR, Dustin ML, Gan WB (2005) ATP mediates rapid microglial response to local brain injury in vivo. Nat Neurosci 8:752–758

Del Rio-Hortega P (1932) Microglia. In: Penfield W (ed) Cytology and cellular pathology of the nervous system. Hoeber, New York, pp 482–534

Del Rio-Hortega P (2012a) Are the glia with very few processes homologous with Schwann cells? by Pio del Rio-Hortega. 1922. Clin Neuropathol 31:460–462

Del Rio-Hortega P (2012b) Studies on neuroglia: glia with very few processes (oligodendroglia) by PA-o del RA-o-Hortega. 1921. Clin Neuropathol 31:440–459

DeLeo JA, Colburn RW, Rickman AJ (1997) Cytokine and growth factor immunohistochemical spinal profiles in two animal models of mononeuropathy. Brain Res 759:50–57

Deloulme JC, Raponi E, Gentil BJ, Bertacchi N, Marks A, Labourdette G, Baudier J (2004) Nuclear expression of S100B in oligodendrocyte progenitor cells correlates with differentiation toward the oligodendroglial lineage and modulates oligodendrocytes maturation. Mol Cell Neurosci 27:453–465

Dinarello CA (1996) Biologic basis for interleukin-1 in disease. Blood 87:2095–2147

Dinarello CA (2007) Historical insights into cytokines. Eur J Immunol 37(Suppl 1):S34–S45

Dray A (2008) New horizons in pharmacologic treatment for rheumatic disease pain. Rheum Dis Clin North Am 34:481–505

Dunn E, Sims JE, Nicklin MJ, O'Neill LA (2001) Annotating genes with potential roles in the immune system: six new members of the IL-1 family. Trends Immunol 22:533–536

Fong AM, Robinson LA, Steeber DA, Tedder TF, Yoshie O, Imai T, Patel DD (1998) Fractalkine and CX3CR1 mediate a novel mechanism of leukocyte capture, firm adhesion, and activation under physiologic flow. J Exp Med 188:1413–1419

Gao YJ, Ji RR (2010) Targeting astrocyte signaling for chronic pain. Neurotherapeutics 7:482–493

Gao YJ, Zhang L, Samad OA, Suter MR, Yasuhiko K, Xu ZZ, Park JY, Lind AL, Ma Q, Ji RR (2009) JNK-induced MCP-1 production in spinal cord astrocytes contributes to central sensitization and neuropathic pain. J Neurosci 29:4096–4108

Gao YJ, Xu ZZ, Liu YC, Wen YR, Decosterd I, Ji RR (2010a) The c-Jun N-terminal kinase 1 (JNK1) in spinal astrocytes is required for the maintenance of bilateral mechanical allodynia under a persistent inflammatory pain condition. Pain 148:309–319

Gao YJ, Zhang L, Ji RR (2010b) Spinal injection of TNF-alpha-activated astrocytes produces persistent pain symptom mechanical allodynia by releasing monocyte chemoattractant protein-1. Glia 58:1871–1880

Garrison CJ, Dougherty PM, Kajander KC, Carlton SM (1991) Staining of glial fibrillary acidic protein (GFAP) in lumbar spinal cord increases following a sciatic nerve constriction injury. Brain Res 565:1–7

George A, Schmidt C, Weishaupt A, Toyka KV, Sommer C (1999) Serial determination of tumor necrosis factor-alpha content in rat sciatic nerve after chronic constriction injury. Exp Neurol 160:124–132

Ginhoux F, Greter M, Leboeuf M, Nandi S, See P, Gokhan S, Mehler MF, Conway SJ, Ng LG, Stanley ER, Samokhvalov IM, Merad M (2010) Fate mapping analysis reveals that adult microglia derive from primitive macrophages. Science 330:841–845

Go VL, Yaksh TL (1987) Release of substance P from the cat spinal cord. J Physiol 391:141–167

Gruber-Schoffnegger D, Drdla-Schutting R, Honigsperger C, Wunderbaldinger G, Gassner M, Sandkuhler J (2013) Induction of thermal hyperalgesia and synaptic long-term potentiation in the spinal cord lamina I by TNF-alpha and IL-1beta is mediated by glial cells. J Neurosci 33:6540–6551

Guo W, Wang H, Watanabe M, Shimizu K, Zou S, LaGraize SC, Wei F, Dubner R, Ren K (2007) Glial-cytokine-neuronal interactions underlying the mechanisms of persistent pain. J Neurosci 27:6006–6018

Hachem S, Aguirre A, Vives V, Marks A, Gallo V, Legraverend C (2005) Spatial and temporal expression of S100B in cells of oligodendrocyte lineage. Glia 51:81–97

Hao S, Mata M, Glorioso JC, Fink DJ (2007) Gene transfer to interfere with TNFalpha signaling in neuropathic pain. Gene Ther 14:1010–1016

Harrison JK, Jiang Y, Chen S, Xia Y, Maciejewski D, McNamara RK, Streit WJ, Salafranca MN, Adhikari S, Thompson DA, Botti P, Bacon KB, Feng L (1998) Role for neuronally derived fractalkine in mediating interactions between neurons and CX3CR1-expressing microglia. Proc Natl Acad Sci U S A 95:10896–10901

Hathway GJ, Vega-Avelaira D, Moss A, Ingram R, Fitzgerald M (2009) Brief, low frequency stimulation of rat peripheral C-fibres evokes prolonged microglial-induced central sensitization in adults but not in neonates. Pain 144:110–118

Hess A, Axmann R, Rech J, Finzel S, Heindl C, Kreitz S, Sergeeva M, Saake M, Garcia M, Kollias G, Straub RH, Sporns O, Doerfler A, Brune K, Schett G (2011) Blockade of TNF-alpha rapidly inhibits pain responses in the central nervous system. Proc Natl Acad Sci U S A 108:3731–3736

Honda S, Sasaki Y, Ohsawa K, Imai Y, Nakamura Y, Inoue K, Kohsaka S (2001) Extracellular ATP or ADP induce chemotaxis of cultured microglia through Gi/o-coupled P2Y receptors. J Neurosci 21:1975–1982

Hughes PM, Botham MS, Frentzel S, Mir A, Perry VH (2002) Expression of fractalkine (CX3CL1) and its receptor, CX3CR1, during acute and chronic inflammation in the rodent CNS. Glia 37:314–327

Hundhausen C, Misztela D, Berkhout TA, Broadway N, Saftig P, Reiss K, Hartmann D, Fahrenholz F, Postina R, Matthews V, Kallen KJ, Rose-John S, Ludwig A (2003) The disintegrin-like metalloproteinase ADAM10 is involved in constitutive cleavage of CX3CL1 (fractalkine) and regulates CX3CL1-mediated cell-cell adhesion. Blood 102:1186–1195

Hundhausen C, Schulte A, Schulz B, Andrzejewski MG, Schwarz N, von Hundelshausen P, Winter U, Paliga K, Reiss K, Saftig P, Weber C, Ludwig A (2007) Regulated shedding of transmembrane chemokines by the disintegrin and metalloproteinase 10 facilitates detachment of adherent leukocytes. J Immunol 178:8064–8072

Imai T, Hieshima K, Haskell C, Baba M, Nagira M, Nishimura M, Kakizaki M, Takagi S, Nomiyama H, Schall TJ, Yoshie O (1997) Identification and molecular characterization of fractalkine receptor CX3CR1, which mediates both leukocyte migration and adhesion. Cell 91:521–530

Imamoto K, Leblond CP (1978) Radioautographic investigation of gliogenesis in the corpus callosum of young rats. II. Origin of microglial cells. J Comp Neurol 180:139–163

Inglis JJ, Notley CA, Essex D, Wilson AW, Feldmann M, Anand P, Williams R (2007) Collagen-induced arthritis as a model of hyperalgesia: functional and cellular analysis of the analgesic actions of tumor necrosis factor blockade. Arthritis Rheum 56:4015–4023

Ji RR, Gereau RW, Malcangio M, Strichartz GR (2009) MAP kinase and pain. Brain Res Rev 60:135–148

Jung S, Aliberti J, Graemmel P, Sunshine MJ, Kreutzberg GW, Sher A, Littman DR (2000) Analysis of fractalkine receptor CX(3)CR1 function by targeted deletion and green fluorescent protein reporter gene insertion. Mol Cell Biol 20:4106–4114

Kawasaki Y, Zhang L, Cheng JK, Ji RR (2008) Cytokine mechanisms of central sensitization: distinct and overlapping role of interleukin-1beta, interleukin-6, and tumor necrosis factor-alpha in regulating synaptic and neuronal activity in the superficial spinal cord. J Neurosci 28:5189–5194

Kettenmann H, Hanisch UK, Noda M, Verkhratsky A (2011) Physiology of microglia. Physiol Rev 91:461–553

Kidd BL, Langford RM, Wodehouse T (2007) Arthritis and pain. Current approaches in the treatment of arthritic pain. Arthritis Res Ther 9:214

Kim DS, Figueroa KW, Li KW, Boroujerdi A, Yolo T, Luo ZD (2009) Profiling of dynamically changed gene expression in dorsal root ganglia post peripheral nerve injury and a critical role of injury-induced glial fibrillary acidic protein in maintenance of pain behaviors [corrected]. Pain 143:114–122

Kim KW, Vallon-Eberhard A, Zigmond E, Farache J, Shezen E, Shakhar G, Ludwig A, Lira SA, Jung S (2011) In vivo structure/function and expression analysis of the CX3C chemokine fractalkine. Blood 118:e156–e167

King IL, Dickendesher TL, Segal BM (2009) Circulating Ly-6C+ myeloid precursors migrate to the CNS and play a pathogenic role during autoimmune demyelinating disease. Blood 113:3190–3197

Koeglsperger T, Li S, Brenneis C, Saulnier JL, Mayo L, Carrier Y, Selkoe DJ, Weiner HL (2013) Impaired glutamate recycling and GluN2B-mediated neuronal calcium overload in mice lacking TGF-beta1 in the CNS. Glia 61:985–1002

Konig C, Zharsky M, Moller C, Schaible HG, Ebersberger A (2014) Involvement of peripheral and spinal tumor necrosis factor alpha in spinal cord hyperexcitability during knee joint inflammation in rats. Arthritis Rheumatol 66:599–609

Kuner R (2010) Central mechanisms of pathological pain. Nat Med 16:1258–1266

Latremoliere A, Woolf CJ (2009) Central sensitization: a generator of pain hypersensitivity by central neural plasticity. J Pain 10:895–926

Lawson LJ, Perry VH, Gordon S (1992) Turnover of resident microglia in the normal adult mouse brain. Neuroscience 48:405–415

Lee KM, Jeon SM, Cho HJ (2009) Tumor necrosis factor receptor 1 induces interleukin-6 upregulation through NF-kappaB in a rat neuropathic pain model. Eur J Pain 13:794–806

Lee KM, Jeon SM, Cho HJ (2010) Interleukin-6 induces microglial CX3CR1 expression in the spinal cord after peripheral nerve injury through the activation of p38 MAPK. Eur J Pain 14:682.e1–e12

Leung L, Cahill CM (2010) TNF-alpha and neuropathic pain—a review. J Neuroinflammation 7:27

Lever IJ, Grant AD, Pezet S, Gerard NP, Brain SD, Malcangio M (2003) Basal and activity-induced release of substance P from primary afferent fibres in NK1 receptor knockout mice: evidence for negative feedback. Neuropharmacology 45:1101–1110

Levison SW, Goldman JE (1993) Both oligodendrocytes and astrocytes develop from progenitors in the subventricular zone of postnatal rat forebrain. Neuron 10:201–212

Levison SW, Goldman JE (1997) Multipotential and lineage restricted precursors coexist in the mammalian perinatal subventricular zone. J Neurosci Res 48:83–94

Levison SW, Chuang C, Abramson BJ, Goldman JE (1993) The migrational patterns and developmental fates of glial precursors in the rat subventricular zone are temporally regulated. Development 119:611–622

Liaw WJ, Stephens RL Jr, Binns BC, Chu Y, Sepkuty JP, Johns RA, Rothstein JD, Tao YX (2005) Spinal glutamate uptake is critical for maintaining normal sensory transmission in rat spinal cord. Pain 115:60–70

Lindia JA, McGowan E, Jochnowitz N, Abbadie C (2005) Induction of CX3CL1 expression in astrocytes and CX3CR1 in microglia in the spinal cord of a rat model of neuropathic pain. J Pain 6:434–438

Ling EA, Penney D, Leblond CP (1980) Use of carbon labeling to demonstrate the role of blood monocytes as precursors of the 'ameboid cells' present in the corpus callosum of postnatal rats. J Comp Neurol 193:631–657

Lister MF, Sharkey J, Sawatzky DA, Hodgkiss JP, Davidson DJ, Rossi AG, Finlayson K (2007) The role of the purinergic P2X7 receptor in inflammation. J Inflamm (Lond) 4:5

Liu YL, Zhou LJ, Hu NW, Xu JT, Wu CY, Zhang T, Li YY, Liu XG (2007) Tumor necrosis factor-alpha induces long-term potentiation of C-fiber evoked field potentials in spinal dorsal horn in rats with nerve injury: the role of NF-kappa B, JNK and p38 MAPK. Neuropharmacology 52:708–715

Liu T, Jiang CY, Fujita T, Luo SW, Kumamoto E (2013) Enhancement by interleukin-1beta of AMPA and NMDA receptor-mediated currents in adult rat spinal superficial dorsal horn neurons. Mol Pain 9:16

Liuzzo JP, Petanceska SS, Devi LA (1999a) Neurotrophic factors regulate cathepsin S in macrophages and microglia: a role in the degradation of myelin basic protein and amyloid beta peptide. Mol Med 5:334–343

Liuzzo JP, Petanceska SS, Moscatelli D, Devi LA (1999b) Inflammatory mediators regulate cathepsin S in macrophages and microglia: a role in attenuating heparan sulfate interactions. Mol Med 5:320–333

Lyons A, Lynch AM, Downer EJ, Hanley R, O'Sullivan JB, Smith A, Lynch MA (2009) Fractalkine-induced activation of the phosphatidylinositol-3 kinase pathway attenuates microglial activation in vivo and in vitro. J Neurochem 110:1547–1556

Malcangio M, Bowery NG (1994) Spinal cord SP release and hyperalgesia in monoarthritic rats: involvement of the GABAB receptor system. Br J Pharmacol 113:1561–1566

Marchand F, Tsantoulas C, Singh D, Grist J, Clark AK, Bradbury EJ, McMahon SB (2009) Effects of Etanercept and Minocycline in a rat model of spinal cord injury. Eur J Pain 13:673–681

Marin-Teva JL, Dusart I, Colin C, Gervais A, Van RN, Mallat M (2004) Microglia promote the death of developing Purkinje cells. Neuron 41:535–547

Marin-Teva JL, Cuadros MA, Martin-Oliva D, Navascues J (2011) Microglia and neuronal cell death. Neuron Glia Biol 7:25–40

Martinon F, Burns K, Tschopp J (2002) The inflammasome: a molecular platform triggering activation of inflammatory caspases and processing of proIL-beta. Mol Cell 10:417–426

McKercher SR, Torbett BE, Anderson KL, Henkel GW, Vestal DJ, Baribault H, Klemsz M, Feeney AJ, Wu GE, Paige CJ, Maki RA (1996) Targeted disruption of the PU.1 gene results in multiple hematopoietic abnormalities. EMBO J 15:5647–5658

McMahon SB, Malcangio M (2009) Current challenges in glia-pain biology. Neuron 64:46–54

Meller ST, Dykstra C, Grzybycki D, Murphy S, Gebhart GF (1994) The possible role of glia in nociceptive processing and hyperalgesia in the spinal cord of the rat. Neuropharmacology 33:1471–1478

Milligan ED, Twining C, Chacur M, Biedenkapp J, O'Connor K, Poole S, Tracey K, Martin D, Maier SF, Watkins LR (2003) Spinal glia and proinflammatory cytokines mediate mirror-image neuropathic pain in rats. J Neurosci 23:1026–1040

Milligan ED, Zapata V, Chacur M, Schoeniger D, Biedenkapp J, O'Connor KA, Verge GM, Chapman G, Green P, Foster AC, Naeve GS, Maier SF, Watkins LR (2004) Evidence that exogenous and endogenous fractalkine can induce spinal nociceptive facilitation in rats. Eur J Neurosci 20:2294–2302

Milligan E, Zapata V, Schoeniger D, Chacur M, Green P, Poole S, Martin D, Maier SF, Watkins LR (2005a) An initial investigation of spinal mechanisms underlying pain enhancement induced by fractalkine, a neuronally released chemokine. Eur J Neurosci 22:2775–2782

Milligan ED, Langer SJ, Sloane EM, He L, Wieseler-Frank J, O'Connor K, Martin D, Forsayeth JR, Maier SF, Johnson K, Chavez RA, Leinwand LA, Watkins LR (2005b) Controlling pathological pain by adenovirally driven spinal production of the anti-inflammatory cytokine, interleukin-10. Eur J Neurosci 21:2136–2148

Milligan ED, Soderquist RG, Malone SM, Mahoney JH, Hughes TS, Langer SJ, Sloane EM, Maier SF, Leinwand LA, Watkins LR, Mahoney MJ (2006) Intrathecal polymer-based interleukin-10 gene delivery for neuropathic pain. Neuron Glia Biol 2:293–308

Miyoshi K, Obata K, Kondo T, Okamura H, Noguchi K (2008) Interleukin-18-mediated microglia/astrocyte interaction in the spinal cord enhances neuropathic pain processing after nerve injury. J Neurosci 28:12775–12787

Mizuno T, Kawanokuchi J, Numata K, Suzumura A (2003) Production and neuroprotective functions of fractalkine in the central nervous system. Brain Res 979:65–70

Nadeau S, Filali M, Zhang J, Kerr BJ, Rivest S, Soulet D, Iwakura Y, de Rivero Vaccari JP, Keane RW, Lacroix S (2011) Functional recovery after peripheral nerve injury is dependent on the pro-inflammatory cytokines IL-1beta and TNF: implications for neuropathic pain. J Neurosci 31:12533–12542

Nimmerjahn A, Kirchhoff F, Helmchen F (2005) Resting microglial cells are highly dynamic surveillants of brain parenchyma in vivo. Science 308:1314–1318

Nishiyori A, Minami M, Ohtani Y, Takami S, Yamamoto J, Kawaguchi N, Kume T, Akaike A, Satoh M (1998) Localization of fractalkine and CX3CR1 mRNAs in rat brain: does fractalkine play a role in signaling from neuron to microglia? FEBS Lett 429:167–172

Ohtori S, Takahashi K, Moriya H, Myers RR (2004) TNF-alpha and TNF-alpha receptor type 1 upregulation in glia and neurons after peripheral nerve injury: studies in murine DRG and spinal cord. Spine (Phila Pa 1976) 29:1082–1088

Okada-Ogawa A, Suzuki I, Sessle BJ, Chiang CY, Salter MW, Dostrovsky JO, Tsuboi Y, Kondo M, Kitagawa J, Kobayashi A, Noma N, Imamura Y, Iwata K (2009) Astroglia in medullary dorsal horn (trigeminal spinal subnucleus caudalis) are involved in trigeminal neuropathic pain mechanisms. J Neurosci 29:11161–11171

Oku R, Satoh M, Fujii N, Otaka A, Yajima H, Takagi H (1987) Calcitonin gene-related peptide promotes mechanical nociception by potentiating release of substance P from the spinal dorsal horn in rats. Brain Res 403:350–354

Old EA, Malcangio M (2012) Chemokine mediated neuron-glia communication and aberrant signalling in neuropathic pain states. Curr Opin Pharmacol 12:67–73

Olechowski CJ, Truong JJ, Kerr BJ (2009) Neuropathic pain behaviours in a chronic-relapsing model of experimental autoimmune encephalomyelitis (EAE). Pain 141:156–164

Pan Y, Lloyd C, Zhou H, Dolich S, Deeds J, Gonzalo JA, Vath J, Gosselin M, Ma J, Dussault B, Woolf E, Alperin G, Culpepper J, Gutierrez-Ramos JC, Gearing D (1997) Neurotactin, a membrane-anchored chemokine upregulated in brain inflammation. Nature 387:611–617

Paolicelli RC, Bolasco G, Pagani F, Maggi L, Scianni M, Panzanelli P, Giustetto M, Ferreira TA, Guiducci E, Dumas L, Ragozzino D, Gross CT (2011) Synaptic pruning by microglia is necessary for normal brain development. Science 333:1456–1458

Park CK, Lu N, Xu ZZ, Liu T, Serhan CN, Ji RR (2011) Resolving TRPV1- and TNF-α-mediated spinal cord synaptic plasticity and inflammatory pain with neuroprotectin D1. J Neurosci 31:15072–15085

Pekny M, Nilsson M (2005) Astrocyte activation and reactive gliosis. Glia 50:427–434

Perry VH, Hume DA, Gordon S (1985) Immunohistochemical localization of macrophages and microglia in the adult and developing mouse brain. Neuroscience 15:313–326

Peters CM, Jimenez-Andrade JM, Kuskowski MA, Ghilardi JR, Mantyh PW (2007) An evolving cellular pathology occurs in dorsal root ganglia, peripheral nerve and spinal cord following intravenous administration of paclitaxel in the rat. Brain Res 1168:46–59

Pezet S, Malcangio M, Lever IJ, Perkinton MS, Thompson SW, Williams RJ, McMahon SB (2002) Noxious stimulation induces Trk receptor and downstream ERK phosphorylation in spinal dorsal horn. Mol Cell Neurosci 21:684–695

Porter JT, McCarthy KD (1997) Astrocytic neurotransmitter receptors in situ and in vivo. Prog Neurobiol 51:439–455

Prinz M, Mildner A (2011) Microglia in the CNS: immigrants from another world. Glia 59:177–187

Raghavendra V, Tanga FY, DeLeo JA (2004) Complete Freunds adjuvant-induced peripheral inflammation evokes glial activation and proinflammatory cytokine expression in the CNS. Eur J Neurosci 20:467–473

Ransohoff RM, Cardona AE (2010) The myeloid cells of the central nervous system parenchyma. Nature 468:253–262

Ransohoff RM, Perry VH (2009) Microglial physiology: unique stimuli, specialized responses. Annu Rev Immunol 27:119–145

Rasley A, Bost KL, Olson JK, Miller SD, Marriott I (2002) Expression of functional NK-1 receptors in murine microglia. Glia 37:258–267

Reeve AJ, Patel S, Fox A, Walker K, Urban L (2000) Intrathecally administered endotoxin or cytokines produce allodynia, hyperalgesia and changes in spinal cord neuronal responses to nociceptive stimuli in the rat. Eur J Pain 4:247–257

Ren K, Torres R (2009) Role of interleukin-1beta during pain and inflammation. Brain Res Rev 60:57–64

Ridet JL, Malhotra SK, Privat A, Gage FH (1997) Reactive astrocytes: cellular and molecular cues to biological function. Trends Neurosci 20:570–577

Romero-Sandoval A, Chai N, Nutile-McMenemy N, DeLeo JA (2008) A comparison of spinal Iba1 and GFAP expression in rodent models of acute and chronic pain. Brain Res 1219:116–126

Rubartelli A, Cozzolino F, Talio M, Sitia R (1990) A novel secretory pathway for interleukin-1 beta, a protein lacking a signal sequence. EMBO J 9:1503–1510

Saijo K, Glass CK (2011) Microglial cell origin and phenotypes in health and disease. Nat Rev Immunol 11:775–787

Salter MW, Beggs S (2014) Sublime microglia: expanding roles for the guardians of the CNS. Cell 158:15–24

Sandkuhler J (2009) Models and mechanisms of hyperalgesia and allodynia. Physiol Rev 89:707–758

Schafer DP, Lehrman EK, Kautzman AG, Koyama R, Mardinly AR, Yamasaki R, Ransohoff RM, Greenberg ME, Barres BA, Stevens B (2012) Microglia sculpt postnatal neural circuits in an activity and complement-dependent manner. Neuron 74:691–705

Schafers M, Svensson CI, Sommer C, Sorkin LS (2003) Tumor necrosis factor-alpha induces mechanical allodynia after spinal nerve ligation by activation of p38 MAPK in primary sensory neurons. J Neurosci 23:2517–2521

Schaible HG, Richter F, Ebersberger A, Boettger MK, Vanegas H, Natura G, Vazquez E, von Segond BG (2009) Joint pain. Exp Brain Res 196:153–162

Schroder K, Tschopp J (2010) The inflammasomes. Cell 140:821–832

Sedel F, Bechade C, Vyas S, Triller A (2004) Macrophage-derived tumor necrosis factor alpha, an early developmental signal for motoneuron death. J Neurosci 24:2236–2246

Seybold VS, McCarson KE, Mermelstein PG, Groth RD, Abrahams LG (2003) Calcitonin gene-related peptide regulates expression of neurokinin1 receptors by rat spinal neurons. J Neurosci 23:1816–1824

Shubayev VI, Myers RR (2000) Upregulation and interaction of TNFalpha and gelatinases A and B in painful peripheral nerve injury. Brain Res 855:83–89

Sofroniew MV, Vinters HV (2010) Astrocytes: biology and pathology. Acta Neuropathol 119:7–35

Sokka T, Kautiainen H, Toloza S, Makinen H, Verstappen SM, Lund HM, Naranjo A, Baecklund E, Herborn G, Rau R, Cazzato M, Gossec L, Skakic V, Gogus F, Sierakowski S, Bresnihan B, Taylor P, McClinton C, Pincus T (2007) QUEST-RA: quantitative clinical assessment of patients with rheumatoid arthritis seen in standard rheumatology care in 15 countries. Ann Rheum Dis 66:1491–1496

Sommer C, Kress M (2004) Recent findings on how proinflammatory cytokines cause pain: peripheral mechanisms in inflammatory and neuropathic hyperalgesia. Neurosci Lett 361:184–187

Sommer C, Schafers M (1998) Painful mononeuropathy in C57BL/Wld mice with delayed Wallerian degeneration: differential effects of cytokine production and nerve regeneration on thermal and mechanical hypersensitivity. Brain Res 784:154–162

Staniland AA, Clark AK, Wodarski R, Sasso O, Maione F, D'Acquisto F, Malcangio M (2010) Reduced inflammatory and neuropathic pain and decreased spinal microglial response in fractalkine receptor (CX3CR1) knockout mice. J Neurochem 114:1143–1157

Steinhauser C, Berger T, Frotscher M, Kettenmann H (1992) Heterogeneity in the membrane current pattern of identified glial cells in the hippocampal slice. Eur J Neurosci 4:472–484

Stevens B, Allen NJ, Vazquez LE, Howell GR, Christopherson KS, Nouri N, Micheva KD, Mehalow AK, Huberman AD, Stafford B, Sher A, Litke AM, Lambris JD, Smith SJ, John SW, Barres BA (2007) The classical complement cascade mediates CNS synapse elimination. Cell 131:1164–1178

Sung CS, Wen ZH, Chang WK, Ho ST, Tsai SK, Chang YC, Wong CS (2004) Intrathecal interleukin-1beta administration induces thermal hyperalgesia by activating inducible nitric oxide synthase expression in the rat spinal cord. Brain Res 1015:145–153

Suter MR, Berta T, Gao YJ, Decosterd I, Ji RR (2009) Large A-fiber activity is required for microglial proliferation and p38 MAPK activation in the spinal cord: different effects of resiniferatoxin and bupivacaine on spinal microglial changes after spared nerve injury. Mol Pain 5:53

Suzuki T, Hide I, Ido K, Kohsaka S, Inoue K, Nakata Y (2004) Production and release of neuroprotective tumor necrosis factor by P2X7 receptor-activated microglia. J Neurosci 24:1–7

Svensson CI, Hua XY, Protter AA, Powell HC, Yaksh TL (2003a) Spinal p38 MAP kinase is necessary for NMDA-induced spinal PGE(2) release and thermal hyperalgesia. Neuroreport 14:1153–1157

Svensson CI, Marsala M, Westerlund A, Calcutt NA, Campana WM, Freshwater JD, Catalano R, Feng Y, Protter AA, Scott B, Yaksh TL (2003b) Activation of p38 mitogen-activated protein kinase in spinal microglia is a critical link in inflammation-induced spinal pain processing. J Neurochem 86:1534–1544

Svensson CI, Fitzsimmons B, Azizi S, Powell HC, Hua XY, Yaksh TL (2005a) Spinal p38beta isoform mediates tissue injury-induced hyperalgesia and spinal sensitization. J Neurochem 92:1508–1520

Svensson CI, Schafers M, Jones TL, Powell H, Sorkin LS (2005b) Spinal blockade of TNF blocks spinal nerve ligation-induced increases in spinal P-p38. Neurosci Lett 379:209–213

Sweitzer SM, Colburn RW, Rutkowski M, DeLeo JA (1999) Acute peripheral inflammation induces moderate glial activation and spinal IL-1beta expression that correlates with pain behavior in the rat. Brain Res 829:209–221

Tanga FY, Raghavendra V, DeLeo JA (2004) Quantitative real-time RT-PCR assessment of spinal microglial and astrocytic activation markers in a rat model of neuropathic pain. Neurochem Int 45:397–407

Tarozzo G, Bortolazzi S, Crochemore C, Chen SC, Lira AS, Abrams JS, Beltramo M (2003) Fractalkine protein localization and gene expression in mouse brain. J Neurosci Res 73:81–88

Tawfik VL, Nutile-McMenemy N, Lacroix-Fralish ML, DeLeo JA (2007) Efficacy of propentofylline, a glial modulating agent, on existing mechanical allodynia following peripheral nerve injury. Brain Behav Immun 21:238–246

Todd AJ (2010) Neuronal circuitry for pain processing in the dorsal horn. Nat Rev Neurosci 11:823–836

Tong N, Perry SW, Zhang Q, James HJ, Guo H, Brooks A, Bal H, Kinnear SA, Fine S, Epstein LG, Dairaghi D, Schall TJ, Gendelman HE, Dewhurst S, Sharer LR, Gelbard HA (2000) Neuronal fractalkine expression in HIV-1 encephalitis: roles for macrophage recruitment and neuroprotection in the central nervous system. J Immunol 164:1333–1339

Trang T, Beggs S, Wan X, Salter MW (2009) P2X4-receptor-mediated synthesis and release of brain-derived neurotrophic factor in microglia is dependent on calcium and p38-mitogen-activated protein kinase activation. J Neurosci 29:3518–3528

Tremblay ME, Lowery RL, Majewska AK (2010) Microglial interactions with synapses are modulated by visual experience. PLoS Biol 8:e1000527

Trentham DE (1982) Collagen arthritis as a relevant model for rheumatoid arthritis. Arthritis Rheum 25:911–916

Tsuda M, Shigemoto-Mogami Y, Koizumi S, Mizokoshi A, Kohsaka S, Salter MW, Inoue K (2003) P2X4 receptors induced in spinal microglia gate tactile allodynia after nerve injury. Nature 424:778–783

Tsuda M, Tozaki-Saitoh H, Inoue K (2012) Purinergic system, microglia and neuropathic pain. Curr Opin Pharmacol 12:74–79

Vazquez E, Kahlenbach J, von Segond BG, Konig C, Schaible HG, Ebersberger A (2012) Spinal interleukin-6 is an amplifier of arthritic pain in the rat. Arthritis Rheum 64:2233–2242

Verge GM, Milligan ED, Maier SF, Watkins LR, Naeve GS, Foster AC (2004) Fractalkine (CX3CL1) and fractalkine receptor (CX3CR1) distribution in spinal cord and dorsal root ganglia under basal and neuropathic pain conditions. Eur J Neurosci 20:1150–1160

Verpoorten N, Claeys KG, Deprez L, Jacobs A, Van Gerwen V, Lagae L, Arts WF, De Meirleir L, Keymolen K, Ceuterick-de GC, De Jonghe P, Timmerman V, Nelis E (2006) Novel frameshift and splice site mutations in the neurotrophic tyrosine kinase receptor type 1 gene (NTRK1) associated with hereditary sensory neuropathy type IV. Neuromuscul Disord 16:19–25

Vincent TL, Williams RO, Maciewicz R, Silman A, Garside P (2012) Mapping pathogenesis of arthritis through small animal models. Rheumatology (Oxford) 51:1931–1941

Viviani B, Bartesaghi S, Gardoni F, Vezzani A, Behrens MM, Bartfai T, Binaglia M, Corsini E, Di LM, Galli CL, Marinovich M (2003) Interleukin-1beta enhances NMDA receptor-mediated intracellular calcium increase through activation of the Src family of kinases. J Neurosci 23:8692–8700

Volterra A, Meldolesi J (2005) Astrocytes, from brain glue to communication elements: the revolution continues. Nat Rev Neurosci 6:626–640

Walsh DA, McWilliams DF (2014) Mechanisms, impact and management of pain in rheumatoid arthritis. Nat Rev Rheumatol 10(10):581–592

Wang DD, Bordey A (2008) The astrocyte odyssey. Prog Neurobiol 86:342–367

Watkins LR, Martin D, Ulrich P, Tracey KJ, Maier SF (1997) Evidence for the involvement of spinal cord glia in subcutaneous formalin induced hyperalgesia in the rat. Pain 71:225–235

Weber A, Wasiliew P, Kracht M (2010) Interleukin-1 (IL-1) pathway. Sci Signal 3:cm1

Wen YR, Suter MR, Kawasaki Y, Huang J, Pertin M, Kohno T, Berde CB, Decosterd I, Ji RR (2007) Nerve conduction blockade in the sciatic nerve prevents but does not reverse the activation of p38 mitogen-activated protein kinase in spinal microglia in the rat spared nerve injury model. Anesthesiology 107:312–321

Weng HR, Chen JH, Cata JP (2006) Inhibition of glutamate uptake in the spinal cord induces hyperalgesia and increased responses of spinal dorsal horn neurons to peripheral afferent stimulation. Neuroscience 138:1351–1360

Williams RO (1998) Rodent models of arthritis: relevance for human disease. Clin Exp Immunol 114:330–332

Williams RO (2004) Collagen-induced arthritis as a model for rheumatoid arthritis. Methods Mol Med 98:207–216

Wolf G, Gabay E, Tal M, Yirmiya R, Shavit Y (2006) Genetic impairment of interleukin-1 signaling attenuates neuropathic pain, autotomy, and spontaneous ectopic neuronal activity, following nerve injury in mice. Pain 120:315–324

Woolf CJ, Mannion RJ (1999) Neuropathic pain: aetiology, symptoms, mechanisms, and management. Lancet 353:1959–1964

Youn DH, Wang H, Jeong SJ (2008) Exogenous tumor necrosis factor-alpha rapidly alters synaptic and sensory transmission in the adult rat spinal cord dorsal horn. J Neurosci Res 86:2867–2875

Zhang J, Shi XQ, Echeverry S, Mogil JS, De KY, Rivest S (2007) Expression of CCR2 in both resident and bone marrow-derived microglia plays a critical role in neuropathic pain. J Neurosci 27:12396–12406

Zhang RX, Li A, Liu B, Wang L, Ren K, Zhang H, Berman BM, Lao L (2008) IL-1ra alleviates inflammatory hyperalgesia through preventing phosphorylation of NMDA receptor NR-1 subunit in rats. Pain 135:232–239

Zhang L, Berta T, Xu ZZ, Liu T, Park JY, Ji RR (2011) TNF-alpha contributes to spinal cord synaptic plasticity and inflammatory pain: distinct role of TNF receptor subtypes 1 and 2. Pain 152:419–427

Zhuang ZY, Wen YR, Zhang DR, Borsello T, Bonny C, Strichartz GR, Decosterd I, Ji RR (2006) A peptide c-Jun N-terminal kinase (JNK) inhibitor blocks mechanical allodynia after spinal nerve ligation: respective roles of JNK activation in primary sensory neurons and spinal astrocytes for neuropathic pain development and maintenance. J Neurosci 26:3551–3560

Zhuang ZY, Kawasaki Y, Tan PH, Wen YR, Huang J, Ji RR (2007) Role of the CX3CR1/p38 MAPK pathway in spinal microglia for the development of neuropathic pain following nerve injury-induced cleavage of fractalkine. Brain Behav Immun 21:642–651

Plasticity of Inhibition in the Spinal Cord

Andrew J. Todd

Contents

1. General Organisation of Spinal Cord Neurons and Circuits That Are Involved in Pain Processing 172
   - 1.1 Primary Afferents 172
   - 1.2 Projection Neurons 173
   - 1.3 Interneurons 174
   - 1.4 Descending Pathways 176
   - 1.5 What We Know About Synaptic/Neuronal Circuits in the Dorsal Horn 176
   - 1.6 Normal Function of Inhibitory Mechanisms 179
2. Plasticity of Inhibition in Neuropathic Pain States 180
   - 2.1 Animal Models of Neuropathic Pain 180
   - 2.2 Reduced Inhibitory Synaptic Transmission in Neuropathic Pain States 181
   - 2.3 Possible Mechanisms for Reduced Inhibition Following Peripheral Nerve Injury 181
3. Conclusions 186

References 186

Abstract

Inhibitory interneurons, which use GABA and/or glycine as their principal transmitter, have numerous roles in regulating the transmission of sensory information through the spinal dorsal horn. These roles are likely to be performed by different populations of interneurons, each with specific locations in the synaptic circuitry of the region. Peripheral nerve injury frequently leads to neuropathic pain, and it is thought that loss of function of inhibitory interneurons in the dorsal horn contributes to this condition. Several mechanisms have been proposed for this disinhibition, including death of inhibitory interneurons,

---

A.J. Todd (✉)
Institute of Neuroscience and Psychology, College of Medical Veterinary and Life Sciences, University of Glasgow, Glasgow G12 8QQ, UK
e-mail: andrew.todd@glasgow.ac.uk

decreased transmitter release, diminished activity of these cells and reduced effectiveness of GABA and glycine as inhibitory transmitters. However, despite numerous studies on this important topic, it is still not clear which (if any) of these mechanisms contributes to neuropathic pain after nerve injury.

Keywords
GABA • Glycine • Neuropathic pain • Inhibitory interneuron

1 General Organisation of Spinal Cord Neurons and Circuits That Are Involved in Pain Processing

In order to understand plastic changes in spinal cord inhibitory mechanisms that can occur during chronic pain states, it is helpful to have a basic knowledge of the circuits that process sensory information in the dorsal horn. Four neuronal components contribute to these circuits: (1) primary afferent axons, (2) projection neurons (cells with axons that travel directly to the brain), (3) interneurons (which can be defined as those neurons with axons that remain within the spinal cord) and (4) axons that descend from the brain.

This chapter therefore begins with a very brief overview of the neuronal components and circuits that are involved in pain processing, with particular emphasis on inhibitory interneurons and their connections. Based on neuronal size and packing density, the dorsal horn can be divided into a series of 6 parallel laminae (Rexed 1952). This account is largely restricted to laminae I–III, since we know considerably more about the neuronal organisation in this region, and it includes the main termination zone for nociceptive primary afferents.

Loss of inhibition in the dorsal horn is thought to contribute to the symptoms and signs of the neuropathic pain that follows nerve injury, and the main part of the chapter will review the various mechanisms that have been proposed to explain this phenomenon.

1.1 Primary Afferents

Primary afferent axons provide sensory input from peripheral tissues and organs and respond to a range of stimulus modalities. They are classified according to their diameter and whether or not they are myelinated. In general, the larger myelinated (Aβ) afferents convey innocuous (tactile and proprioceptive) mechanical information, while most fine myelinated (Aδ) and unmyelinated (C) fibres function as thermoreceptors or nociceptors. Their central projections are arranged in a very specific way, with Aδ and C fibres arborising mainly in the superficial dorsal horn (laminae I and II of Rexed), while Aβ tactile afferents terminate in the deeper laminae (III–V).

Nociceptive unmyelinated afferents can be divided into two broad groups, based on their neurochemical properties. One group consists of axons that contain neuropeptides, such as calcitonin gene-related peptide (CGRP), substance P and galanin. In the rat, CGRP appears to be present in all of these afferents, and since they are the only source of the peptide in the dorsal horn, CGRP is a convenient marker for their central terminals. Axons in the other major class lack neuropeptides and can be identified by their ability to bind the plant lectin IB4 and by their expression of Mas-related G protein-coupled receptor D (Mrgd) (Zylka et al. 2005; Todd and Koerber 2012). These two classes of afferent differ in both their central and peripheral terminations. Peptidergic nociceptors innervate most tissues of the body and project mainly to lamina I and the outer part of lamina II (IIo). In contrast, the non-peptidergic (Mrgd$^+$) C fibres innervate the skin and project to a region that overlaps the inner and outer parts of lamina II.

Two types of myelinated nociceptors have been identified. One group projects to lamina I, with limited extension into lamina IIo, and at least some of these can be identified in anatomical studies by their ability to transport cholera toxin B subunit (CTb) that has been injected into a peripheral nerve. The other class projects diffusely to the whole of laminae I–V, and these afferents are not labelled by transganglionic transport of CTb (Light et al. 1979; Woodbury and Koerber 2003).

1.2 Projection Neurons

Projection neurons are concentrated in lamina I, virtually absent from lamina II and scattered through the deeper laminae (III–VI). Many of these cells have axons that cross the midline and ascend in the ventrolateral white matter forming the anterolateral tract (ALT). The targets of this tract include the thalamus, periaqueductal grey matter, lateral parabrachial area and several medullary nuclei. Most (if not all) of these cells have axons that project to more than one of these regions.

The majority of ALT projection neurons in lamina I, together with a distinctive population of those in lamina III, express the neurokinin 1 receptor (NK1r), which is a target for substance P (Todd 2010). Among the lamina I ALT neurons that lack the NK1r, we have identified a population of giant cells, which can be recognised because of the very high density of excitatory and inhibitory synapses that coat their cell bodies and dendrites. While the dendrites of lamina I projection neurons remain within the lamina, those of the lamina III ALT cells have a more widespread distribution, and these cells give rise to prominent dorsally directed dendrites that penetrate the superficial laminae. Virtually all of these projection neurons respond to noxious stimulation (Bester et al. 2000; Polgár et al. 2007), and they presumably provide a major source of nociceptive input to the brain.

1.3 Interneurons

The great majority of the neurons in laminae I–III have axons that remain in the spinal cord and are therefore defined as interneurons. It is thought that all of these cells give rise to locally arborising axons, while a significant proportion also have long intersegmental axonal projections. It is clear that there is a diverse array of interneuron types, each of which is likely to have specific functions (Graham et al. 2007; Todd 2010; Zeilhofer et al. 2012). There have therefore been numerous attempts to classify the interneurons into discrete populations, based on a variety of morphological, electrophysiological, neurochemical and/or developmental criteria. A fundamental distinction can be made between inhibitory interneurons, which use GABA and/or glycine, and excitatory (glutamatergic) interneurons.

1.3.1 Inhibitory Interneurons

The inhibitory cells can be recognised with antibodies against GABA or glycine and account for around one-third of all neurons in laminae I–II and 40 % of those in lamina III (Polgár et al. 2003, 2013a) (Fig. 1). Since projection neurons in the dorsal horn are glutamatergic, it is generally assumed that all of these inhibitory cells are interneurons. Within laminae I–III, virtually all neurons that show high levels of glycine are also GABA-immunoreactive, and this suggests that inhibitory interneurons in this region are all GABAergic, with some using glycine as a co-transmitter (Todd 2010). However, electrophysiological studies have suggested that most inhibitory synapses in this region use one or other transmitter, and it is likely that this results from differential distribution of the corresponding receptors at these synapses (Zeilhofer et al. 2012).

Interneurons in lamina II have been studied most extensively and have been divided into four main morphological classes: islet, central, vertical and radial cells (Grudt and Perl 2002; Yasaka et al. 2007). However, even in this lamina, not all cells can be assigned to one of these classes, and the relationship between morphology and neurotransmitter type is not entirely straightforward. A correlative electrophysiological and anatomical study of ~60 lamina II neurons showed that all islet cells (which are defined by their highly elongated rostrocaudal dendritic trees) were inhibitory but that the remaining inhibitory interneurons did not show any characteristic morphological properties (Yasaka et al. 2010).

An alternative approach to classifying the inhibitory interneurons has been to use neurochemical markers. In this way, we have identified four nonoverlapping populations of inhibitory cells in laminae I–III of the rat, based on expression of neuropeptide Y (NPY), galanin, neuronal nitric oxide synthase (nNOS) or parvalbumin (Tiong et al. 2011). Between them, these cells account for over half of the inhibitory interneurons in laminae I–II and a smaller proportion of those in lamina III. The parvalbumin cells correspond to some of the lamina II islet cells

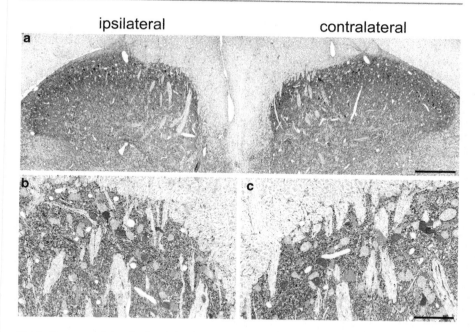

Fig. 1 Immunostaining for GABA in a semithin (0.5 μm) section from the L4 segment of a rat that had undergone chronic constriction injury of the left sciatic nerve 2 weeks previously. Panel (**a**) shows a low magnification view through the dorsal horn on both sides, while panels (**b**) and (**c**) show higher magnification views of the medial parts of the ipsilateral and contralateral dorsal horns, respectively. Numerous immunoreactive and non-immunoreactive cell bodies can be seen in the dorsal horn on each side, and there is no detectable difference between the two sides. Stereological analysis showed that the proportion of cells in each of laminae I, II and III that were GABA-immunoreactive did not differ between ipsilateral and contralateral sides and did not differ from that seen in naïve or sham-operated animals. Scale bars: **a** = 200 μm; **b**, **c** = 50 μm. Reproduced with permission from Polgár et al. (2003)

(and have similar morphology in lamina III), while the other three neurochemical types are not islet cells and are morphologically heterogeneous (AJ Todd, N Iwagaki and RP Ganley, unpublished data).

There are several mouse lines in which subsets of inhibitory interneurons express green fluorescent protein (GFP), and these have been used to investigate the functions of these cells. These include the GIN mouse (Oliva et al. 2000), in which around one-third of GABAergic neurons in lamina II express GFP; the PrP-GFP mouse (Hantman et al. 2004), in which GFP labels a subset of the nNOS- and galanin-containing GABAergic neurons (Iwagaki et al. 2013); and the GlyT2-EGFP mouse, in which glycinergic neurons are labelled.

1.3.2 Excitatory Interneurons

All of the neurons that are not projection cells and lack GABA and glycine-immunoreactivity are thought to be glutamatergic (excitatory) interneurons, and these account for 60–70 % of the neuronal population in laminae I–III. Initially, it was difficult to be certain of their neurotransmitter phenotype, but this can now be confirmed by their expression of vesicular glutamate transporter 2 (VGLUT2) (Maxwell et al. 2007; Schneider and Walker 2007; Yasaka et al. 2010). Some of these cells belong to the vertical or radial classes, but many cannot be assigned to any morphological class. Although there are several neurochemical markers (e.g. certain neuropeptides, the calcium-binding proteins calbindin and calretinin, protein kinase Cγ) that are largely restricted to excitatory interneurons, little is yet known about whether these represent discrete functional populations (Todd 2010).

1.4 Descending Pathways

The main descending projections from the brain are the monoaminergic systems, which originate in the raphe nuclei of the medulla (serotonergic) and the locus coeruleus and adjacent regions of the pons (norepinephrinergic). However, there is also a GABAergic/glycinergic projection from the ventromedial medulla that arborises throughout the dorsal horn (Antal et al. 1996).

1.5 What We Know About Synaptic/Neuronal Circuits in the Dorsal Horn

Although our knowledge concerning the spinal cord circuits that process somatosensory information is still extremely limited, there has been some progress in identifying specific pathways (Fig. 2). It is likely that most (if not all) dorsal horn neurons receive glutamatergic synapses from both primary afferents and excitatory interneurons, as well as GABAergic, glycinergic or mixed synapses from local inhibitory interneurons. However, there is emerging evidence that these are organised in a selective way, with certain types of primary afferent and interneuron preferentially innervating specific types of dorsal horn neuron (Todd 2010). The inhibitory synapses described above are formed by the axons of inhibitory interneurons and the dendrites or cell bodies of other neurons (axodendritic or axosomatic synapses), and these underlie postsynaptic inhibition, which is the major form of inhibition in the dorsal horn. However, some GABAergic axon terminals form synapses onto primary afferent axons (axoaxonic synapses), and these generate presynaptic inhibition, which can control specific types of sensory input. In addition, there are dendrodendritic (and dendroaxonic) synapses, in which the presynaptic element is the dendrite of a local GABAergic neuron. The functions of these synapses are not well understood.

As stated above, the majority of ALT projection neurons in laminae I and III express the NK1r. These cells are densely innervated by substance P-containing primary afferent nociceptors, which provide around half of the glutamatergic synapses on their cell bodies and dendrites (Polgár et al. 2010; Baseer et al. 2012). This generates a powerful input and is likely to underlie the nociceptive

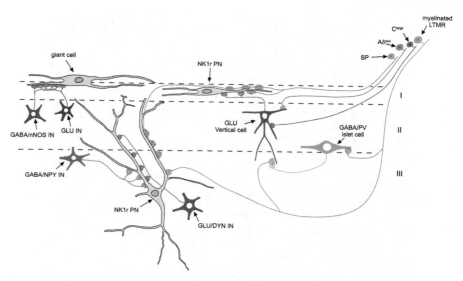

Fig. 2 A diagram illustrating some of the synaptic connections that have been identified in laminae I–III of the rodent dorsal horn. Three anterolateral tract projection neurons are indicated: a lamina I giant cell and projection neurons (PN) in laminae I and III that express the neurokinin 1 receptor (NK1r). Both lamina I and lamina III NK1r-expressing cells receive numerous synapses from peptidergic primary afferents that contain substance P (SP). The NK1r-expressing lamina I PN receives excitatory synapses from glutamatergic (GLU) vertical cells in lamina II, which are thought to be innervated by Aδ nociceptors ($A\delta^{noci}$), non-peptidergic C fibre nociceptors (C^{Mrgd}) and myelinated low-threshold mechanoreceptors (LTMR). The myelinated LTMRs also innervate GABAergic islet cells that contain parvalbumin (PV), and they receive axoaxonic synapses from the PV cells. The lamina III PNs are selectively innervated by two distinct classes of interneuron: inhibitory cells that express neuropeptide Y (NPY) and excitatory (glutamatergic) cells that express dynorphin. The giant lamina I projection neurons appear to receive little or no direct primary afferent input but are densely innervated by excitatory and inhibitory interneurons. Many of the latter contain neuronal nitric oxide synthase (nNOS). For further details, see text

responses of these cells. The remaining glutamatergic synapses on the projection neurons are presumably derived mainly from excitatory interneurons, and these are thought to include the vertical cells in lamina II (Lu and Perl 2005; Cordero-Erausquin et al. 2009). One of the functions of the excitatory interneurons that innervate lamina I projection neuron is to provide polysynaptic input from Aβ low-threshold mechanoreceptors (LTMRs). This contributes low-threshold components to the receptive fields of some of the projection cells (Bester et al. 2000; Andrew 2009), and loss of inhibition is thought to strengthen this indirect low-threshold pathway, leading to tactile allodynia in chronic pain states (Torsney and MacDermott 2006; Keller et al. 2007; Lu et al. 2013) (see below). We have recently found that lamina II vertical cells receive numerous contacts from myelinated LTMRs, which suggests that they may provide a disynaptic link between these afferents and lamina I ALT neurons (Yasaka et al 2014).

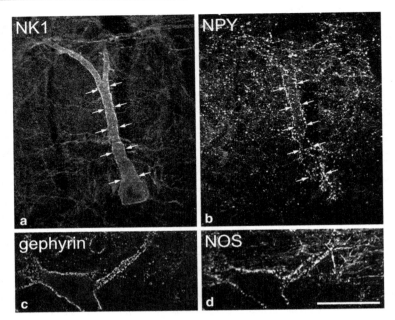

Fig. 3 Selective innervation of projection neurons by different types of inhibitory interneuron. Panels (**a**, **b**) show part of a parasagittal section of rat dorsal horn scanned to reveal the neurokinin 1 receptor (NK1) and neuropeptide Y (NPY). (**a**) The cell body and dorsal dendrite of a large NK1 receptor-immunoreactive lamina III projection neuron are visible. (**b**) NPY-containing axons, which are derived from local inhibitory interneurons, form a plexus in laminae I and II, and there is a dense cluster of these axons that contact the projection neuron. *Arrows* indicate corresponding locations in the two images. Panels (**c**, **d**) show part of a horizontal section through lamina I stained for the inhibitory synapse-associated protein gephyrin and the neuronal form of nitric oxide synthase (NOS). Gephyrin puncta (corresponding to inhibitory synapses) are scattered throughout lamina I and outline giant projection cells, one of which is seen here. The neuron is surrounded by numerous NOS-containing axons, again derived from local inhibitory interneurons. Scale bar = 50 μm for all parts (Parts **a** and **b** modified from Polgár et al. (1999), with permission from the Society of Neuroscience; parts **c** and **d** modified from Puskár et al. (2001), with permission from Elsevier)

Unlike the NK1r-expressing projection neurons, the giant cells in lamina I apparently receive little or no primary afferent input, but these cells also respond to noxious stimuli, which are probably transmitted by polysynaptic pathways involving excitatory interneurons (Polgár et al. 2008).

There is evidence that different classes of ALT projection neuron are selectively innervated by specific populations of interneurons. For example, the large lamina III ALT neurons with prominent dorsal dendrites receive around one-third of their inhibitory synapses from NPY-containing GABAergic interneurons, whereas NPY is present in between 5 and 15 % of all GABAergic axons in laminae I–III (Polgar et al. 2011) (Fig. 3). These inhibitory inputs are apparently derived from a small subset of NPY-containing inhibitory interneurons, as both the size of the boutons and the intensity of NPY-immunoreactivity are considerably higher than those in the general population of NPY axons. In addition, the lamina III ALT neurons are also targeted by excitatory interneurons that express the opioid peptide dynorphin (Baseer et al. 2012). The giant cells in lamina I are also selectively innervated by inhibitory interneurons but this time by GABAergic cells that contain nNOS (Puskár et al. 2001). At present, less is known about the inhibitory inputs to the NK1r$^+$ lamina I ALT neurons, although we have found that in the mouse, some of these cells are also innervated by nNOS-containing GABAergic axons (N Baseer and AJ Todd, unpublished data).

It is likely that presynaptic inhibitory circuits are also arranged in a highly selective way. Peptidergic primary afferents receive few axoaxonic synapses, whereas these are found in moderate numbers on non-peptidergic nociceptors and are frequently associated with myelinated LTMRs. The axons that form synapses with the non-peptidergic nociceptors are enriched with GABA but not glycine, while those associated with Aδ D-hair afferents contain high levels of both GABA and glycine, indicating that they originate from different populations of inhibitory interneurons (Todd 1996). It has recently been shown that the parvalbumin-containing interneurons selectively innervate myelinated LTMRs (Hughes et al. 2012), since three-quarters of parvalbumin-immunoreactive axons in the inner part of lamina II (IIi) formed axoaxonic synapses on the central terminals of type II synaptic glomeruli (Ribeiro-da-Silva and Coimbra 1982), which are thought to be derived from Aδ D-hair afferents. It is not yet known which cells are responsible for presynaptic inhibition of the non-peptidergic C nociceptors.

1.6 Normal Function of Inhibitory Mechanisms

Inhibitory interneurons in the dorsal horn are thought to perform several different functions. For example, Sandkuhler (2009) has identified four specific mechanisms that are involved in the control of pain: (1) **attenuation** of the nociceptive inputs to dorsal horn neurons to achieve an appropriate level of activation in response to painful stimuli; (2) **muting**, to prevent spontaneous activity in neurons (including projection cells) that are driven by nociceptors; (3) **separating** different modalities, in order to prevent crosstalk that might lead to allodynia; and (4) **limiting** the spatial spread of sensory inputs, in order to restrict sensation to somatotopically appropriate body regions. Failure of each of these mechanisms would be expected to lead to the various symptoms that are seen in chronic pain states: hyperalgesia, spontaneous pain, allodynia and radiating/referred pain, respectively. Additional roles include the inhibition of itch, in response to counter-stimuli such as scratching (Davidson et al. 2009; Ross et al. 2010; Akiyama et al. 2011), and the sharpening of tactile acuity, by surround inhibition of LTMR afferents.

Since there are apparently many different populations of inhibitory interneurons in the dorsal horn, it is not likely that each of these functions is performed by a single population. However, different interneuron populations probably have a specific range of functions. For example, since many nNOS- and NPY-containing

GABAergic neurons respond to noxious stimuli (Polgar et al. 2013b), those that innervate projection neurons in laminae I and III are likely to have a role in attenuating nociceptive inputs. In contrast, the parvalbumin interneurons, which generate presynaptic inhibition of myelinated LTMR inputs, may be involved in maintaining tactile acuity and preventing tactile allodynia (Hughes et al. 2012). Mice lacking the transcription factor Bhlhb5 show exaggerated itch that is associated with loss of inhibitory interneurons from the dorsal horn (Ross et al. 2010). We have recently found that there is a highly selective loss of nNOS- and galanin-containing inhibitory interneurons from these mice, which suggests that one or both of these populations is responsible for scratch-mediated inhibition of itch (Kardon et al 2014).

2 Plasticity of Inhibition in Neuropathic Pain States

Peripheral nerve injuries frequently give rise to neuropathic pain, which is characterised by spontaneous pain, allodynia and hyperalgesia. There are two lines of evidence to suggest that loss of inhibition in the spinal dorsal horn contributes to neuropathic pain. Firstly, suppressing inhibition by intrathecal administration of antagonists acting at $GABA_A$ or glycine receptors produces signs of tactile allodynia and hyperalgesia, which resemble those seen after nerve injury (Yaksh 1989; Sivilotti and Woolf 1994; Miraucourt et al. 2007; Lu et al. 2013). Secondly, a few studies have provided direct evidence for a loss of inhibitory synaptic transmission, by showing reduced inhibitory postsynaptic currents or potentials (IPSCs, IPSPs) after nerve injury (Moore et al. 2002; Scholz et al. 2005; Yowtak et al. 2011; Lu et al. 2013) (see below).

2.1 Animal Models of Neuropathic Pain

Before discussing the evidence for plasticity of inhibition in neuropathic pain, it is first necessary to give a brief account of the animal models that have been used to investigate this issue. Complete transection of a peripheral nerve in humans generally leads to an area of anaesthesia, whereas partial nerve injuries are more likely to give rise to symptoms and signs of neuropathic pain. Based on this observation, several partial nerve injury models have been developed in rodents, and some of the most commonly used ones will be mentioned here. Bennett and Xie (1988) reported that loose ligation of the sciatic nerve led to oedema and subsequent self-strangulation of the nerve, similar to an entrapment neuropathy. Animals that have undergone this procedure, which is known as **chronic constriction injury** (**CCI**), gradually develop thermal and mechanical hyperalgesia, together with tactile allodynia. These are maximal at ~2 weeks after surgery and last for around 2 months (Attal et al. 1990). A disadvantage of the method is the potential variability that results from unavoidable differences in the tightness of the ligatures, and application of a polyethylene cuff around the nerve has therefore been used as an alternative. Kim and Chung (1992) developed the **spinal nerve ligation (SNL)** model, which involves tight ligation of the L5 (and optionally the L6) spinal nerve(s). This leads to a rapid and long-lasting tactile allodynia and thermal hyperalgesia. Another technique, **spared nerve injury (SNI)**, involves tight ligation of two of the three major branches of the sciatic nerve (tibial and common peroneal), leaving the remaining branch (sural) intact, and results in a rapid onset of tactile allodynia, with cold allodynia and a moderate degree of heat hyperalgesia (Decosterd and Woolf 2000).

2.2 Reduced Inhibitory Synaptic Transmission in Neuropathic Pain States

Electrical stimulation of dorsal roots in spinal cord slice preparations normally evokes (polysynaptic) IPSCs in dorsal horn neurons, due to activation of inhibitory interneurons that are presynaptic to the recorded cell. Moore et al. (2002) reported that the proportion of lamina II neurons showing these evoked IPSCs (eIPSCs) was reduced in CCI and SNI models but was unchanged in slices from rats that had undergone complete transection of the sciatic nerve (when compared to slices from unoperated animals). In addition, the amplitude and duration of eIPSCs was reduced in both CCI and SNI models. There was also a reduction in the frequency of miniature IPSCs (mIPSCs; i.e. those recorded in the presence of tetrodotoxin to block synaptic activity) in both neuropathic pain models. Similar results were reported by the same group in a subsequent study (Scholz et al. 2005), while Yowtak et al. (2011) found a reduction in mIPSC frequency (but not amplitude) in the SNL model.

The reduced frequency of mIPSCs indicates a presynaptic mechanism involving the inhibitory interneurons, and this is consistent with immunocytochemical evidence that there is no reduction of $GABA_A$ receptors in the superficial dorsal horn (Moore et al. 2002; Polgár and Todd 2008). Although it has been reported that there is a loss of $GABA_B$ receptors from the L5 dorsal root ganglion after SNL (Engle et al. 2012), this would have no effect on IPSCs in the dorsal horn.

2.3 Possible Mechanisms for Reduced Inhibition Following Peripheral Nerve Injury

Several different mechanisms have been proposed to explain loss of inhibition in the superficial dorsal horn after nerve injury. Some of these are consistent with the reduced mIPSC frequency, for example, death of inhibitory interneurons and reduced neurotransmitter release. Other mechanisms that have been suggested would not account for the change in mIPSCs, although these could still contribute to neuropathic pain. These will be discussed in the following sections.

2.3.1 Loss of Inhibitory Interneurons

One mechanism that could result in loss of inhibition would be death of inhibitory interneurons, and numerous studies have addressed this issue. There has been considerable controversy over whether or not there is significant neuronal death in the dorsal horn after peripheral nerve injury and, if so, whether this affects inhibitory interneurons. Three approaches have been used to investigate these questions: (1) detection of apoptotic cell death, (2) assessment of the number of neurons in each lamina after nerve injury and (3) analysis of the numbers of GABAergic neurons, by using immunocytochemistry, in situ hybridisation for glutamate decarboxylase (GAD, the enzyme that synthesises GABA) or mouse lines in which these GABAergic neurons express GFP. It should be noted that there are two isoforms of GAD, named from their molecular weights: GAD65 and GAD67, and both of these are present in the dorsal horn (Mackie et al. 2003).

Several studies have reported apoptosis in the dorsal horn following peripheral nerve injury, based on staining with the terminal deoxynucleotidyl transferase-mediated biotinylated UTP nick end labelling (TUNEL) method (Kawamura et al. 1997; Azkue et al. 1998; Whiteside and Munglani 2001; Moore et al. 2002; Polgár et al. 2005; Scholz et al. 2005) or detection of the activated form of caspase-3 (Scholz et al. 2005), which is thought to lead to inevitable apoptotic cell death. In some of these studies, no attempt was made to identify whether the apoptotic nuclei belonged to neurons, but Azkue et al. (1998) concluded that TUNEL-positive cells were neuronal, based on expression of a cytoskeletal marker, while Moore et al. (2002) reported that 10 % of them were immunoreactive for the neuronal marker NeuN. In addition, Scholz et al. (2005) observed extensive co-localisation of activated caspase-3 and NeuN after SNI. Polgár et al. (2005) also looked for evidence of neuronal apoptosis in the SNI model but found that although there were numerous TUNEL-positive cells in the ipsilateral dorsal horn, these were virtually all associated with the calcium-binding protein Iba-1, a marker for microglia. They also found no coexistence between NeuN and either TUNEL staining or activated caspase-3 in the SNI model, despite the fact that NeuN co-localised with both these markers in the developing olfactory bulb, where neuronal apoptosis is known to occur. This indicates that failure to detect coexistence in the spinal cord was unlikely to result from early loss of NeuN, making neurons undetectable. Polgár et al. (2005) also observed that while TUNEL-positive nuclei were present in the ipsilateral dorsal horn, they were also found in relatively large numbers in the adjacent white matter, including the dorsal columns, where neurons are seldom present.

There has also been disagreement as to whether the numbers of neurons in the dorsal horn are altered after nerve injury. Scholz et al. (2005) observed a ~20 % loss of neurons from laminae I to III 4 weeks after SNI, whereas two other studies observed no loss of neurons from this region in either the SNI or CCI models (Polgár et al. 2004, 2005). All of these studies used stereological methods to assess the packing density of NeuN-immunoreactive profiles, and it is therefore difficult to explain the discrepancy.

There has also been considerable controversy over whether the numbers of GABAergic neurons are reduced after peripheral nerve injury. Two early studies reported a dramatic (~80–100 %) loss of GABA-immunostaining after CCI (Ibuki et al. 1997; Eaton et al. 1998). Surprisingly, both of these studies also observed a substantial loss of GABA on the contralateral side, even though signs of neuropathic pain are generally not present on this side. Scholz et al. (2005) reported a 25 % reduction in the number of cells with GAD67 mRNA after SNI, although this is not consistent with the earlier report from this group that GAD67 protein and mRNA levels were not altered in this model (Moore et al. 2002). In complete contrast, Polgár et al. (2003) found no alteration in immunostaining for either GABA (Fig. 1) or glycine in the ipsilateral or contralateral dorsal horn of animals that had undergone CCI and which showed clear signs of thermal hyperalgesia. Specifically, the proportions of neurons that were GABA- and/or glycine-immunoreactive on either side of CCI rats did not differ from those in naïve animals. Again, this discrepancy is difficult to interpret. However, immunocytochemistry for the amino acids is technically challenging, requiring rapid and efficient fixation with a relatively high concentration of glutaraldehyde. Technical issues associated with retention of GABA are likely to underlie the differences between the immunocytochemical studies.

A recent study by Yowtak et al. (2013) examined the effects of SNL on the GIN mouse (see above) and reported a reduction of ~30 % in the number of GFP$^+$ cells in the lateral part of lamina II in the L5 segment. Since around a third of GABAergic cells are GFP$^+$ in this line (Heinke et al. 2004), they interpreted this result as indicating a loss of GABAergic neurons. However, it is possible that this reflects downregulation of GFP, rather than cell loss. In addition, the significance of this finding is difficult to interpret, since the signs of neuropathic pain in the SNL model depend on inputs from the intact L4 spinal nerve (Todd 2012), which does not innervate the lateral part of the L5 segment (Shehab et al. 2008).

Taken together, these studies indicate that there is a great deal of controversy over whether there is any neuronal loss after peripheral nerve injury and, if so, whether this involves inhibitory interneurons. Importantly, in the studies that found no neuronal apoptosis or loss of GABAergic neurons, there were clear signs of neuropathic pain (Polgár et al. 2003, 2004, 2005), similar to those found in the original reports of these models. This strongly suggests that loss of GABAergic neurons is not required for the development of neuropathic pain after nerve injury.

2.3.2 Depletion of Transmitter from the Axons of GABAergic Neurons

An alternative explanation for loss of inhibitory function is that the GABAergic neurons in the affected dorsal horn have lower levels of GABA (e.g. due to decreased synthesis), leading to reduced transmitter release. In support of this suggestion, Moore et al. (2002), who used both immunocytochemistry and Western blots, reported a 20–40 % depletion of GAD65, but no change in the levels of GAD67, after CCI and SNI.

Polgár and Todd (2008) looked for direct evidence of transmitter depletion in GABAergic axon terminals in lamina II, by quantifying post-embedding immunogold labelling for GABA with electron microscopy, in rats that had undergone SNI. In order to identify GABAergic boutons, they used pre-embedding immunocytochemistry with antibody against the β3 subunit of the $GABA_A$ receptor, which showed no change after the nerve injury. Because absolute values of immunogold labelling vary from section to section, they compared GABAergic boutons (i.e. those presynaptic at a $GABA_A{}^+$ synapse) on the sides ipsilateral and contralateral to the nerve injury and found that there was no difference in the density of immunogold labelling for GABA between boutons on the two sides. They also found no change in the level of immunostaining for the vesicular GABA transporter (VGAT), which suggests that there was no loss of GABAergic axons in this model.

The available evidence therefore suggests that while GAD65 may be depleted, this does not lead to a significant reduction in the amount of GABA in inhibitory boutons.

2.3.3 Reduced Excitation of Inhibitory Interneurons

One potential mechanism that has received relatively little attention until recently is reduced activation of inhibitory interneurons, either due to changes in their intrinsic properties or due to diminished excitatory drive. Schoffnegger et al. (2006) reported that the passive and active membrane properties, as well as the firing patterns, of GFP-labelled neurons in the GIN mouse were similar in animals that had undergone CCI or a sham operation. They also found that while the proportion of GFP neurons receiving monosynaptic primary afferent input was slightly lower following CCI, the pattern of input from different types of afferent was similar, when compared to mice that had undergone sham operation.

Recently, Leitner et al. (2013) reported that the frequency of miniature excitatory postsynaptic currents (EPSCs) recorded from GFP cells in this mouse line was significantly lower in animals that had undergone CCI, compared to sham-operated animals, indicating a reduced excitatory drive to this subset of GABAergic neurons. They proposed that this was due to reduced release probability at excitatory synapses, rather than due to a reduction in the number of excitatory synapses that the cells received. This conclusion was based on three lines of analysis: (1) the number of dendritic spines (which are sites for excitatory synapses); (2) immunostaining for GAD67, PSD-95 and synaptophysin; and (3) assessment of paired-pulse ratios (PPRs). No change was detected in the number of dendritic spines following CCI, and this was taken as evidence against loss of excitatory synapses. However, dendritic spines are likely to account for only part of the excitatory synaptic input to lamina II neurons, as many of these cells have few such spines. In addition, the number of spines varies enormously between neurons, making it difficult to detect subtle changes. Immunostaining revealed no change in the number of contacts between PSD-95 puncta and GAD67-immunoreactive profiles. However, GAD67 is not present at detectable levels in cell bodies or dendrites of GABAergic neurons in lamina II, and it is therefore difficult to interpret this finding. Specifically, the PSD-95 puncta adjacent to GAD67-immunoreactive profiles would not represent excitatory synapses onto GABAergic neurons. There was a significant increase in PPR for monosynaptic Aδ and C inputs following CCI, and this provides evidence of a reduced release probability at synapses formed by these afferents. Leittner et al. also found direct evidence for a reduced nociceptive drive to inhibitory interneurons following CCI, since the percentage of GFP cells that expressed the transcription factor Fos (a marker for neuronal activation) after noxious heat stimulation was significantly reduced.

These findings are of considerable interest, as they suggest that reduced afferent input to inhibitory interneurons may underlie the loss of inhibition after peripheral nerve injury. The results of Leittner et al. clearly suggest that reduced release probability at excitatory synapses on inhibitory interneurons contributes to this phenomenon. However, a loss of excitatory synapses should not be ruled out for the reasons outlined above. In addition, it is thought that there is a substantial loss of the central terminals of non-peptidergic C nociceptors following nerve injury (Castro-Lopes et al. 1990; Molander et al. 1996), and the postsynaptic targets of these afferents are known to include the dendrites of local inhibitory interneurons (Todd 1996). It is therefore very likely that there is some loss of primary afferent synapses on these cells.

2.3.4 Reduced Effectiveness of Inhibitory Transmission

An alternative mechanism involving an altered postsynaptic effect of GABA and glycine has been proposed by Coull et al. (2003), who reported that the potassium-chloride co-transporter KCC2 was downregulated in lamina I neurons, leading to a dramatic rise in intracellular chloride ion concentrations. This meant that opening of $GABA_A$ and glycine receptors caused a smaller hyperpolarisation than normal, and this could even reverse to a depolarisation. These authors subsequently provided evidence that brain-derived neurotrophic factor (BDNF) released from activated microglia was responsible for the alteration in KCC2 and, therefore, the reduced effectiveness of inhibitory neurotransmitters (Coull et al. 2005).

While this is an attractive hypothesis, there are some reasons for caution in accepting that alterations in KCC2 expression contribute to neuropathic pain. Firstly, intrathecal administration of $GABA_A$ agonists can reverse thermal hyperalgesia and tactile allodynia in rats that have undergone SNL (Hwang and Yaksh 1997; Malan et al. 2002), indicating that GABA retains an anti-nociceptive role in neuropathic pain states. Secondly, the proposed alteration in the reversal potential for chloride would not explain the reduction of mIPSC frequency that has been reported after peripheral nerve injury (Moore et al. 2002; Scholz et al. 2005; Yowtak et al. 2011).

2.3.5 Role of Glycinergic Circuits

While many of the studies described above have concentrated on loss of GABAergic function, there is also evidence that disrupted glycinergic transmission could contribute to tactile allodynia in neuropathic pain states. Miraucourt et al. (2007) demonstrated that blocking glycine receptors in the spinal trigeminal nucleus leads to a form of dynamic allodynia and can result in brush-evoked activation of presumed nociceptive neurons in lamina I. This mechanism appears to involve PKCγ-expressing excitatory interneurons in lamina IIi, which are directly activated by myelinated LTMRs (Peirs et al. 2014). Lu et al. (2013) have recently provided evidence to suggest that there is a polysynaptic pathway involving PKCγ+ interneurons and two other classes of excitatory interneuron (transient central cells and vertical cells) that link Aβ LTMRs with lamina I projection neurons. They proposed that feedforward inhibition from glycinergic lamina III neurons to the PKCγ interneurons normally reduced their responses to Aβ activation and prevented transmission through this pathway. The inhibitory synaptic connection between glycinergic lamina III cells and PKCγ interneurons was weakened after SNL, and this allowed tactile information to reach lamina I cells. However, while this effect was seen in slices from the L5 segment, it was not observed in those from L4. This suggests that the disinhibition only affects inputs from damaged primary afferents (i.e. those in the L5 root). Since the sensory inputs that are perceived as painful in the SNL model are transmitted by the intact L4 spinal nerve, it is not clear how this mechanism could contribute to the allodynia seen in this model.

3 Conclusions

The studies described above demonstrate that there is a wealth of information concerning spinal inhibitory mechanisms in various neuropathic pain states, much of it contradictory. Therefore, despite over 20 years of investigation into the mechanisms that underlie neuropathic pain, there are still many questions that remain to be answered. Although it seems likely that loss of inhibition in the dorsal horn plays a major role, we still do not know precisely how important this is for the different symptoms that occur in neuropathic pain states.

An important recent finding is that there are several distinct populations of inhibitory interneuron, which are likely to perform specific functions. It is very likely that these populations are differentially affected by peripheral nerve injury. Future studies will need to investigate the roles of these populations in controlling different aspects of somatic sensation and the involvement of each of these populations in different forms of neuropathic pain.

References

Akiyama T, Iodi Carstens M, Carstens E (2011) Transmitters and pathways mediating inhibition of spinal itch-signaling neurons by scratching and other counter stimuli. PLoS One 6:e22665

Andrew D (2009) Sensitization of lamina I spinoparabrachial neurons parallels heat hyperalgesia in the chronic constriction injury model of neuropathic pain. J Physiol 587:2005–2017

Antal M, Petko M, Polgar E, Heizmann CW, Storm-Mathisen J (1996) Direct evidence of an extensive GABAergic innervation of the spinal dorsal horn by fibres descending from the rostral ventromedial medulla. Neuroscience 73:509–518

Attal N, Jazat F, Kayser V, Guilbaud G (1990) Further evidence for 'pain-related' behaviours in a model of unilateral peripheral mononeuropathy. Pain 41:235–251

Azkue JJ, Zimmermann M, Hsieh TF, Herdegen T (1998) Peripheral nerve insult induces NMDA receptor-mediated, delayed degeneration in spinal neurons. Eur J Neurosci 10:2204–2206

Baseer N, Polgar E, Watanabe M, Furuta T, Kaneko T, Todd AJ (2012) Projection neurons in lamina III of the rat spinal cord are selectively innervated by local dynorphin-containing excitatory neurons. J Neurosci 32:11854–11863

Bennett GJ, Xie YK (1988) A peripheral mononeuropathy in rat that produces disorders of pain sensation like those seen in man. Pain 33:87–107

Bester H, Chapman V, Besson JM, Bernard JF (2000) Physiological properties of the lamina I spinoparabrachial neurons in the rat. J Neurophysiol 83:2239–2259

Castro-Lopes JM, Coimbra A, Grant G, Arvidsson J (1990) Ultrastructural changes of the central scalloped (C1) primary afferent endings of synaptic glomeruli in the substantia gelatinosa Rolandi of the rat after peripheral neurotomy. J Neurocytol 19:329–337

Cordero-Erausquin M, Allard S, Dolique T, Bachand K, Ribeiro-da-Silva A, De Koninck Y (2009) Dorsal horn neurons presynaptic to lamina I spinoparabrachial neurons revealed by transynaptic labeling. J Comp Neurol 517:601–615

Coull JA, Boudreau D, Bachand K et al (2003) Trans-synaptic shift in anion gradient in spinal lamina I neurons as a mechanism of neuropathic pain. Nature 424:938–942

Coull JA, Beggs S, Boudreau D et al (2005) BDNF from microglia causes the shift in neuronal anion gradient underlying neuropathic pain. Nature 438:1017–1021

Davidson S, Zhang X, Khasabov SG, Simone DA, Giesler GJ Jr (2009) Relief of itch by scratching: state-dependent inhibition of primate spinothalamic tract neurons. Nat Neurosci 12:544–546

Decosterd I, Woolf CJ (2000) Spared nerve injury: an animal model of persistent peripheral neuropathic pain. Pain 87:149–158

Eaton MJ, Plunkett JA, Karmally S, Martinez MA, Montanez K (1998) Changes in GAD- and GABA-immunoreactivity in the spinal dorsal horn after peripheral nerve injury and promotion of recovery by lumbar transplant of immortalized serotonergic precursors. J Chem Neuroanat 16:57–72

Engle MP, Merrill MA, Marquez De Prado B, Hammond DL (2012) Spinal nerve ligation decreases gamma-aminobutyric acid B receptors on specific populations of immunohistochemically identified neurons in L5 dorsal root ganglion of the rat. J Comp Neurol 520:1663–1677

Graham BA, Brichta AM, Callister RJ (2007) Moving from an averaged to specific view of spinal cord pain processing circuits. J Neurophysiol 98:1057–1063

Grudt TJ, Perl ER (2002) Correlations between neuronal morphology and electrophysiological features in the rodent superficial dorsal horn. J Physiol 540:189–207

Hantman AW, van den Pol AN, Perl ER (2004) Morphological and physiological features of a set of spinal substantia gelatinosa neurons defined by green fluorescent protein expression. J Neurosci 24:836–842

Heinke B, Ruscheweyh R, Forsthuber L, Wunderbaldinger G, Sandkuhler J (2004) Physiological, neurochemical and morphological properties of a subgroup of GABAergic spinal lamina II neurones identified by expression of green fluorescent protein in mice. J Physiol 560:249–266

Hughes DI, Sikander S, Kinnon CM, Boyle KA, Watanabe M, Callister RJ, Graham BA (2012) Morphological, neurochemical and electrophysiological features of parvalbumin-expressing cells: a likely source of axo-axonic inputs in the mouse spinal dorsal horn. J Physiol 590:3927–3951

Hwang JH, Yaksh TL (1997) The effect of spinal GABA receptor agonists on tactile allodynia in a surgically-induced neuropathic pain model in the rat. Pain 70:15–22

Ibuki T, Hama AT, Wang XT, Pappas GD, Sagen J (1997) Loss of GABA-immunoreactivity in the spinal dorsal horn of rats with peripheral nerve injury and promotion of recovery by adrenal medullary grafts. Neuroscience 76:845–858

Iwagaki N, Garzillo F, Polgar E, Riddell JS, Todd AJ (2013) Neurochemical characterisation of lamina II inhibitory interneurons that express GFP in the PrP-GFP mouse. Mol Pain 9:56

Kardon AP, Polgár E, Hachisuka J et al (2014) Dynorphin acts as a neuromodulator to inhibit itch in the dorsal horn of the spinal cord. Neuron 82:573–586

Kawamura T, Akira T, Watanabe M, Kagitani Y (1997) Prostaglandin E1 prevents apoptotic cell death in superficial dorsal horn of rat spinal cord. Neuropharmacology 36:1023–1030

Keller AF, Beggs S, Salter MW, De Koninck Y (2007) Transformation of the output of spinal lamina I neurons after nerve injury and microglia stimulation underlying neuropathic pain. Mol Pain 3:27

Kim SH, Chung JM (1992) An experimental model for peripheral neuropathy produced by segmental spinal nerve ligation in the rat. Pain 50:355–363

Leitner J, Westerholz S, Heinke B et al (2013) Impaired excitatory drive to spinal GABAergic neurons of neuropathic mice. PLoS One 8:e73370

Light AR, Trevino DL, Perl ER (1979) Morphological features of functionally defined neurons in the marginal zone and substantia gelatinosa of the spinal dorsal horn. J Comp Neurol 186:151–171

Lu Y, Perl ER (2005) Modular organization of excitatory circuits between neurons of the spinal superficial dorsal horn (laminae I and II). J Neurosci 25:3900–3907

Lu Y, Dong H, Gao Y et al (2013) A feed-forward spinal cord glycinergic neural circuit gates mechanical allodynia. J Clin Invest 123:4050–4062

Mackie M, Hughes DI, Maxwell DJ, Tillakaratne NJ, Todd AJ (2003) Distribution and colocalisation of glutamate decarboxylase isoforms in the rat spinal cord. Neuroscience 119:461–472

Malan TP, Mata HP, Porreca F (2002) Spinal GABA(A) and GABA(B) receptor pharmacology in a rat model of neuropathic pain. Anesthesiology 96:1161–1167

Maxwell DJ, Belle MD, Cheunsuang O, Stewart A, Morris R (2007) Morphology of inhibitory and excitatory interneurons in superficial laminae of the rat dorsal horn. J Physiol 584:521–533

Miraucourt LS, Dallel R, Voisin DL (2007) Glycine inhibitory dysfunction turns touch into pain through PKCgamma interneurons. PLoS One 2:e1116

Molander C, Wang HF, Rivero-Melian C, Grant G (1996) Early decline and late restoration of spinal cord binding and transganglionic transport of isolectin B4 from Griffonia simplicifolia I after peripheral nerve transection or crush. Restor Neurol Neurosci 10:123–133

Moore KA, Kohno T, Karchewski LA, Scholz J, Baba H, Woolf CJ (2002) Partial peripheral nerve injury promotes a selective loss of GABAergic inhibition in the superficial dorsal horn of the spinal cord. J Neurosci 22:6724–6731

Oliva AA Jr, Jiang M, Lam T, Smith KL, Swann JW (2000) Novel hippocampal interneuronal subtypes identified using transgenic mice that express green fluorescent protein in GABAergic interneurons. J Neurosci 20:3354–3368

Peirs C, Patil S, Bouali-Benazzouz R, Artola A, Landry M, Dallel R (2014) Protein kinase C gamma interneurons in the rat medullary dorsal horn: distribution and synaptic inputs to these neurons, and subcellular localization of the enzyme. J Comp Neurol 522:393–413

Polgár E, Todd AJ (2008) Tactile allodynia can occur in the spared nerve injury model in the rat without selective loss of GABA or GABA(A) receptors from synapses in laminae I-II of the ipsilateral spinal dorsal horn. Neuroscience 156:193–202

Polgár E, Shehab SAS, Watt C, Todd AJ (1999) GABAergic neurons that contain neuropeptide Y selectively target cells with the Neurokinin 1 receptor in laminae III and IV of the rat spinal cord. J Neurosci 19:2637–2646

Polgár E, Hughes DI, Riddell JS, Maxwell DJ, Puskar Z, Todd AJ (2003) Selective loss of spinal GABAergic or glycinergic neurons is not necessary for development of thermal hyperalgesia in the chronic constriction injury model of neuropathic pain. Pain 104:229–239

Polgár E, Gray S, Riddell JS, Todd AJ (2004) Lack of evidence for significant neuronal loss in laminae I-III of the spinal dorsal horn of the rat in the chronic constriction injury model. Pain 111:144–150

Polgár E, Hughes DI, Arham AZ, Todd AJ (2005) Loss of neurons from laminas I-III of the spinal dorsal horn is not required for development of tactile allodynia in the spared nerve injury model of neuropathic pain. J Neurosci 25:6658–6666

Polgár E, Campbell AD, MacIntyre LM, Watanabe M, Todd AJ (2007) Phosphorylation of ERK in neurokinin 1 receptor-expressing neurons in laminae III and IV of the rat spinal dorsal horn following noxious stimulation. Mol Pain 3:4

Polgár E, Al-Khater KM, Shehab S, Watanabe M, Todd AJ (2008) Large projection neurons in lamina I of the rat spinal cord that lack the neurokinin 1 receptor are densely innervated by VGLUT2-containing axons and possess GluR4-containing AMPA receptors. J Neurosci 28:13150–13160

Polgár E, Al Ghamdi KS, Todd AJ (2010) Two populations of neurokinin 1 receptor-expressing projection neurons in lamina I of the rat spinal cord that differ in AMPA receptor subunit composition and density of excitatory synaptic input. Neuroscience 167:1192–1204

Polgar E, Sardella TC, Watanabe M, Todd AJ (2011) Quantitative study of NPY-expressing GABAergic neurons and axons in rat spinal dorsal horn. J Comp Neurol 519:1007–1023

Polgar E, Durrieux C, Hughes DI, Todd AJ (2013a) A quantitative study of inhibitory interneurons in laminae I-III of the mouse spinal dorsal horn. PLoS One 8:e78309

Polgar E, Sardella TC, Tiong SY, Locke S, Watanabe M, Todd AJ (2013b) Functional differences between neurochemically defined populations of inhibitory interneurons in the rat spinal dorsal horn. Pain 154:2606–2615

Puskár Z, Polgár E, Todd AJ (2001) A population of large lamina I projection neurons with selective inhibitory input in rat spinal cord. Neuroscience 102:167–176

Rexed B (1952) The cytoarchitectonic organization of the spinal cord in the cat. J Comp Neurol 96:414–495

Ribeiro-da-Silva A, Coimbra A (1982) Two types of synaptic glomeruli and their distribution in laminae I-III of the rat spinal cord. J Comp Neurol 209:176–186

Ross SE, Mardinly AR, McCord AE et al (2010) Loss of inhibitory interneurons in the dorsal spinal cord and elevated itch in Bhlhb5 mutant mice. Neuron 65:886–898

Sandkuhler J (2009) Models and mechanisms of hyperalgesia and allodynia. Physiol Rev 89:707–758

Schneider SP, Walker TM (2007) Morphology and electrophysiological properties of hamster spinal dorsal horn neurons that express VGLUT2 and enkephalin. J Comp Neurol 501:790–809

Schoffnegger D, Heinke B, Sommer C, Sandkuhler J (2006) Physiological properties of spinal lamina II GABAergic neurons in mice following peripheral nerve injury. J Physiol 577:869–878

Scholz J, Broom DC, Youn DH et al (2005) Blocking caspase activity prevents transsynaptic neuronal apoptosis and the loss of inhibition in lamina II of the dorsal horn after peripheral nerve injury. J Neurosci 25:7317–7323

Shehab SA, Al-Marashda K, Al-Zahmi A, Abdul-Kareem A, Al-Sultan MA (2008) Unmyelinated primary afferents from adjacent spinal nerves intermingle in the spinal dorsal horn: a possible mechanism contributing to neuropathic pain. Brain Res 1208:111–119

Sivilotti L, Woolf CJ (1994) The contribution of GABAA and glycine receptors to central sensitization: disinhibition and touch-evoked allodynia in the spinal cord. J Neurophysiol 72:169–179

Tiong SY, Polgar E, van Kralingen JC, Watanabe M, Todd AJ (2011) Galanin-immunoreactivity identifies a distinct population of inhibitory interneurons in laminae I-III of the rat spinal cord. Mol Pain 7:36

Todd AJ (1996) GABA and glycine in synaptic glomeruli of the rat spinal dorsal horn. Eur J Neurosci 8:2492–2498

Todd AJ (2010) Neuronal circuitry for pain processing in the dorsal horn. Nat Rev Neurosci 11:823–836

Todd AJ (2012) How to recognise collateral damage in partial nerve injury models of neuropathic pain. Pain 153:11–12

Todd AJ, Koerber HR (2012) Neuroanatomical substrates of spinal nociception. In: McMahon S, Koltzenburg M, Tracey I, DC T (eds) Wall and Melzack's textbook of pain, 6th edn. Elsevier, Edinburgh, pp 73–90

Torsney C, MacDermott AB (2006) Disinhibition opens the gate to pathological pain signaling in superficial neurokinin 1 receptor-expressing neurons in rat spinal cord. J Neurosci 26:1833–1843

Whiteside GT, Munglani R (2001) Cell death in the superficial dorsal horn in a model of neuropathic pain. J Neurosci Res 64:168–173

Woodbury CJ, Koerber HR (2003) Widespread projections from myelinated nociceptors throughout the substantia gelatinosa provide novel insights into neonatal hypersensitivity. J Neurosci 23:601–610

Yaksh TL (1989) Behavioral and autonomic correlates of the tactile evoked allodynia produced by spinal glycine inhibition: effects of modulatory receptor systems and excitatory amino acid antagonists. Pain 37:111–123

Yasaka T, Kato G, Furue H et al (2007) Cell-type-specific excitatory and inhibitory circuits involving primary afferents in the substantia gelatinosa of the rat spinal dorsal horn in vitro. J Physiol 581:603–618

Yasaka T, Tiong SYX, Hughes DI, Riddell JS, Todd AJ (2010) Populations of inhibitory and excitatory interneurons in lamina II of the adult rat spinal dorsal horn revealed by a combined electrophysiological and anatomical approach. Pain 151:475–488

Yasaka T, Tiong SYX, Polgár E, Watanabe M, Kumamoto E, Riddell JS, Todd AJ (2014) A putative relay circuit providing low-threshold mechanoreceptive input to lamina I projection neurons via vertical cells in lamina II of the rat dorsal horn. Mol Pain 10:3

Yowtak J, Lee KY, Kim HY, Wang J, Kim HK, Chung K, Chung JM (2011) Reactive oxygen species contribute to neuropathic pain by reducing spinal GABA release. Pain 152:844–852

Yowtak J, Wang J, Kim HY, Lu Y, Chung K, Chung JM (2013) Effect of antioxidant treatment on spinal GABA neurons in a neuropathic pain model in the mouse. Pain 154:2469–2476

Zeilhofer HU, Wildner H, Yevenes GE (2012) Fast synaptic inhibition in spinal sensory processing and pain control. Physiol Rev 92:193–235

Zylka MJ, Rice FL, Anderson DJ (2005) Topographically distinct epidermal nociceptive circuits revealed by axonal tracers targeted to Mrgprd. Neuron 45:17–25

Modulation of Peripheral Inflammation by the Spinal Cord

Linda S. Sorkin

Contents

1	The Dorsal Root Reflex	193
2	Acute Inflammatory Models	195
	2.1 Monoarthritis-Kaolin/Carrageenan Knee	195
3	Joint Inflammation: Immuno-Active Agents	196
4	Acute Cutaneous Inflammation	197
5	Chronic Models of Inflammation (Arthritis)	199
6	Sympathetic Effects on Peripheral Inflammation Are Biphasic	200
7	Spinovagal Circuitry	201
References		202

Abstract

The central nervous system, and the spinal cord in particular, is involved in multiple mechanisms that influence peripheral inflammation. Both pro- and anti-inflammatory feedback loops can involve just the peripheral nerves and spinal cord or can also include more complex, supraspinal structures such as the vagal nuclei and the hypothalamic-pituitary axis. Analysis is complicated by the fact that inflammation encompasses a constellation of end points from simple edema to changes in immune cell infiltration and pathology. Whether or not any of these individual elements is altered by any potential mechanism is determined by a complex algorithm including, but not limited to, chronicity of the inflammation, tissue type, instigating stimulus, and state/tone of the immune system. Accordingly, the pharmacology and anatomical substrate of spinal cord modulation of peripheral inflammation are discussed with regard to peripheral tissue type, inflammatory insult (initiating stimulus), and duration of the inflammation.

L.S. Sorkin (✉)
Department of Anesthesiology, University of California, San Diego, La Jolla, CA, USA
e-mail: lsorkin@ucsd.edu

Keywords
Neurogenic inflammation • Dorsal root reflex • Adenosine • Sympathetic nervous system • Arthritis

Inflammation, much like acute pain, can be thought of as a series of protective measures in the face of tissue-damaging stimuli such as pathogens and irritants, as well as debris and by-products from damaged cells. Acute inflammation is primarily a response of the vasculature in conjunction with the innate immune system; together they form the inflammatory response. This manifests clinically as the traditional cardinal signs of inflammation: rubor and calor, redness and heat due to increased blood flow and vasodilation; tumor, edema secondary to plasma extravasation; and dolor, pain as a result of nociceptor stimulation and loss of function. The sequential infiltration of immuno-active cells (neutrophils, macrophages, etc.) is an integral part of inflammation and, through release of pro-inflammatory and chemotactic agents, contributes mightily to the process. Lastly, there are longer-term changes specific to tissue type, such as bone reabsorption and joint remodeling that must also be considered part of the inflammatory disease. There is no doubt that the central nervous system is involved in several positive and negative feedback loops that influence the manifestation of these elements individually or collectively; the tone, sign, and impact of any particular neural influence can vary depending on differences in the severity and stage of the inflammatory disease. Accordingly, this discussion of spinal cord modulation of peripheral inflammation will be broken down both by tissue type, inflammatory model (initiating stimulus), and duration of the inflammation.

Following section of the innervating nerve, capsaicin-induced increases in cutaneous blood flow (flare) were greatly ameliorated compared to the control condition (Lin et al. 1999). Rhizotomy of lumbar sensory roots is fundamentally different than nerve section in that it maintains connectivity between the dorsal root ganglia (DRG) and the peripheral terminals, and thus, the peripheral terminals remain intact and sympathetic efferent fibers are also spared. After rhizotomy, induced increases in cutaneous blood flow and paw thickness are also reduced in comparison to the intact animal (Lin et al. 1999) stressing that the central connection to the spinal cord is of prime importance (Fig. 1). Thus, complete manifestation of inflammation requires a connection to the spinal cord via the dorsal root to enhance this release. Local production of pro-inflammatory cytokines and the resultant infiltration of immune cells can further enhance this process.

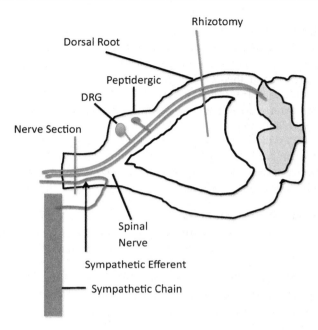

Fig. 1 Transection of the dorsal root, e.g., rhizotomy, separates the spinal cord from the DRG. All central terminals of sensory afferent fibers degenerate; peripheral terminals and sympathetic efferent fibers remain intact. Relationships and interactions between the sympathetic and somatic peripheral terminals are unchanged. Intrathecal capsaicin has a similar effect, but the central fiber loss is specific for the peptidergic, primarily C, nociceptors. Nerve section also separates the spinal cord from the periphery; however, in this instance the central, but not the peripheral primary afferent, terminals remain intact. There is also loss of sympathetic efferent fibers

1 The Dorsal Root Reflex

The most widely accepted explanation for the inflammation-induced spinofugal positive feedback loop resulting in enhanced peripheral neurotransmitter and peptide release is based on dorsal root reflex (DRR) involvement (Cervero and Laird 1996b; Schmidt 1971; Willis 1999). This theory posits that the maintained, increased afferent activity towards the spinal cord engendered by peripheral inflammation triggers a primary afferent depolarization (PAD) in the central terminals of nociceptive primary afferent fibers. Low levels of PAD result in presynaptic inhibition of afferent activity; this is thought to occur, in part, through low-threshold afferent fiber (large diameter) activation of γ-aminobutyric acid (GABA)-containing inhibitory interneurons that in turn have collaterals ending on the nociceptive afferent terminal. We now know that the primary afferent fibers and DRG neurons are slightly more positive (less polarized) than neurons in the central nervous system. When $GABA_A$ receptors on the central terminals of primary afferent nociceptors are activated, Cl^- channels open as they do at all $GABA_A$ receptors, but the Cl^- flux is out of the terminal rather than inward, resulting in a small PAD and decreased central release of afferent neurotransmitters, i.e., PAD. This large fiber inhibition of small-fiber throughput is the central tenet of Melzack and Wall's gate control theory (Melzack and Wall 1965). Following sustained tissue injury, or a high-frequency afferent barrage in nociceptive afferent fibers, sensitized nociceptors can activate the inhibitory interneurons and hijack the PAD pathway (Cervero and Laird 1996b). This hypothesis is substantiated by the anatomy, which shows that GABAergic synapses are often postsynaptic as well as presynaptic to primary afferent terminals (Barber et al. 1978; Carlton and Hayes 1990), many of which are identified or presumptive nociceptors (Alvarez et al. 1992; Bernardi et al. 1995). In addition, tissue inflammation induces an increase in spinal dorsal horn GABA release (Castro-Lopes et al. 1992, 1994). This enhanced presynaptic GABAergic activity on primary afferent terminals leads to a greater level of depolarization, which, if threshold is reached, results in antidromic activity in the afferent fibers. These peripherally directed action potentials are known as DRRs (Schmidt 1971). In addition, at this point orthodromic nociceptive throughput is restored due to the loss of the inhibitory PAD, resulting in enhanced activation of the ascending system (Cervero and Laird 1996a, b; Willis 1999). Thus, drugs or manipulations that alter DRRs elicit covariant changes in inflammation and inflammatory pain. It would be expected that agents that cause presynaptic inhibition via hyperpolarization of the terminal or blockade of neurotransmitter release would reduce both the DRRs and nociception (Cervero and Laird 1996b). In rodent skin, DRR firing in response to intradermal capsaicin has been demonstrated in Aδ and C, but not in Aβ fibers (Lin et al. 2000; Wang et al. 2004; Willis 1999). This antidromic activity results in primary afferent neurotransmitter release in the skin, which extends beyond the injection site. However, in joint afferents of cats and monkeys, DRRs have been demonstrated in group II, III, and IV fibers, indicating that they are generated in large myelinated fibers as well (Sluka et al. 1995). This difference is thought to be due to variations between cutaneous and joint fiber systems rather than to a species difference, although the specific underlying dorsal horn anatomy is not known and the intensity/specificity of the afferent stimuli (kaolin/carrageenan for the joint and capsaicin for the skin) may also be factors. These data conflict with those of an earlier study indicating that electrical stimulation of the distal end of a sectioned posterior articular nerve at group IV, but not group III fiber strength, was sufficient to elicit plasma extravasation and blood cell infiltration into the knee synovia (Ferrell and Russell 1986).

2 Acute Inflammatory Models

2.1 Monoarthritis-Kaolin/Carrageenan Knee

Pharmacology: According to the above scheme, spinal blockade of either the monosynaptic non-NMDA glutamatergic receptor on the inhibitory interneuron or alternatively of the presynaptic GABA$_A$ receptor located on the nociceptive terminal should block DRRs and modulate the peripheral signs of inflammation. Utilizing an acute (several hours) arthritis model generated by kaolin/carrageenan injection into the knee capsule, Westlund, Sluka, and colleagues demonstrated this to be the case. First, they showed that local spinal administration of non-NMDA, but importantly not NMDA glutamate receptor antagonists, significantly blocked manifestations of peripheral inflammation in the awake behaving rat; this was done using both joint size (edema) and temperature as output measures of inflammation (Sluka et al. 1994a; Sluka and Westlund 1993). These data are consistent with involvement of a non-NMDA-mediated monosynaptic afferent drive being an essential part of the pro-inflammatory spinal influence on peripheral inflammation. The data argue against a requirement for spinal glia, whose activation is more closely linked to NMDA and metabotropic glutamatergic receptor-mediated processes, playing a prominent role in this reflexive activity. Second, spinal administration of GABA$_A$, but not GABA$_B$ receptor antagonists, also reduced the signs of knee joint inflammation (Sluka et al. 1993). Both of these findings are consistent with the DRR hypothesis. Unexpectedly, spinal posttreatment with the antagonists partially reversed already established knee joint swelling, but did not reduce the elevated joint temperature (Sluka et al. 1994a), indicating perhaps that different aspects of the inflammatory response are under separate controls and that specifics of these paradigms change during the course of the inflammation. Importantly, if after injection of irritants into the knee capsule, recordings were taken from the proximal stump of the medial articular nerve, which innervates the knee joint, parallel results were observed in that spinal non-NMDA and GABA$_A$, but not NMDA or GABA$_B$ receptor antagonists, blocked DRRs (Rees et al. 1995). Interestingly, when Willis, working in the same lab, later examined the pharmacology of the spinal component contributing to enhanced cutaneous blood flow and edema in the zone of secondary hyperalgesia following intradermal capsaicin, he confirmed the involvement of GABA$_A$ and non-NMDA glutamate receptors, but saw an additional requirement for NMDA receptor activation (Lin et al. 1999). It is unknown if this difference is due to the location of the insult (joint vs. skin) or to some other variant in experimental design. If it is the former, the variation in pharmacology might be a reflection of the different sensory afferent fiber types involved. An alternate hypothesis proposing an anti-inflammatory outcome of spinal GABAergic activation has been proposed; this posits that GABAergic inhibition leads to a reduction of spinal p38 activation (see below), which results in decreased peripheral signs of inflammation (Kelley et al. 2008) perhaps through ascending connections to the vagus. Inflammation-activated spinal p38 implies a dorsal horn microglial link (Boyle et al. 2006; Svensson et al. 2003); however, this spinal microglial to vagus linkage has not yet been fully substantiated.

Pathways: The DRR theory requires the involvement of only the sensory fibers running through the nerve trunk and dorsal root and the inhibitory interneurons for both the afferent and efferent legs of the reflex and does not require participation of the autonomic nervous systems, glia, supraspinal centers, or of the adreno-hypothalamic (HPA) axis. This is supported by experiments showing that peripheral signs of inflammation and DRRs in the acute inflammatory phase of the kaolin/carrageenan model are unaffected by sympathectomy and high spinal cord transection but are readily blocked by rhizotomy, lidocaine application to the peripheral nerve, and nerve crush (Rees et al. 1994; Sluka et al. 1994b).

3 Joint Inflammation: Immuno-Active Agents

Pharmacology: A second class of acute joint inflammation models is initiated by systemic immunization followed in a few weeks by injection of an immuno-active agent into the knee capsule resulting in a monoarthritis. The model has been tested with a variety of pharmacological treatments applied to the spinal cord prior to knee injection. In the acute phase (up to 6 h after the knee injection), spinal administration of the NMDA receptor antagonist ketamine prevented vascular leakage (edema), infiltration of mononuclear leukocytes, and preserved joint integrity (Boettger et al. 2010a). Spinal pretreatment with tumor necrosis factor (TNF) antagonists (TNF-neutralizing antibody or thalidomide, which restricts TNF synthesis by increasing its mRNA degradation (Moreira et al. 1993)) reduced vascular leakage and infiltration of polymorphonuclear leukocytes into the synovial fluid (Bressan et al. 2010, 2012) and helped to maintain joint integrity (Boettger et al. 2010a). The TNF-neutralizing antibody, but not thalidomide, also blocked infiltration of mononuclear leukocytes into the knee (Bressan et al. 2010, 2012). In the acute phase, i.t. administration of the TNF-neutralizing antibody not only reduced joint swelling but also blocked the shift to a sympathetically dominated autonomic nervous system as measured by increased heart rate and heart rate variability (Boettger et al. 2010b). Spinal pretreatment with furosemide, which inhibits DRRs, or with glial (fluorocitrate and minocycline) or fractalkine inhibitors also successfully blocked edema and leukocyte infiltration into knee synovial fluid during the acute phase of this model (Bressan et al. 2012). These results were unaffected by corticosteroid synthesis inhibitors, indicating a lack of participation of the HPA axis. Coadministration of furosemide with either glial inhibitor produced no additional effect, indicating, perhaps, a common pathway. Thus, as seen for the kaolin/carrageenan model of knee inflammation, DRRs appear to be a necessary element in the efferent leg of spinal modulation of peripheral inflammation. Taken together the pharmacological results indicate that, unlike the kaolin/carrageenan model of joint inflammation, there is an intraspinal cord NMDA and glial dependence of the DRRs in the early stages of adjuvant-induced arthritis (AA). Spinal administration of morphine prevented vascular leakage (edema), infiltration of mononuclear leukocytes, and preserved joint integrity (Boettger et al. 2010a). This last result could be due to the opiate-induced reduction of primary afferent glutamate release, which would block the incoming signal and reduce activation of postsynaptic neurons and glia. Alternatively, presynaptic actions of the opiate could reduce primary afferent excitability, thus, affecting a reduction in DRR magnitude (Cervero and Laird 1996b).

4 Acute Cutaneous Inflammation

Pharmacology: Another experimental variant designed to demonstrate spinal contributions to modulation of the acute inflammatory response utilizes an intradermal or subcutaneous carrageenan paradigm that measures either vascular leakage (edema) or the mechanistically different neutrophil infiltration (measured as myeloperoxidase activity, MPO) as outcome measures indicative of inflammation. Intrathecal pretreatment with indomethacin, a prostaglandin synthesis inhibitor, reduces the edema in this model (Daher and Tonussi 2003) as does i.t. morphine (Brock and Tonussi 2008). Unexpectedly, neutrophil infiltration is not blocked following i.t. morphine. If this result holds, it points to different spinal feedback mechanisms in the control of plasma extravasation and immune cell chemotaxis as i.t. morphine should undoubtedly have blocked the DRRs. The morphine effect on edema apparently involves a spinal nitric oxide/cyclic-guanosine monophosphate (c-GMP) pathway (Brock and Tonussi 2008), which is likely to be within the afferent terminals. In similar experiments, spinal pretreatment with either 5HT1 receptor agonists or 5HT2 receptor antagonists also reduced paw swelling (Daher and Tonussi 2003). Significantly, spinal administration of adenosine A1, but not adenosine A2-specific agonists, dose-dependently inhibited dermal neutrophil accumulation due to intradermal carrageenan (Bong et al. 1996; Sorkin et al. 2003). The adenosine effect was mimicked by a spinal NMDA antagonist and reversed by NMDA agonists. This latter experiment demonstrated that the NMDA linkage is downstream of the adenosine effect. What was unexpected, given the data on the kaolin/carrageenan knee joint model, was that spinal non-NMDA receptor antagonists had no affect on the intradermal carrageenan-induced neutrophil accumulation (Bong et al. 1996). Unfortunately, spinal GABAergic agents and glial antagonists were not tested in this model (Pinter et al. 2002).

Paw carrageenan elicited a massive reduction in peripheral adenosine levels, which was temporally linked with the neutrophil infiltration (Bong et al. 1996). This adenosine loss, as well as the increase in neutrophil accumulation, is blocked by spinal pretreatment with an NMDA antagonist or an adenosine A1 agonist. Basal levels of peripheral adenosine in the tissue act on A2 receptors to inhibit peripheral neutrophil infiltration; lower concentrations of adenosine, in contrast, preferentially activate A1 receptors, which enhance neutrophil chemotaxis (Cronstein et al. 1992; Nolte et al. 1992). Thus, maintenance of basal peripheral adenosine concentration is thought to be the basis of the anti-allodynic activity of both the spinal adenosine A1 agonists and NMDA receptor antagonists. Surprisingly, although both substance P and TNF increased at the site of inflammation, neither of the spinal pretreatments that reduced neutrophil infiltration in this model affected either of these agents. Thus, these data appear to conflict with the well-known chemotactic effects of substance P and pro-inflammatory cytokines. There have to be other, as yet, unidentified factors that can block neutrophil infiltration despite the permissive influence of the tachykinin and cytokine. Levels of sympathetic nervous system mediators were not investigated in these studies. Taken together, these data indicate that intradermal carrageenan-induced inflammation, like that induced by intradermal capsaicin, is enhanced by a spinal NMDA-mediated linkage (Sorkin et al. 2003).

Pathways: Chemical elimination of the capsaicin-sensitive small afferent sensory fibers had no effect on neutrophil infiltration in the skin carrageenan model or on the ability of intrathecal adenosine A1 agonists to modify the inflammatory response (Sorkin et al. 2003). In agreement with the negative capsaicin results, Chen et al. (2007) recently observed no effect of capsaicin pretreatment and resulting loss of the peptidergic sensory afferents on acute paw swelling induced by paw carrageenan or complete Freund's adjuvant, but did find that loss of peptidergic innervation reduced increases in paw thickness induced by injection of bee venom. These authors argue that the role of small peptidergic afferents varies according to unidentified aspects of the initiating inflammatory stimulus. Chemical sympathectomy resulted in a strong, but nonsignificant, trend towards decreased carrageenan-induced neutrophil infiltration. However, despite the reduction of basal inflammation due to loss of the sympathetic terminals, intrathecal adenosine A1 agonists resulted in a further reduction in the number of neutrophils found in the injected skin (Sorkin et al. 2003). This implies that the spinal adenosine mediated modulatory effect did not require engagement of the sympathetic nervous system.

It is curious that chemical sympathectomy was ineffective in altering neutrophil infiltration as the sympathetic neurotransmitters norepinephrine and adenosine are known to change in acute inflammation and are well known as leukocyte chemotactic agents (Rose et al. 1988; Straub et al. 2000), although loss of the sympathetic efferents also had no effect on the DRR-dependent knee joint inflammation (Sluka et al. 1994b). The most surprising result of the intradermal carrageenan study was that bilateral carrageenan injections in animals with unilateral rhizotomies resulted in symmetrical infiltration of neutrophils into the injection sites (Sorkin et al. 2003). These data strongly suggest that unlike the kaolin/carrageenan knee joint model, neutrophil trafficking in skin inflammation is independent of endogenous spinal influences. Loss of spinal adenosine modulation of the neutrophil infiltration following rhizotomy points to a role for sensory fibers in this process. Given the previous lack of effect of spinal adenosine agonists on peripheral substance P in the inflamed tissues (Bong et al. 1996), small peptidergic afferents are not the relevant ones, neither are the sympathetic efferents (Sorkin et al. 2003). Thus, the spinal adenosine modulation of the peripheral inflammation is mediated by some combination of large myelinated fibers, non-peptidergic small fibers, or an ascending system with an efferent leg not found within the peripheral nerve.

5 Chronic Models of Inflammation (Arthritis)

Animal models of chronic inflammation that last for weeks are used to study spinal cord control of peripheral inflammation. These model rheumatoid arthritis (RA) and include adjuvant-induced arthritis (AA) and, less frequently, collagen-induced arthritis (CIA).

Pharmacology: Continuous i.t. administration of cyclohexyladenosine (CHA), an adenosine A1 agonist, greatly ameliorated the symptoms of AA including joint swelling, bone changes (demineralization, erosions, and heterotopic bone formation), cartilage destruction, synovial integrity/narrowing of the joint space, hyperemia, and expression of activator protein 1 (AP-1) in the joint. This spinal treatment was efficacious if it began as late as 8 days postimmunization when animals first presented with clinical signs, but had a much smaller, nonsignificant effect on paw swelling when the course of treatment began after clinical signs were well established (day 14) (Boyle et al. 2002). Despite the more than 80 % decrease in AA-induced presentation of clinical symptoms and paw volume that was observed with i.t. CHA treatment, CHA only reduced the AA-induced c-Fos expression in the superficial dorsal horn by 22 % (Boyle et al. 2002). Reduction of c-Fos expression in other spinal laminae was unaffected by the A1 agonist. Similar anti-inflammatory results were observed with either i.t. administration of a TNF-neutralizing antibody or SB203580, a p38 inhibitor, with the addition that these agents also suppressed synovial infiltration of immune cells and expression of the pro-inflammatory cytokines interleukin (IL)-1β, IL-6, and TNF and MMP3, a key gene involved with extracellular matrix degradation (Boettger et al. 2010b; Boyle et al. 2006). In addition, i.t. etanercept blocked the arthritis-induced shift in the autonomic nervous system towards sympathetic dominance (Boettger et al. 2010b).

Interestingly, the p38 antagonist did not affect peripheral T cell proliferation. Continuous administration of morphine or of the NMDA antagonist ketamine throughout a 3-week course of AA caused a major decrease in joint swelling and infiltration of the synovia by pro-inflammatory immune cells (Boettger et al. 2010a). This is likely due to a loss of DRRs due to presynaptic actions on μ-opiate and NMDA receptors, respectively. The opiate-associated reduction in swelling was maintained over the entire period and did not develop tolerance. Rats with continuous administration of a catecholamine reuptake inhibitor or a β2-adrenergic receptor antagonist that began prior to disease induction also showed delayed presentation of clinical signs and less severe joint damage than controls (Levine et al. 1988). Smaller, yet significant, protective effects of these agents on joint injury were obtained when treatment was confined to the period either before or after presentation of clinical disease indicating that reduction in endogenous catecholamines was beneficial throughout the entire 28-day time course. Subsequent studies from other groups have confirmed the role of early β2 receptor activation as contributory to joint damage and report that administration of the β2-agonist salbutamol after disease onset was of clinical benefit and reduced inflammation and joint damage (Lubahn et al. 2004).

Pathways: Much as in the acute inflammatory models, section of major lumbar nerve trunks, greatly diminished the development of arthritis in the formerly innervated limb (Courtright and Kuzell 1965; Kane et al. 2005). In AA, dorsal rhizotomy in lumbar segments decreased the time to presentation of clinical symptoms, while in cervical segments magnitude of the radiographic symptoms was greatly exacerbated (Levine et al. 1986). Thus, the more limited sensory lesion has opposite effects from severing the entire nerve trunk including the sensory terminals and the sympathetic efferents. When i.t. capsaicin pretreatment was used to eliminate only the central peptidergic sensory fibers, rather than severing the entire dorsal root, severity of the inflammation was consistently ameliorated (Colpaert et al. 1983; Cruwys et al. 1995; Hood et al. 2001; Levine et al. 1986). The obvious differences between rhizotomy and central capsaicin are the maintained connection provided to the spinal cord via the non-capsaicin-sensitive afferents (myelinated and non-peptidergic C fibers). Both treatments preserve the sympathetic efferent fibers, and it is thought that both rhizotomy and intrathecal capsaicin preserve the DRGs and peripheral peptidergic fibers (Holzer 1991). Interestingly, capsaicin pretreatment also reduced AA-induced T cell infiltration into the synovium (Hood et al. 2001). In a combined lesion study, Levine pretreated animals with capsaicin, to produce a bilateral loss of the peptidergic afferent connection in conjunction with a unilateral rhizotomy. The rats displayed increased disease severity compared to the capsaicin only side, but not compared to control arthritic animals (Levine et al. 1986). The results outlined above point to the complexity of the wiring and the necessity to involve additional autonomic and endocrine feedback loops to explain the system.

6 Sympathetic Effects on Peripheral Inflammation Are Biphasic

Release of neurotransmitters/mediators from sympathetic terminals in the joint (or other inflamed tissue) during inflammation can alter blood flow, vascular endothelium permeability, and immune cell chemotaxis. During inflammation these processes contribute to increases in joint and tissue swelling in addition to clearance of injury products from the tissue. It follows that sympathectomy prior to induction of AA or CIA, with a loss of the sympathetic terminals in the inflamed joint, results in amelioration of bone damage and a delay in onset of the clinical symptoms of arthritis (Harle et al. 2008; Levine et al. 1985, 1986). If the sympathetic lesion is limited to the lymph nodes that drain the hind limbs, swelling and joint damage are exacerbated (Lorton et al. 1999) highlighting the involvement of the immune system in the generation of the inflammatory response. Integrity of the peripheral sympathetic terminal has been proposed to participate in inflammation aside from its function as part of an efferent leg of a feedback loop and may be adrenergic independent. Local synovial release of inflammatory agents such as bradykinin acts directly on sympathetic terminal varicosities to release prostaglandins and adenosine (Green et al. 1998). Boettger and colleagues

(2010b) have reported decreases in heart rate variability and complexity, which are indicative of "a shift in the autonomic balance towards a sympathetically dominated state" during early phases of the disease or possibly a change in the sympathetic to parasympathetic tone (Waldburger and Firestein 2010). Follow-up studies indicate that the pro-inflammatory sympathetic effect is mediated in the inflamed tissue by peripheral β2 adrenergic and adenosine A2 receptors (Green et al. 1991). In contrast, high-dose epinephrine acting at α2 adrenergic receptors reduced the severity and onset of the inflammatory disease (Coderre et al. 1991).

More recently, studies have shown that timing of the sympathectomy relative to the induction of AIA plays an important role (Harle et al. 2008; Lubahn et al. 2004) with neutral, or even pro-inflammatory, results when the sympathetic nervous system was interrupted after induction and early phases of inflammation (Ebbinghaus et al. 2012; Harle et al. 2005). This change in sign may be due to a functional loss of sympathetic innervation of the inflamed tissue over time; this loss has been observed in rheumatoid arthritis patients and in animal models of RA (Donnerer et al. 1991; Mapp et al. 1990; Straub and Harle 2005). Indeed, there is an inverse relationship between tyrosine hydroxylase (TH)-positive nerve fibers and both the inflammation index and released IL-6 in synovial tissue of RA patients (Miller et al. 2000). This reduction is specific for the sympathetic nerve endings and may be due to increased secretion of mediators of axon repulsion, such as semaphorin 3c; pertidergic sensory nerve fibers are unaffected (Miller et al. 2004). The biphasic nature of the sympathetic system is most prominent in the polyarthritic RA models where swelling and inflammatory signs are present throughout the full duration experiment, usually about 2 months. In the acute monoarthritis model mentioned previously, joint swelling presents on day 1 and resolves in 7–10 days (Ebbinghaus et al. 2012). As in the polyarthritis model, sympathectomy prior to symptom presentation in the monoarthritis model ameliorated the disease, and posttreatment was without effect. However, if a second mBSA injection is made in the knee after resolution, preinjection sympathectomy is once again effective. This implies that in this model and, perhaps, in clinical RA, the sympathetic terminals come back and become functional between flares once the inflammation goes down. In the newly "sympathectomized" joint, the lower levels of norepinephrine mediate their effects predominantly by the higher-affinity α-adrenergic receptors that are pro-inflammatory (Straub and Harle 2005).

7 Spinovagal Circuitry

Activation of the vagus nerve is proposed to have anti-inflammatory actions throughout the entire course of several inflammatory diseases including endotoxic shock and inflammatory bowel disease via a negative feedback loop (Borovikova et al. 2000a, b). The effector mechanism is release of acetylcholine (Ach) from vagal peripheral terminals; Ach then activates α7-nicotinic receptors on macrophages and fibroblast like synoviocytes. This results in decreased synthesis/expression of a host of pro-inflammatory cytokines and other inflammatory mediators. Details for vagal control of joint and cutaneous inflammation are less well defined. Electrical stimulation of the peripheral vagus nerve prevents the edema expected after intraplantar carrageenan injection. This anti-inflammatory action is mimicked by i.c.v. injection of the mitogen-activated protein (MAP) kinase and pro-inflammatory mediator synthesis inhibitor, CNI-1493 (Bernik et al. 2002). Bilateral cervical vagotomy and systemic atropine both antagonize the CNI-1493 effect, thus, solidly implicating the vagus nerve as an anti-inflammatory effector (Borovikova et al. 2000a). In animals without inflammation, the same dose of the p38 MAP kinase inhibitor SB203580 that reduced peripheral inflammation in AIA animals when given into the lumbar i.t. space acutely increases vagal outflow and the resultant peripheral cholinergic activity (Waldburger et al. 2008a). Thus, it has been proposed that spinal p38 activation triggers an ascending signal to the brainstem indicative of peripheral injury, which contributes to peripheral inflammation as well as contributing to the shift in the autonomic balance towards sympathetic dominance. When this pathway is inhibited, as by spinal administration of SB203580, nociception and peripheral inflammation are reduced. The specifics of this pathway are, as yet, not fully defined and include activation of vagal efferent fibers and the subsequent release of Ach from vagal terminals (Waldburger et al. 2008b; Wang et al. 2003) that results in a diminution of the sympathetic shift.

The spinal cord contributes to peripheral inflammation via a multitude of mechanisms. Some, like the DRR and control of peripheral adenosine levels, appear to be intrinsic to the spinal cord. Others, like sympathetic nervous system modulation of peripheral inflammation, can do so via a spinal cord linkage via interactions of sympathetic efferent terminal varicosities with nociceptive nerve terminals or by the local release of norepinephrine and other neurotransmitters. In the main, these are part of positive feedback loops and are pro-inflammatory. The sympathetic nervous system can further alter peripheral inflammation by modulation of B and T cell function (Pongratz and Straub 2010). Parasympathetic influences are anti-inflammatory and always involve a supraspinal linkage encompassing the vagal nuclei. Analysis is complicated by the fact that inflammation encompasses a constellation of end points from simple edema to changes in immune cell infiltration and pathology. Whether or not any of these individual elements is altered by each potential mechanism is determined by a complex algorithm including, but not limited to, chronicity of the inflammation, tissue type, instigating stimulus, and state/tone of the immune system.

References

Alvarez FJ, Kavookjian AM, Light AR (1992) Synaptic interactions between GABA-immunoreactive profiles and the terminals of functionally defined myelinated nociceptors in the monkey and cat spinal cord. J Neurosci 12(8):2901–2917

Barber RP, Vaughn JE, Saito K, McLaughlin BJ, Roberts E (1978) GABAergic terminals are presynaptic to primary afferent terminals in the substantia gelatinosa of the rat spinal cord. Brain Res 141:35–55

Bernardi PS, Valtschanoff JG, Weinberg RJ, Schmidt HH, Rustioni A (1995) Synaptic interactions between primary afferent terminals and GABA and nitric oxide-synthesizing neurons in superficial laminae of the rat spinal cord. J Neurosci 15(2):1363–1371

Bernik TR, Friedman SG, Ochani M, DiRaimo R, Ulloa L, Yang H, Sudan S, Czura CJ, Ivanova SM, Tracey KJ (2002) Pharmacological stimulation of the cholinergic antiinflammatory pathway. J Exp Med 195(6):781–788

Boettger MK, Weber K, Gajda M, Brauer R, Schaible HG (2010a) Spinally applied ketamine or morphine attenuate peripheral inflammation and hyperalgesia in acute and chronic phases of experimental arthritis. Brain Behav Immun 24(3):474–485

Boettger MK, Weber K, Grossmann D, Gajda M, Bauer R, Bar KJ, Schulz S, Voss A, Geis C, Brauer R, Schaible HG (2010b) Spinal tumor necrosis factor alpha neutralization reduces peripheral inflammation and hyperalgesia and suppresses autonomic responses in experimental arthritis: a role for spinal tumor necrosis factor alpha during induction and maintenance of peripheral inflammation. Arthritis Rheum 62(5):1308–1318

Bong GW, Rosengren S, Firestein GS (1996) Spinal cord adenosine receptor stimulation in rats inhibits peripheral neutrophil accumulation. The role of N-methyl-D-aspartate receptors. J Clin Invest 98(12):2779–2785

Borovikova LV, Ivanova S, Nardi D, Zhang M, Yang H, Ombrellino M, Tracey KJ (2000a) Role of vagus nerve signaling in CNI-1493-mediated suppression of acute inflammation. Auton Neurosci 85(1–3):141–147

Borovikova LV, Ivanova S, Zhang M, Yang H, Botchkina GI, Watkins LR, Wang H, Abumrad N, Eaton JW, Tracey KJ (2000b) Vagus nerve stimulation attenuates the systemic inflammatory response to endotoxin. Nature 405(6785):458–462

Boyle DL, Moore J, Yang L, Sorkin LS, Firestein GS (2002) Stimulation of spinal adenosine (ADO) receptors inhibits inflammation and joint destruction in rat adjuvant arthritis. Arthritis Rheum 46(11):3076–3082

Boyle DL, Jones TL, Hammaker D, Svensson CI, Rosengren S, Albani S, Sorkin L, Firestein GS (2006) Regulation of peripheral inflammation by spinal p38 MAP kinase in rats. PLoS Med 3(9):e338

Bressan E, Mitkovski M, Tonussi CR (2010) LPS-induced knee-joint reactive arthritis and spinal cord glial activation were reduced after intrathecal thalidomide injection in rats. Life Sci 87(15–16):481–489

Bressan E, Peres KC, Tonussi CR (2012) Evidence that LPS-reactive arthritis in rats depends on the glial activity and the fractalkine-TNF-alpha signaling in the spinal cord. Neuropharmacology 62(2):947–958

Brock SC, Tonussi CR (2008) Intrathecally injected morphine inhibits inflammatory paw edema: the involvement of nitric oxide and cyclic-guanosine monophosphate. Anesth Analg 106(3):965–971, table of contents

Carlton SM, Hayes ES (1990) Light microscopic and ultrastructural analysis of GABA-immunoreactive profiles in the monkey spinal cord. J Comp Neurol 300(2):162–182

Castro-Lopes JM, Tavares I, Tolle TR, Coito A, Coimbra A (1992) Increase in GABAergic cells and GABA levels in the spinal cord in unilateral inflammation of the hindlimb in the rat. Eur J Neurosci 4(4):296–301

Castro-Lopes JM, Tavares I, Tölle TR, Coimbra A (1994) Carrageenan-induced inflammation of the hind foot provokes a rise of GABA-immunoreactive cells in the rat spinal cord that is prevented by peripheral neurectomy or neonatal capsaicin treatment. Pain 56(2):193–201

Cervero F, Laird JM (1996a) Mechanisms of allodynia: interactions between sensitive mechanoreceptors and nociceptors. Neuroreport 7(2):526–528

Cervero F, Laird JM (1996b) Mechanisms of touch-evoked pain (allodynia): a new model. Pain 68(1):13–23

Chen HS, He X, Wang Y, Wen WW, You HJ, Arendt-Nielsen L (2007) Roles of capsaicin-sensitive primary afferents in differential rat models of inflammatory pain: a systematic comparative study in conscious rats. Exp Neurol 204(1):244–251

Coderre TJ, Basbaum AI, Helms C, Levine JD (1991) High-dose epinephrine acts at alpha 2-adrenoceptors to suppress experimental arthritis. Brain Res 544(2):325–328

Colpaert FC, Donnerer J, Lembeck F (1983) Effects of capsaicin on inflammation and on the substance P content of nervous tissues in rats with adjuvant arthritis. Life Sci 32(16):1827–1834

Courtright LJ, Kuzell WC (1965) Sparing effect of neurological deficit and trauma on the course of adjuvant arthritis in the rat. Ann Rheum Dis 24(4):360–368

Cronstein BN, Levin RI, Philips M, Hirschhorn R, Abramson SB, Weissmann G (1992) Neutrophil adherence to endothelium is enhanced via adenosine A1 receptors and inhibited via adenosine A2 receptors. J Immunol 148(7):2201–2206

Cruwys SC, Garrett NE, Kidd BL (1995) Sensory denervation with capsaicin attenuates inflammation and nociception in arthritic rats. Neurosci Lett 193(3):205–207

Daher JB, Tonussi CR (2003) A spinal mechanism for the peripheral anti-inflammatory action of indomethacin. Brain Res 962(1–2):207–212

Donnerer J, Amann R, Lembeck F (1991) Neurogenic and non-neurogenic inflammation in the rat paw following chemical sympathectomy. Neuroscience 45(3):761–765

Ebbinghaus M, Gajda M, Boettger MK, Schaible HG, Brauer R (2012) The anti-inflammatory effects of sympathectomy in murine antigen-induced arthritis are associated with a reduction of Th1 and Th17 responses. Ann Rheum Dis 71(2):253–261

Ferrell WR, Russell NJ (1986) Extravasation in the knee induced by antidromic stimulation of articular C fibre afferents of the anaesthetized cat. J Physiol 379:407–416

Green PG, Basbaum AI, Helms C, Levine JD (1991) Purinergic regulation of bradykinin-induced plasma extravasation and adjuvant-induced arthritis in the rat. Proc Natl Acad Sci U S A 88(10):4162–4165

Green PG, Miao FJ, Strausbaugh H, Heller P, Janig W, Levine JD (1998) Endocrine and vagal controls of sympathetically dependent neurogenic inflammation. Ann N Y Acad Sci 840:282–288

Harle P, Mobius D, Carr DJ, Scholmerich J, Straub RH (2005) An opposing time-dependent immune-modulating effect of the sympathetic nervous system conferred by altering the cytokine profile in the local lymph nodes and spleen of mice with type II collagen-induced arthritis. Arthritis Rheum 52(4):1305–1313

Harle P, Pongratz G, Albrecht J, Tarner IH, Straub RH (2008) An early sympathetic nervous system influence exacerbates collagen-induced arthritis via CD4+CD25+ cells. Arthritis Rheum 58(8):2347–2355

Holzer P (1991) Capsaicin: cellular targets, mechanisms of action, and selectivity for thin sensory neurons. Pharmacol Rev 43:144–201

Hood VC, Cruwys SC, Urban L, Kidd BL (2001) The neurogenic contribution to synovial leucocyte infiltration and other outcome measures in a guinea pig model of arthritis. Neurosci Lett 299(3):201–204

Kane D, Lockhart JC, Balint PV, Mann C, Ferrell WR, McInnes IB (2005) Protective effect of sensory denervation in inflammatory arthritis (evidence of regulatory neuroimmune pathways in the arthritic joint). Ann Rheum Dis 64(2):325–327

Kelley JM, Hughes LB, Bridges SL Jr (2008) Does gamma-aminobutyric acid (GABA) influence the development of chronic inflammation in rheumatoid arthritis? J Neuroinflammation 5:1

Levine JD, Moskowitz MA, Basbaum AI (1985) The contribution of neurogenic inflammation in experimental arthritis. J Immunol 135(2 Suppl):843s–847s

Levine JD, Dardick SJ, Roizen MF, Helms C, Basbaum AI (1986) Contribution of sensory afferents and sympathetic efferents to joint injury in experimental arthritis. J Neurosci 6:3423–3429

Levine JD, Coderre TJ, Helms C, Basbaum AI (1988) Beta 2-adrenergic mechanisms in experimental arthritis. Proc Natl Acad Sci U S A 85(12):4553–4556

Lin Q, Wu J, Willis WD (1999) Dorsal root reflexes and cutaneous neurogenic inflammation after intradermal injection of capsaicin in rats. J Neurophysiol 82(5):2602–2611

Lin Q, Zou X, Willis WD (2000) Adelta and C primary afferents convey dorsal root reflexes after intradermal injection of capsaicin in rats. J Neurophysiol 84(5):2695–2698

Lorton D, Lubahn C, Klein N, Schaller J, Bellinger DL (1999) Dual role for noradrenergic innervation of lymphoid tissue and arthritic joints in adjuvant-induced arthritis. Brain Behav Immun 13(4):315–334

Lubahn CL, Schaller JA, Bellinger DL, Sweeney S, Lorton D (2004) The importance of timing of adrenergic drug delivery in relation to the induction and onset of adjuvant-induced arthritis. Brain Behav Immun 18(6):563–571

Mapp PI, Kidd BL, Gibson SJ, Terry JM, Revell PA, Ibrahim NB, Blake DR, Polak JM (1990) Substance P-, calcitonin gene-related peptide- and C-flanking peptide of neuropeptide Y-immunoreactive fibres are present in normal synovium but depleted in patients with rheumatoid arthritis. Neuroscience 37(1):143–153

Melzack R, Wall PD (1965) Pain mechanisms: a new theory. Science 150:7

Miller LE, Justen HP, Scholmerich J, Straub RH (2000) The loss of sympathetic nerve fibers in the synovial tissue of patients with rheumatoid arthritis is accompanied by increased norepinephrine release from synovial macrophages. FASEB J 14(13):2097–2107

Miller LE, Weidler C, Falk W, Angele P, Schaumburger J, Scholmerich J, Straub RH (2004) Increased prevalence of semaphorin 3C, a repellent of sympathetic nerve fibers, in the synovial tissue of patients with rheumatoid arthritis. Arthritis Rheum 50(4):1156–1163

Moreira AL, Sampaio EP, Zmuidzinas A, Frindt P, Smith KA, Kaplan G (1993) Thalidomide exerts its inhibitory action on tumor necrosis factor alpha by enhancing mRNA degradation. J Exp Med 177(6):1675–1680

Nolte D, Lorenzen A, Lehr HA, Zimmer FJ, Klotz KN, Messmer K (1992) Reduction of postischemic leukocyte-endothelium interaction by adenosine via A2 receptor. Naunyn Schmiedebergs Arch Pharmacol 346(2):234–237

Pinter E, Than M, Chu DQ, Fogg C, Brain SD (2002) Interaction between interleukin 1beta and endogenous neurokinin 1 receptor agonists in mediating plasma extravasation and neutrophil accumulation in the cutaneous microvasculature of the rat. Neurosci Lett 318(1):13–16

Pongratz G, Straub RH (2010) The B cell, arthritis, and the sympathetic nervous system. Brain Behav Immun 24(2):186–192

Rees H, Sluka KA, Westlund KN, Willis WD (1994) Do dorsal root reflexes augment peripheral inflammation? Neuroreport 5(7):821–824

Rees H, Sluka KA, Westlund KN, Willis WD (1995) The role of glutamate and GABA receptors in the generation of dorsal root reflexes by acute arthritis in the anaesthetized rat. J Physiol 484 (Pt 2):437–445

Rose FR, Hirschhorn R, Weissmann G, Cronstein BN (1988) Adenosine promotes neutrophil chemotaxis. J Exp Med 167(3):1186–1194

Schmidt RF (1971) Presynaptic inhibition in the vertebrate nervous system. Rev Physiol Biochem Pharmacol 63:21–101

Sluka KA, Westlund KN (1993) Centrally administered non-NMDA but not NMDA receptor antagonists block peripheral knee joint inflammation. Pain 55(2):217–225

Sluka KA, Willis WD, Westlund KN (1993) Joint inflammation and hyperalgesia are reduced by spinal bicuculline. Neuroreport 5(2):109–112

Sluka KA, Jordan HH, Westlund KN (1994a) Reduction in joint swelling and hyperalgesia following post-treatment with a non-NMDA glutamate receptor antagonist. Pain 59(1):95–100

Sluka KA, Lawand NB, Westlund KN (1994b) Joint inflammation is reduced by dorsal rhizotomy and not by sympathectomy or spinal cord transection. Ann Rheum Dis 53(5):309–314

Sluka KA, Rees H, Westlund KN, Willis WD (1995) Fiber types contributing to dorsal root reflexes induced by joint inflammation in cats and monkeys. J Neurophysiol 74(3):981–989

Sorkin LS, Moore J, Boyle DL, Yang L, Firestein GS (2003) Regulation of peripheral inflammation by spinal adenosine: role of somatic afferent fibers. Exp Neurol 184(1):162–168

Straub RH, Harle P (2005) Sympathetic neurotransmitters in joint inflammation. Rheum Dis Clin North Am 31(1):43–59, viii

Straub RH, Mayer M, Kreutz M, Leeb S, Scholmerich J, Falk W (2000) Neurotransmitters of the sympathetic nerve terminal are powerful chemoattractants for monocytes. J Leukoc Biol 67(4):553–558

Svensson CI, Marsala M, Westerlund A, Calcutt NA, Campana WM, Freshwater JD, Catalano R, Feng Y, Protter AA, Scott B, Yaksh TL (2003) Activation of p38 mitogen-activated protein kinase in spinal microglia is a critical link in inflammation-induced spinal pain processing. J Neurochem 86(6):1534–1544

Waldburger JM, Firestein GS (2010) Regulation of peripheral inflammation by the central nervous system. Curr Rheumatol Rep 12(5):370–378

Waldburger JM, Boyle DL, Edgar M, Sorkin LS, Levine YA, Pavlov VA, Tracey K, Firestein GS (2008a) Spinal p38 MAP kinase regulates peripheral cholinergic outflow. Arthritis Rheum 58(9):2919–2921

Waldburger JM, Boyle DL, Pavlov VA, Tracey KJ, Firestein GS (2008b) Acetylcholine regulation of synoviocyte cytokine expression by the alpha7 nicotinic receptor. Arthritis Rheum 58(11):3439–3449

Wang H, Yu M, Ochani M, Amella CA, Tanovic M, Susarla S, Li JH, Yang H, Ulloa L, Al-Abed Y, Czura CJ, Tracey KJ (2003) Nicotinic acetylcholine receptor alpha7 subunit is an essential regulator of inflammation. Nature 421(6921):384–388

Wang J, Ren Y, Zou X, Fang L, Willis WD, Lin Q (2004) Sympathetic influence on capsaicin-evoked enhancement of dorsal root reflexes in rats. J Neurophysiol 92(4):2017–2026

Willis WD Jr (1999) Dorsal root potentials and dorsal root reflexes: a double-edged sword. Exp Brain Res 124(4):395–421

The Relationship Between Opioids and Immune Signalling in the Spinal Cord

Jacob Thomas, Sanam Mustafa, Jacinta Johnson, Lauren Nicotra, and Mark Hutchinson

Contents

1	Introduction	208
2	Opioid-Induced Initiation of Non-neuronal Cell Intracellular Signalling in the Central Nervous System	209
3	Non-neuronal Central Immune Cells	209
	3.1 Microglia	210
	3.2 Astrocytes	210
	3.3 Other Cell Types	211
	3.4 Central Immune Synergy	211
4	Involvement of Immunocompetent Cells in Opioid Pharmacodynamics	212
5	Stereoselective and Non-stereoselective Receptor Binding	212
6	Non-stereoselective Activation of Central Immune Cells	213
7	Soluble Contributors to Opioid Analgesia Opposition	213
	7.1 Cytokines	213
	7.2 Proinflammatory Cytokine-Mediated Neuronal Excitation	215
	7.3 Chemokines	216
	7.4 Cholecystokinin	217
	7.5 ATP	219
	7.6 Nitric Oxide	220
	7.7 Sphingomyelins	220
8	Understanding the Molecular Mechanisms of Receptor Crosstalk	222
9	Immediate Clinical Implications of Opioid-Induced Cytokine Signalling	225
10	Sex Differences in Analgesics	227
11	Conclusion	228
References		229

J. Thomas (✉) • J. Johnson • L. Nicotra
Discipline of Pharmacology, School of Medical Sciences, University of Adelaide, Adelaide, Australia
e-mail: jacob.thomas@adelaide.edu.au

S. Mustafa • M. Hutchinson
Discipline of Physiology, School of Medical Sciences, University of Adelaide, Adelaide, Australia

© Springer-Verlag Berlin Heidelberg 2015
H.-G. Schaible (ed.), *Pain Control*, Handbook of Experimental Pharmacology 227,
DOI 10.1007/978-3-662-46450-2_11

Abstract

Opioids are considered the gold standard for the treatment of moderate to severe pain. However, heterogeneity in analgesic efficacy, poor potency and side effects are associated with opioid use, resulting in dose limitations and suboptimal pain management. Traditionally thought to exhibit their analgesic actions via the activation of the neuronal G-protein-coupled opioid receptors, it is now widely accepted that neuronal activity of opioids cannot fully explain the initiation and maintenance of opioid tolerance, hyperalgesia and allodynia. In this review we will highlight the evidence supporting the role of non-neuronal mechanisms in opioid signalling, paying particular attention to the relationship of opioids and immune signalling.

Keywords

Opioid • Analgesia • Tolerance • Immune signalling • Cytokines • Chemokines • Glia • Non-stereoselectivity

1   Introduction

Opioids not only suppress pain, they also activate endogenous counter-regulatory mechanisms that, for example, actively oppose opioid-induced pain suppression, enhance analgesic tolerance wherein repeated opioid doses lose their ability to suppress pain, and enhance dependence as continued opioid exposure is required to stave off drug withdrawal (Watkins et al. 2009). Despite the continual clinical use of opioids over several millennia, and intense scientific research in the past century, a complete understanding of opioid action remains elusive. Of particular importance to this review are those opioid-induced systems that counter-regulate the beneficial and wanted opioid actions. For example, hypertrophy of the cyclic AMP system, enhancement of N-methyl-D-aspartate (NMDA) receptor activity, upregulation of P-glycoprotein and hetero-dimerisation and trafficking of μ-opioid/δ-opioid receptors have been shown to contribute to opioid tolerance and addiction. However, in recent years immune signalling within the central nervous system (CNS), such as that derived from non-neuronal cells, has become the focus of many groups (Johnston et al. 2004; Shavit et al. 2005; Hutchinson et al. 2008a). This 'central immune signalling' profoundly affects all types of cells within the CNS, contributing to the development of the negative side effects of opioids, such as tolerance and enhanced pain states.

2 Opioid-Induced Initiation of Non-neuronal Cell Intracellular Signalling in the Central Nervous System

It is now thought that exclusively considering neuronal activity provides an incomplete understanding of the initiation and maintenance of opioid tolerance, hyperalgesia and allodynia. In recent times, one of the most prominently reported cascades in the CNS influenced by opioid exposure is the mitogen-activated protein kinase (MAPK) signalling pathway. The MAPK pathway comprises a collection of secondary messengers that are recruited by cell surface receptors in response to extracellular stimuli to elicit various cellular responses, such as gene expression. Phosphorylation of the three key kinases of this pathway, p38, JNK, and ERK, results in an active functional signalling complex. Morphine has been shown to activate both p38 and ERK within microglia, which can be prevented by administration of (−)-naloxone (an opioid antagonist) and minocycline (a glial attenuator) (Cui et al. 2006; Xie et al. 2010). The role of JNK is less clear; Guo and co-workers have reported morphine-mediated phosphorylation of JNK in astrocytes in an NMDA receptor-dependent fashion (Guo et al. 2009), whereas others have reported it to be unaffected (Wang et al. 2009). In addition to the MAPK pathway, the IP3/Akt pathway is also activated by opioid exposure and appears to be involved in activation of microglial ERK (Takayama and Ueda 2005; Horvath and DeLeo 2009). The common downstream consequence of MAPK and related pathway signalling is activation of NF-κB, which is responsible for the transcriptional activation of a large number of immune factors, such as IL-1β, IL-6 and TNF-α (Baeuerle and Henkel 1994). Classical G-protein-coupled opioid receptors play a fundamental role in opioid pharmacology; however, as discussed below, a key role for nonclassical opioid sites has been for the most part overlooked.

3 Non-neuronal Central Immune Cells

It has recently been recognised that non-neuronal immunocompetent cells (glia–astrocytes and microglia ['glia'] and endothelial cells) of the CNS and brain play a powerful modulatory role in pain and opioid pharmacodynamics (Hutchinson et al. 2011). Activation of these immunocompetent cells is thought to enhance spinal nociceptive transmission and behavioural responsiveness via the release of central immune signals such as cytokines, chemokines, ATP, nitric oxide and excitatory amino acids (Watkins et al. 2005, 2009; Hutchinson et al. 2008a). It has been suggested that μ-opioid receptor (MOR) agonists are responsible for the glial activation. However, findings by Kao and colleagues indicate that MOR expression is absent from spinal cord astrocytes and microglia, suggesting that these cell types are indirectly activated by MOR agonists under chronic opioid tolerance conditions (Kao et al. 2012).

3.1 Microglia

Microglia are a subset of glial cells within the brain and CNS that make up 5–12 % of all cells and 5–10 % of all glia. Microglia are the resident immune cells of the CNS and under basal conditions scan their microenvironment, performing immune surveillance (Raivich 2005). Microglia are believed to be the most reactive and mobile cells of the CNS and a shift to an activated state can occur within minutes (Morioka et al. 1991). Given their immunological roles, it is no surprise microglia share many of the same immune signalling and response systems as peripheral immune cells. Critical to sensing their microenvironment, microglia express key innate-immune receptors and accompanying response pathways such as the innate-immune pattern recognition toll-like receptors (TLRs). The sensing capabilities of the innate-immune receptors are key in activating early response pathways to molecules such as endogenous danger signals (e.g. heat shock proteins) and xenobiotics (e.g. opioids) (Buchanan et al. 2010). Activation of microglia results in changes in morphology, rapid proliferation, upregulated receptor expression (e.g. complement receptors, TLRs) and changes in function (e.g. migration to sites of damage, phagocytosis and release of proinflammatory mediators). After the activation stimulus has resolved, microglia can either return to their basal state or enter a 'primed' state. Primed microglia do not constitutively produce proinflammatory mediators but may overrespond to new challenges, both in speed and magnitude of release of proinflammatory mediators (Perry et al. 1985; Watkins et al. 2007).

3.2 Astrocytes

Astrocytes are the most abundant glial cell, significantly outnumber neurons, populate all regions of the CNS and, for a long time, were considered exclusively as the metabolic supporting cells for neurons. Astrocytes have highly dynamic processes and are well suited to share synaptic functions with neurons due to their location, organisation and the morphology of their processes that, in combination with microglia, are capable of completely encapsulating neuronal synapses to form the tetrapartite synapse (De Leo et al. 2006; Watkins et al. 2009). The intimate contact astrocytes and microglia have with neurons allows these cells to directly modulate neuron-to-neuron synaptic communication. Astrocytes are known to play an important structural and metabolic role in the homeostasis of the extracellular environment, providing the required conditions for the function of neurons and synapses. Some key roles include forming the blood–brain barrier; metabolic support to neurons, supplying nutrients and neurotransmitters; maintenance of the extracellular environment such as uptake and release of neurotransmitters; regulation of ion concentrations; and detection of neuronal damage (Johnston et al. 2004; Shavit et al. 2005; Hutchinson et al. 2008a, 2011; Araque and Navarrete 2010; Smith 2010). Like microglia, upon stimulation, astrocytes are capable of changing from their basal but active state to elicit a proinflammatory response profile characterised by changes in morphology, proliferation and expression of inflammatory factors such as cytokines and chemokines (Cui et al. 2006; Ben Achour and Pascual 2010; Xie et al. 2010). The immune signalling can further activate other nearby glia and ultimately leads to altered homeostatic balance resulting in the release of various soluble factors involved in neuronal hyperexcitability and the development of abnormal pain.

3.3 Other Cell Types

Microglia and astrocytes are not likely to be the only non-neuronal cell types able to elicit a proinflammatory response profile. Endothelial cells, fibroblasts, oligodendrocytes and other cell types in both the spinal cord and overlying meninges can also produce many of the same neuroexcitatory substances as astrocytes and microglia. For example, it has recently been hypothesised that following the release of central immune signals, the tight junctions of endothelial cells of the blood–brain barrier become leaky, exposing the CNS to peripheral immune signals (Guo et al. 2009; Grace et al. 2011). However, blood–brain barrier endothelial cells remain a significant, yet largely uncharacterised, source of central immune signalling and contributor to altered neuronal function (Quan et al. 2003; Wang et al. 2009).

3.4 Central Immune Synergy

Glia and their released products can work in synergy resulting in an enhanced state of activation and further release of central immune products. With regard to activation, microglia release substances that induce astrocyte activation, expression of adhesion molecules and release of glutamate, TNF, IL-1β and nitric oxide. Astrocytes in return can release substances that stimulate microglial activation, proliferation and production of IL-1β, TNF-α, IL-6 and nitric oxide. The release of these central immune products can synergise and induce the release of others. For example, proinflammatory cytokines can synergise with each other as well as with neurotransmitters and neuromodulators, such as norepinephrine, prostaglandin E2 (PGE2) and nitric oxide. The synergy of TNF and IL-1β with ATP can enhance PGE2 release (Loredo and Benton 1998; Takayama and Ueda 2005; Horvath and DeLeo 2009). Nitric oxide potentiates IL-1β, which can induce PGE2 production and substance P release from sensory afferent terminals in the spinal cord. Substance P can potentiate IL-1β-induced release of IL-6 and PGE2 from human spinal cord astrocytes (Baeuerle and Henkel 1994; Morioka et al. 2002; Watkins et al. 2007). The synergy of these products can therefore lead to the production and release of substances that further activate central immune cells within the CNS leading to enhanced modulation of excitatory neurotransmission.

4 Involvement of Immunocompetent Cells in Opioid Pharmacodynamics

It is now widely accepted that neuronal activity of opioids cannot fully explain the initiation and maintenance of opioid tolerance, hyperalgesia and allodynia. A greater understanding of the role of non-neuronal immunocompetent cells of the CNS and brain is required to fully understand the intricate mechanisms behind pain and opioid pharmacodynamics (Hutchinson et al. 2011). Increased astrocyte activation in the spinal cord following chronic systemic morphine administration was the first report linking glia to morphine tolerance (Song and Zhao 2001; Watkins et al. 2005, 2009; Hutchinson et al. 2008a). Importantly, co-administration of fluorocitrate (a glial metabolic inhibitor) with morphine significantly attenuated not only glial activation but also morphine tolerance. Further work has demonstrated that morphine activates both microglia and astrocytes (Song and Zhao 2001; Cui et al. 2008; Bland et al. 2009; Kao et al. 2012) which is associated with the upregulation and release of proinflammatory cytokines (Raghavendra et al. 2002, 2004; Johnston et al. 2004; Raivich 2005; Hutchinson et al. 2008a, 2009). This opioid-induced non-neuronal cell-mediated anti-analgesia is significantly reduced by co-administration with either the glial attenuators minocycline, ibudilast or fluorocitrate (Morioka et al. 1991; Song and Zhao 2001; Cui et al. 2008; Hutchinson et al. 2008a, b, 2009), or by directly blocking proinflammatory cytokine actions (Fairbanks and Wilcox 2000; Shavit et al. 2005; Hutchinson et al. 2008a, c; Buchanan et al. 2010).

5 Stereoselective and Non-stereoselective Receptor Binding

In the infancy of opioid research, attention was focused directly towards the stereoselective receptors that were shown to be critical for opioid analgesic responses. Classical opioid receptors are stereoselective, as they bind (−)-opioid isomers but not (+)-opioid isomers. Intriguingly, nonclassical opioid actions were observed in one of the first studies that used synthesised inactive opioid stereoisomers (Takagi et al. 1960; Perry et al. 1985; Watkins et al. 2007). Takagi et al. (1960) demonstrated that co-administration of (+)-morphine with (−)-morphine gave rise to naïve tolerance with significantly reduced (−)-morphine-induced analgesia. Further evidence for nonclassical opioid receptors was established in early opioid binding studies conducted by Goldstein et al. (1971). Goldstein demonstrated that the classical stereoselective opioid receptors only made up a small portion of total opioid binding (2 %). Non-stereoselective nonclassical opioid binding sites were responsible for more than half (53 %), while 46 % was attributed to non-specific trapped and dissolved binding. Since then, several studies have reported that (+)-opioid agonists suppress (−)-opioid analgesia (Wu et al. 2007), an effect attributed to glial activation (Wu et al. 2005) which is independent of classical MORs (Wu et al. 2006). To further highlight the involvement of nonclassical opioid receptors, continuous infusion of morphine, oxymorphone or fentanyl administered to opioid receptor knockout mice initiated immediate and steady declines in nociceptive thresholds culminating in several days of unremitting hyperalgesia (Juni et al. 2007; Waxman et al. 2009). Not only did this suggest an involvement of nonclassical opioid receptors but also indicated such receptors are likely responsible for activating endogenous counter-regulatory mechanisms that actively oppose opioid-induced pain suppression and enhance analgesic tolerance and opioid dependence.

6 Non-stereoselective Activation of Central Immune Cells

Recent work by Hutchinson and colleagues has suggested that the innate-immune toll-like receptor-4 (TLR4) is involved in the non-stereoselective binding of opioids (Watkins et al. 2009; Hutchinson et al. 2010; Wang et al. 2012). In vivo, in vitro and in silico approaches provided converging lines of evidence that members of each structural class of opioids activate TLR4 and that opioid antagonists such as naloxone and naltrexone non-stereoselectively block TLR4 signalling (Hutchinson et al. 2010). It was demonstrated that acute pharmacological blockade of TLR4, genetic knockout of TLR4 or blockade of TLR4 downstream signalling leads to a marked potentiation of the magnitude and duration of opioid analgesia, with TLR4 modulation of opioid actions in wild-type animals occurring within minutes (Hutchinson et al. 2010). Furthermore, Wang et al. (2012) demonstrated that morphine binds the human TLR4 accessory protein, MD-2, inducing TLR4 oligomerisation to activate TLR4 signalling. Within the CNS, TLR4 is predominantly expressed by microglia and astrocytes, but expression has also been demonstrated on other non-neuronal cells such as endothelial cells (Tanga et al. 2005; Wang et al. 2010).

7 Soluble Contributors to Opioid Analgesia Opposition

As outlined below, there have been numerous studies looking at soluble factors that either reduce or contribute to opioid-induced pain enhancement. This suggests that, for many situations of abnormal pain, it is not generally just one mediator contributing to the initiation and maintenance but rather a combination thereof. Here, we will highlight the evidence that suggests acute opioid analgesia is substantially modified by the rapid opioid-induced initiation of central immune signalling and that, upon repeated opioid exposure, continued central immune signalling leads to analgesic tolerance and enhanced pain states.

7.1 Cytokines

Cytokines are proteins involved in paracrine and autocrine communication. Cytokines bind to specific receptors on the surface of target cells, which generally activate intracellular signalling and second messenger cascades. One of the key features of immune-derived cytokines is their ability to trigger the feedforward release of more proinflammatory cytokines, which is an important feature of inflammation. In the CNS, cytokines are an effective means for inducing physiological responses to stress, immunological challenges and pathological conditions, but cytokine signalling can also have detrimental effects on neuronal signalling. Within the CNS, astrocytes, microglia, oligodendrocytes and endothelial cells are robust producers of cytokines, in particular, IL-1β, IL-6 and TNF-α. These proinflammatory cytokines are known to be involved in initiating and maintaining states of enhanced pain in pathologies such as neuropathic pain (Milligan et al. 2003; Milligan and Watkins 2009). Importantly however, the levels of cytokine signalling required to elicit a behavioural response in the CNS is far below the quantitative threshold considered as a classical inflammatory response. Hence, the concepts of central immune signalling and homeostatic immune signalling have been proposed (Hutchinson et al. 2011). A recent major development in opioid pharmacology was the demonstration that these key molecules are also significantly involved in opioid analgesia (Johnston et al. 2004; Shavit et al. 2005; Hutchinson et al. 2008a).

Proinflammatory cytokines are gaining traction in the involvement in opioid-induced tolerance and enhanced pain states. Of particular interest is the evidence that proinflammatory cytokines can substantially modify opioid analgesia following a single administration. Blockade of the key inflammatory cytokine IL-1β using an exogenous IL-1 receptor antagonist (IL-1ra) potentiated both acute and chronic, systemic and intrathecal morphine analgesia (Fairbanks and Wilcox 2000; Johnston et al. 2004; Shavit et al. 2005; Hutchinson et al. 2008a). These results have been confirmed using three separate genetically modified strains of mice lacking IL-β function: a transgenic knock-in of IL-1ra such that IL-1ra is over-expressed, IL-1 receptor knockout leaving IL-1β without its cognate receptor and IL-1 receptor accessory protein knockout preventing IL-1 receptor signalling due to the lack of an intracellular link to the associated toll/IL-1 receptor signalling cascade (Shavit et al. 2005). In each case, morphine analgesia was significantly potentiated and prolonged (Shavit et al. 2005). The acute morphine-induced IL-1β signalling caused nearly an eightfold decrease in morphine analgesic potency as demonstrated by a profound leftward shift in the morphine dose–response curve in the presence of IL-1ra (Hutchinson et al. 2008a). IL-1β not only reduces the potency of morphine, it also reduces the duration of effect. Administration of IL-1ra after the normal analgesic response returned to pre-drug baseline unmasked significant analgesia (Shavit et al. 2005; Hutchinson et al. 2008a). This proinflammatory mediated anti-analgesic effect is not limited to IL-1β, as unmasking and/or potentiation of morphine analgesia is also observed by blocking the action of IL-6 and TNF-α (Johnston et al. 2004; Hutchinson et al. 2008a).

Potentiating/unmasking opioid analgesia can also be achieved using less-specific pharmacological interventions such as minocycline or ibudilast that globally disrupt glial cell activation and subsequently the release of inflammatory mediators (Hutchinson et al. 2008a, 2009). Attenuation of glial activation with ibudilast resulted in a fivefold increase in analgesic potency (Hutchinson et al. 2009). This induction of anti-analgesic central immune signalling is not a phenomenon limited to morphine, as oxycodone analgesia was also potentiated threefold by ibudilast (Hutchinson et al. 2009). In addition, the activation of endogenous anti-inflammatory systems that result in elevations of IL-10 or administration of exogenous IL-10 was also capable of potentiating acute morphine analgesia (Fairbanks and Wilcox 2000; Johnston et al. 2004; Hutchinson et al. 2008a). Thus, acute opioid-induced proinflammatory central immune signalling can be pharmacologically modified to enhance acute opioid analgesia.

While numerous studies have demonstrated clear opioid-induced activation of central immune signalling responses, repeated attempts to quantify short-term transcriptional and/or translational events of these proinflammatory central immune signals, after acute in vivo opioid administration, have failed (Johnston et al. 2004; Hutchinson et al. 2008a). Cytokine receptors and their ligands exhibit very high affinity and potency; thus, very low quantities of opioid-induced cytokine release (at levels undetectable by current cytokine quantification techniques) can potentially cause a significant biological effect. Moreover, it is possible that these short-term effects result from the activation of stored immature protein and therefore do not require transcription and translation.

7.2 Proinflammatory Cytokine-Mediated Neuronal Excitation

The release of proinflammatory products can result in enhanced neuronal excitation in the dorsal horn of the spinal cord (Kawasaki et al. 2008) and actively oppose opioid analgesia (Johnston et al. 2004; Shavit et al. 2005; Hutchinson et al. 2008a). Neurons of the spinal cord express receptors for proinflammatory cytokines and chemokines and exhibit increased neuronal excitability in response to these immune signals (Oka et al. 1994; Dame and Juul 2000; Holmes et al. 2004; Kawasaki et al. 2008). These immune-derived cytokines contribute to a phenomenon called central sensitisation, which has been well studied in the area of neuropathic pain (Kawasaki et al. 2008). IL-1β release has been shown to induce the phosphorylation of NMDA receptors on neurons which leads to an increase in calcium conductivity (Viviani et al. 2003; Broom et al. 2004). TNF-α increases α-amino-3-hydroxy-5-methyl-4-isoxazolepropionic acid (AMPA) receptor conductivity while also increasing spontaneous neurotransmitter release from neuronal presynaptic terminals (De et al. 2003). IL-1β and TNF-α synergistically upregulate neuronal cell surface expression of both NMDA and AMPA receptors while downregulating gamma-aminobutyric acid (GABA) receptors (Stellwagen et al. 2005). TNF-α also enhances neuroexcitability in response to glutamate (Emch et al. 2001), and IL-1β induces the release of the neuroexcitant ATP via an NMDA-mediated mechanism (Sperlágh et al. 2004). Beyond these actions, proinflammatory cytokines also lead to the release of a host of neuroexcitatory substances, including more proinflammatory cytokines, nitric oxide, prostaglandins, nerve growth factors, reactive oxygen species, proinflammatory chemokines (e.g. CCL2/MCP-1, CXCL8/IL-8, CXCL10/IP-10) and BDNF (Watkins et al. 1999; Samad et al. 2001; Inoue 2006). Proinflammatory cytokines can also indirectly lead to elevations in extracellular glutamate levels via downregulation of glial and neuronal glutamate transporters (Tawfik et al. 2006). Thus, taken together, opioid-induced release of proinflammatory cytokines exert multiple effects resulting in increased neuronal excitatory tone, which is, in part, the basis behind enhancing the development of hyperalgesia and tolerance.

7.3 Chemokines

Chemokines are a family of small proteins characterised as chemotactic cytokines, involved in cellular migration and intercellular communication. Chemokine receptors are members of the G-protein-coupled receptor (GPCR) superfamily. Chemokines and their receptors have four subclasses of families: C, CC, CXC and CX3C (Murphy et al. 2000). Some chemokines are considered to be proinflammatory and released during an immune response to recruit cells of the immune system to specific sites, while other chemokines are considered homeostatic and control the migration of cells during normal processes of tissue maintenance and development. Some key CNS and brain-derived chemokines include, but are not limited to, CCL2/MCP-1, CCL3/MIP-1α, CCL5/RANTES, CXCL12/SDF-1, CX3CL1/fractalkine and CXCL10. The distributions of chemokines within the CNS are heterogeneous where many have an established involvement in the modulation of pain and opioid pharmacodynamics. There is now a large amount of data indicating that chemokines and their receptors can influence both the acute and chronic stages of pain and opioid analgesia (Szabo et al. 2002; Johnston et al. 2004; Chen et al. 2007a; Triantafilou et al. 2008).

Fractalkine is unique among the typically promiscuous chemokines in that it only binds one known receptor, CX3C receptor-1 (CX3CR1), and this receptor binds only fractalkine (Murphy et al. 2000). Spinal fractalkine is expressed and tethered to the extracellular surface of sensory afferents and intrinsic neurons (Asensio and Campbell 1999), while the fractalkine receptor is predominantly expressed by microglia (Verge et al. 2004). Fractalkine can be cleaved and released, forming a diffusible neuron-to-microglial signal, where binding of fractalkine to its receptor results in activation of microglia (Chapman et al. 2000). Fractalkine binding to its receptor leads to NFκB and p38 MAPK activation (Stievano et al. 2004) followed by production of proinflammatory cytokines and chemokines (Stievano et al. 2004; Johnston et al. 2004). An intrathecal injection of exogenous fractalkine produces both thermal hyperalgesia and mechanical allodynia (Milligan et al. 2004, 2005) and does so via the actions of activated glia. This suggests that fractalkine can facilitate nociception independent of opioid receptor desensitisation and that microglia play a large role in chemokine-mediated pain facilitation.

The involvement of fractalkine in opioid pharmacology was highlighted with experiments demonstrating that co-administration of morphine with an intrathecal fractalkine receptor neutralising antibody potentiated acute morphine analgesia and attenuated the development of tolerance, hyperalgesia and allodynia (Johnston et al. 2004). Similarly, Hutchinson et al. (2008a) also demonstrated fractalkine's ability to oppose acute morphine analgesia. Hutchinson et al. (2008a) further examined opioid-induced release of fractalkine. Analysis of both lumbar dorsal spinal cord sections and cerebrospinal fluid demonstrated that chronic treatment with morphine or methadone caused significant elevations of fractalkine (Hutchinson et al. 2008a). However, fractalkine is not the only chemokine that can oppose opioid analgesia. A study by Szabo et al. (2002) found that pretreatment with CCL5/RANTES (the ligand for CCR1 and CCR5) or SDF-1/CXCL12 (the ligand for CXCR4) followed by opioid administration into the periaqueductal grey (PAG) matter of the brain resulted in a significantly reduced antinociceptive effect.

Another chemokine that is likely to be important in the pharmacodynamics of opioid analgesia is CCL2 (formerly monocyte chemoattractant protein-1, MCP-1). A growing body of evidence ranging from in vitro molecular profiling studies in dorsal root ganglia and spinal cord to data from in vivo assessment, including studies in knockout mice, indicated that CCL2 and its receptor CCR2 contribute to enhanced pain. Studies using chronic morphine or the selective μ-opioid agonist [D-Ala2, N-MePhe4, Gly-ol]-enkephalin (DAMGO) have shown an increase in expression of CCL2 (Rock et al. 2006). Intrathecal administration of CCL2 induces microglial activation, which is abolished in CCR2 (CCL2 receptor) knockout mice (Zhang et al. 2007). Given that opioids induce CCL2 upregulation in the spinal cord, which closely precedes microglial reactivity, it is likely that CCL2 may play a role in initiating a neuron–glial central immune signalling process causing microglial reactivity leading to counter-regulation of opioid-induced analgesia.

Further studies have been conducted implicating chemokines in opposing acute opioid analgesia. For example, the antinociceptive actions of μ-, δ- and κ-opioid receptor agonists are blocked or significantly reduced when the chemokines RANTES/CCL5; the ligand for CCR1, CCR5 or SDF-1α/CXCL12; and the ligand for CXCR4 or CX3CL1/fractalkine are either administered into the PAG of rat 30 min before or co-administered with the opioid agonists (Szabo et al. 2002; Chen et al. 2007a, b). While these behavioural studies do not directly implicate heterologous desensitisation (see Sect. 8), a study by Zhang et al. (2004) demonstrated that proinflammatory chemokines are capable of desensitising MORs on peripheral sensory neurons.

7.4 Cholecystokinin

Cholecystokinin (CCK) was first recognised as a major gastrointestinal hormone responsible for gallbladder contraction and pancreatic enzyme secretion. It has since been discovered in the brain and spinal cord where it is thought to function as a neurotransmitter. CCK exerts its physiological effects via two different GPCRs, CCKB receptor, the predominant receptor found in the brain and in the terminals of neurons, and CCKA receptors, which are abundant in the peripheral tissues (Raiteri and Paudice 1993).

Several forms of CCK have been detected; however, sulphated octapeptide C terminal, CCK8, is the most predominant form of CCK in the CNS (Vanderhaeghen et al. 1975). Within the CNS, the distribution of CCK is heterogeneous. Under normal conditions, CCK is not found in either the DRG or terminals of primary afferents but has been found in the superficial laminae of the spinal cord (Wiesenfeld-Hallin and Xu 1996; Ossipov et al. 2003). However, the peptide is particularly concentrated in regions involved in nociceptive transmission such as the PAG, thalamus, raphe nuclei and the medullary reticular formation (Raiteri and Paudice 1993; Wiesenfeld-Hallin and Xu 1996). In addition, CCK-containing projections from the RVM to the spinal cord have been identified (Mantyh and Hunt 1984). Interestingly, the areas where CCK neurons are located are involved in the mediation of the supraspinal and spinal analgesic effect of morphine. In these areas the actions of CCK have been shown to counteract opioid antinociception (Xie et al. 2005).

The spinal and supraspinal administration of CCK produces behavioural signs of hyperalgesia and enhanced activity of dorsal horn neurons consistent with a pronociceptive role. Systemic or perispinal CCK potently antagonised opioid analgesia produced by foot shock and morphine (Itoh et al. 1982; Faris et al. 1983; Li and Han 1989). CCK antagonists produce significant enhancements of exogenous and endogenous opioid analgesia and, interestingly, could slow or prevent the development of opioid tolerance in some paradigms (Watkins et al. 1985a, b; Dourish et al. 1990; Rezayat et al. 1994; Chapman et al. 1995). In addition, antisense oligodeoxynucleotide 'knock-down' of the CCKB receptor also enhances morphine antinociception (Vanderah et al. 1994).

During the extensive studies on the behavioural effects of CCK-opioid interactions, it was found that regions with well-documented roles in analgesia, such as the PAG and the RVM, were involved in CCK-mediated opposition of analgesia. For example, opioid tolerance induced by repeated microinjections of morphine into the PAG or via systemic morphine injections was reversed by PAG microinjection of proglumide, a CCK receptor antagonist (Vanderah et al. 1994). Within the RVM, only one group of neurons, OFF cells, is activated by μ-opioid agonists (Fields et al. 1983; Heinricher et al. 1994), while ON-cell firing is directly inhibited by opioid agonists (Bederson et al. 1990; Pan et al. 1990; Fields 1992; Heinricher et al. 1992). This is sufficient to produce behaviourally measurable antinociception (Heinricher et al. 1994; Heinricher and Tortorici 1994). However, low dose of CCK microinjected into the RVM blocked the analgesic effect of systemically administered morphine by preventing activation of OFF cells (Heinricher et al. 2001). At this dose, CCK had no effect on the spontaneous activity of these neurons or on the activity of ON cells. The same research group later demonstrated that microinjection of a higher dose of CCK into the RVM selectively activated ON cells and produced behavioural hyperalgesia (Heinricher 2004). This indicates that the pronociceptive actions of the peptide are mediated by neural elements distinct from those mediating the anti-opioid effects (Heinricher 2004).

There is considerable evidence that while CCK modulates the antinociceptive activity of opioids, the opioids in turn promote CCK release. In vivo microdialysis demonstrated that systemic and spinal administration of morphine increased cerebrospinal levels of CCK (de Araujo Lucas et al. 1998). Following a single systemic injection of morphine, CCK mRNA was significantly increased in the hypothalamus and spinal cord (Ding and Bayer 1993). Prolonged exposure to morphine resulted in an accelerated increase in CCK mRNA and CCK peptide (Zhou et al. 1992; Ding and Bayer 1993). For example, after 1, 3 and 6 days of exposure to morphine, whole brain levels of pro-CCK mRNA increased by 52 %, 62 % and 97 %, respectively (Zhou et al. 1992). Since opioids are known to induce the release of endogenous CCK, it is thought that this could be sufficient to activate both the anti-opioid (OFF-cell inhibition) and pronociceptive (ON-cell activation) circuits in parallel (Ossipov et al. 2003).

7.5 ATP

Of all the known neurotransmitters involved in enhanced pain modulation, ATP and a subset of spinal cord ATP receptors, termed purinergic P2X receptors, have gained focus in the facilitation of pain. The activation of the purinergic P2X4 receptor in microglia by ATP results in phosphorylation of p38 MAPK (Trang et al. 2009), which has been shown to be critical for microglial signalling and neuropathic pain sensitisation (Ji et al. 2009). Activation of p38 by phosphorylation leads to the synthesis and release of several glial products, such as the proinflammatory cytokines IL-1β and TNF-α (Ji et al. 2009; McMahon and Malcangio 2009) and the neurotrophin BDNF (Trang et al. 2009). As reviewed earlier, these mediators have been shown to modulate both excitatory and inhibitory synaptic transmission in the spinal cord nociceptive circuitry, leading to an increase in pain sensitivity.

The Horvath group has presented several lines of compelling evidence that demonstrate a critical role of P2X4 in morphine tolerance. Within minutes, morphine administration increased microglial migration via a novel interaction between μ-opioid and P2X4 receptors, which is dependent upon PI3K/Akt pathway activation (Horvath and DeLeo 2009). Persistent morphine infusion in rats induced a marked increase in the expression of spinal P2X4 receptors, the microglial surface marker CD11b and astrocytic GFAP levels (Horvath and DeLeo 2009; Horvath et al. 2010). Intrathecal injections of P2X4 antisense oligodeoxynucleotides inhibit morphine-induced P2X4 receptor expression. Importantly, the antisense oligodeoxynucleotide treatment almost completely prevents the development of antinociceptive tolerance to systemically administered morphine (Horvath et al. 2010). This suggests spinal cord microglial P2X4 signalling modulates the spinal cord neuronal plasticity underlying morphine tolerance. The activation of P2X4 receptor may elicit morphine tolerance by producing glial mediators TNF-α, IL-1β and BDNF via p38 activation; intrathecal infusion of a p38 inhibitor has also been shown to prevent morphine tolerance (Cui et al. 2006; Chen and Sommer 2009).

7.6 Nitric Oxide

Nitric oxide (NO) is a free radical that, among other functions, behaves as an intracellular and intercellular messenger in the nervous system (Snyder 1992). It is synthesised by nitric oxide synthase (NOS), of which three isoforms have been characterised. The neuronal and endothelial isoforms are constitutively expressed in the CNS, whereas the third form is inducible and found in macrophages and inflammatory cells (González-Hernández and Rustioni 1999). Previous studies have suggested a possible role for nitric oxide in acute nociception as well as the development of chronic pain (Tao et al. 2003). For instance, persistent thermal hyperalgesia induced by sciatic nerve injury can be reversed by the administration of the NOS inhibitor NG-nitro-L-arginine methyl ester (L-NAME) (Meller et al. 1992). Moreover, the NOS/nitric oxide system also participates in the development of opioid tolerance and withdrawal. NOS activity is increased in chronic morphine-treated mouse brains and the NOS mRNA level is greater in morphine-tolerant rat spinal cords (Machelska et al. 1997) where the inducible nitric oxide synthase (iNOS) isoform is likely to be the key enzyme responsible for increased NO production (Célérier et al. 2006).

In a study by Kolesnikov et al. (1993), the co-administration of the NOS inhibitor NG-nitro-L-arginine (L-NOARG) with morphine slowly reverses established morphine tolerance over 5 days despite the continued administration of morphine. In addition, a single dose of L-NOARG was shown to retard the development of morphine tolerance for several days. Similarly, intrathecal co-administration of the NOS inhibitor L-NAME with morphine significantly potentiated acute tail-flick and hind paw analgesia compared to morphine alone (Hutchinson et al. 2008a). It has been shown that glial activation can occur via reactive oxygen species, including NO (Meller and Gebhart 1993; Freeman et al. 2008), which suggests that opioid-induced NO release could indirectly modulate opioid-induced analgesia via glial activation and further release of proinflammatory mediators (Holguin et al. 2004).

7.7 Sphingomyelins

Sphingomyelins are a class of membrane sphingolipids found largely in the brain and nervous tissue (Bryan et al. 2008). The sphingomyelin degradation pathway produces ceramide, which is broken down into sphingosine and ceramide-1-phosphate. Sphingosine is further phosphorylated into sphingosine-1-phosphate (S1P) by the action of isoenzymes, sphingosine kinases (SphK) 1 and 2 (Pyne et al. 2009). S1P can act as a second messenger intracellularly and as a ligand for GPCRs (S1P1, S1P2, S1P3, S1P4, S1P5). In the CNS, SphK1/S1P signalling plays a key role in neuron-specific functions such as the regulation of neurotransmitter release from neurons and in the proliferation and survival of neurons and glia (Okada et al. 2009). However, it has recently been shown that under chronic morphine conditions, ceramide and its metabolic pathway contribute to morphine tolerance and hyperalgesia. In a study by Muscoli et al. (2010), chronic morphine was shown to upregulate both the sphingolipid ceramide in spinal astrocytes and microglia, but not neurons, or spinal S1P, the end product of ceramide metabolism. Co-administering morphine with intrathecal administration of pharmacological inhibitors of ceramide and S1P attenuated the development of hyperalgesia and tolerance and blocked increased formation of glial-related proinflammatory cytokines such as TNF-α, IL-1β and IL-6 which, as discussed before, are known modulators of neuronal excitability (Muscoli et al. 2010). SphK1, a key enzyme of the sphingolipid metabolic pathway, can alter the expression and production of proinflammatory cytokines and nitric oxide in microglia. LPS treatment was shown to increase SphK1 mRNA and protein expression, while suppression of SphK1 by its inhibitor, N,N-dimethylsphingosine (DMS), resulted in decreased mRNA expression of TNF-α, IL-1β and iNOS and release of TNF-α and NO in LPS-activated microglia. The addition of S1P increased the expression levels of TNF-α, IL-1β and iNOS and production of TNF-α and NO in activated microglia suggesting that suppression of SphK1 in activated microglia inhibits the production of proinflammatory cytokines and NO (Nayak et al. 2010).

In addition to playing a role in the production of proinflammatory cytokines, ceramide is involved in the production of reactive nitroxidative species, including superoxide, nitric oxide and peroxynitrite (Muscoli et al. 2007). These species can increase steady-state concentrations of ceramide by activating sphingomyelinases and by increasing the degradation of ceramidases, the enzymes responsible for the degradation of ceramide (Pautz et al. 2002). Peroxynitrite can nitrate mitochondrial manganese superoxide dismutase (MnSOD) (an enzyme responsible for regulating concentrations of superoxide dismutase (SOD)) to inactivate the enzyme. This results in an increase in superoxide levels, thereby favouring peroxynitrite formation (Muscoli et al. 2007). In addition to MnSOD, it is thought that peroxynitrite inactivates proteins of central importance in glutamate homeostasis, including glutamate transporters and glutamine synthase. Loss of the transport function leads to increased glutamate levels in the synaptic cleft, overstimulation of NMDA receptor and neurotoxicity. The involvement of peroxynitrite in the development of morphine tolerance and hyperalgesia appears to be of importance as co-administration of morphine with the peroxynitrite decomposition catalyst, Fe (III) 5,10,15,20-tetrakis (N-methylpyridinium-4-yl) porphyrin, blocked protein nitration and prevented the development of tolerance in a dose-dependent manner (Muscoli et al. 2007, 2010). In addition, it is now thought that the neuroprotective actions of minocycline are mediated by direct and specific scavenging of peroxynitrite. A study by Schildknecht et al. (2011) demonstrated that minocycline acts as a highly selective scavenger of peroxynitrite at submicromolar concentrations in various cellular models, including human neurons. This could be of particular importance as in addition to potentiating opioid analgesia, minocycline has been shown to have neuroprotective properties in a variety of chronic neurodegenerative diseases such as Alzheimer disease, Parkinson disease and amyotrophic lateral sclerosis (Schildknecht et al. 2011).

The connection between chronic administration of morphine and the activation of the ceramide metabolic pathway is hypothesised to be via TLR4. LPS activation of TLR4 receptors expressed on monocytes and macrophages activates enzymes in the de novo and sphingomyelinase pathways, leading to increased production of ceramide that, in turn, activates NF-κB and MAPKs to increase the production of proinflammatory products discussed previously (Muscoli et al. 2010).

8 Understanding the Molecular Mechanisms of Receptor Crosstalk

As mentioned earlier, opioids, cytokines and chemokines mediate their biological effects via their cognate GPCRs. GPCRs are the largest family of cell surface receptors implicated in signal transduction. Historically it was understood that these seven-transmembrane (7TM) receptors existed and functioned as monomeric units, acting like 'on and off' switches to transduce extracellular signals in a linear G-protein-dependent manner. However, it is now widely accepted that GPCRs can influence the signalling outcomes, and hence the biological response, of other unrelated receptors at multiple levels and this is often referred to as receptor 'crosstalk'. Receptor crosstalk can be achieved through diverse mechanisms which, although not fully understood, offer the tantalising opportunity for developing highly selective pharmaceutical drugs.

One such mechanism is 'heterologous desensitisation'. Desensitisation is a regulatory mechanism which can completely or partially abolish signal transduction. It has evolved to prevent overstimulation of GPCRs in the presence of continuous agonist stimulation and is important in both physiological and pharmacological settings. Desensitisation can be classified as either homologous or heterologous in nature. Homologous desensitisation occurs when a given GPCR is activated by its cognate ligand and is then desensitised to prevent further signal transduction. Heterologous desensitisation however describes the situation where the activation of one GPCR can lead to the desensitisation of other unrelated and often inactivated GPCRs. An in-depth review of desensitisation mechanisms is beyond the scope of this chapter; please see reviews (Freedman and Lefkowitz 1996; Gainetdinov et al. 2004). However, in general, it is believed that second messenger-dependent protein kinases such as cAMP-dependent protein kinase A (PKA) and protein kinase C (PKC) are primarily responsible for heterologous desensitisation.

The observation that morphine and heroin administration in patients often results in decreased resistance to infections, taken together with the overlapping expression patterns of opioid and chemokine/cytokine receptors and their ligands, has led to much interest in the identification of crosstalk between these receptors. Indeed it has been well documented that the μ- and δ-opioid receptor agonists exert an inhibitory effect on both antibody and cellular immune response (Pellis et al. 1986; Taub et al. 1991) and cytokine expression (Peterson et al. 1987; Chao et al. 1993; Belkowski et al. 1995). Studies conducted by Liu and co-workers in the early 1990s concluded that opioid pretreatment results in the inhibition of the complement-derived chemotactic factor-dependent chemotactic response of leukocytes (Liu et al. 1992). These findings were further supported by the work of Grimm and researchers, extending this inhibitory effect of opioid pretreatment to the responses mediated by the chemokines CCL3, CCL5, CCL2 and CXCL8 (Grimm et al. 1998). Interestingly, with the use of selective μ- or δ-opioid receptor agonists, these inhibitory effects were attributed to the activation of only the μ- and δ-opioid receptors and not the κ-opioid receptor. Furthermore, Grimm and colleagues demonstrated that the chemokine receptors CXCR1 and CXCR2 could be phosphorylated by opioid treatment (Grimm et al. 1998). These findings provided strong evidence of heterologous desensitisation of the chemokine CXCR1 and CXCR2 receptors by opioid receptor activation and a potential mechanism by which opioids may exhibit their immunosuppressive effects.

Not surprisingly, there is evidence in literature supporting the hypothesis that desensitisation is bidirectional and chemokines can influence the perception of pain and inhibit opioid-induced analgesia. Szabo and co-workers investigated the effects of chemokines CCL5 (CCR1 and CCR5 ligand) and CXCL12 (CXCR4 ligand) pretreatment on opioid-induced analgesia (Szabo et al. 2002). In the tail-flick test, rats that were pretreated with CXCL12 followed by DAMGO treatment (MOR agonist) exhibited a dose-dependent reduction in analgesic responses compared to control rats pretreated with saline. When receptor phosphorylation was investigated, CCL5 treatment resulted in the phosphorylation of the MOR at a similar level to that induced by DAMGO treatment. It is important to note the desensitisation of the μ- or δ-opioid receptor was due to the activation of CCR5, CCR2, CCR7 and CXCR4 receptors but not CXCR1 or CXCR2 (Szabo et al. 2002). This suggests that heterologous desensitisation is not indiscriminate but provides another level of highly precise regulation.

It has been previously reported that in vitro activation of the MOR increases the expression of the neuroprotective chemokine CCL5 (Wetzel et al. 2000; Avdoshina et al. 2010). In a recent study conducted by Campbell and researchers, it was determined that naltrexone, an opioid receptor antagonist, could block this morphine-mediated increase of CCL5 (Campbell et al. 2013). The exact mechanism behind this is yet unknown but it is clear that it is mediated by opioid receptors. The inability of morphine to activate glia in the absence of CCR5, the receptor for CCL5 (El-Hage et al. 2008), strongly suggests that complex signalling mechanisms beyond heterologous desensitisation are in place.

Receptors can also crosstalk via a phenomenon known as heteromerisation. Defined as a 'macromolecular complex composed of at least two (functional) receptor units with biochemical properties that are demonstrably different from those of its individual components' (Ferré et al. 2009), it is an elaborate mechanism by which GPCRs can influence the signalling outcomes of one or more receptors (please see reference for an example, Mustafa et al. 2012). In keeping with the theme of this chapter, heteromerisation may also offer an explanation for early observations that opioids can directly act as chemoattractants (Simpkins et al. 1984; van Epps and Saland 1984). Sophisticated techniques have been developed to identify and investigate heteromerisation and these have been summarised elsewhere (Mustafa and Pfleger 2011).

Based on overlapping expression patterns and the findings described above, Heinisch and colleagues conducted studies to identify co-localisation, and hence interactions, between the chemokine receptors, CXCR4 and CX3CR1, and the MOR. The findings from these studies demonstrated that the MOR co-localised with both the CXCR4 and CX3CR1 receptors on individual neurons in several regions of the brain including cingulate cortex (Heinisch et al. 2011). Interestingly, in whole-cell patch-clamp recordings of periaqueductal grey neurons in a rat brain slice preparation, morphine-induced membrane hyperpolarisation was either blocked or reduced in the presence of CXCL12 (CXCR4 ligand) or CX3CL1 (CX3CR1 ligand), respectively (Heinisch et al. 2011). Heteromerisation of the MOR with CXCR4 or CX3CR1, hence close proximity of these receptors, may explain the CXCR4 or CX3CR1-induced heterologous desensitisation of the MOR and therefore potentially also shed light on the limited benefits of opioid analgesics for the treatment of inflammatory pain.

As different cell types or tissues will express a unique combination of receptors, heteromerisation may indeed explain why heterologous desensitisation is not indiscriminate but occurs only between specific receptor types. It also provides an opportunity to target heteromers in a tissue-specific or even temporal manner with the tantalising prospect of understanding and reducing 'off-target' effects experienced with current pharmaceuticals. The identification and development of heteromer-selective ligands, which only activate a specific heteromer combination, is fast becoming an important research objective (Mustafa et al. 2010). Another approach to selectively activate heteromers is through the synthesis of bivalent ligands, which combine the pharmacophores of ligands for the respective constituent receptor units.

Multiple publications have supported the interactions between the MOR and chemokine receptor CCR5 (Suzuki et al. 2002; Szabo et al. 2002, 2003; Chen et al. 2004). In order to further understand the implications of these interactions, Yuan and colleagues synthesised a bivalent ligand incorporating the pharmacophores of naltrexone (MOR antagonist) and maraviroc (a CCR5 antagonist) (Yuan et al. 2013). In a study designed to investigate HIV-1 entry into human astrocytes, the bivalent ligand was effective in significantly inhibiting viral entry when compared to maraviroc treatment alone (Yuan et al. 2013). Naltrexone treatment did not have any effect. This example highlights the importance of identifying physiological relevant heteromers and investigating their function, as these receptor complexes may be the real pharmacological target for many pathophysiological states.

Although GPCR–GPCR interactions are likely to play an important role in the actions of opioids and chemokines, recent evidence suggesting that morphine may also activate TLR4 to mediate proinflammatory response (Wang et al. 2012), analgesia (Hutchinson et al. 2010) and opioid drug reward (Thomas and Hutchinson 2012) should not be overlooked but investigated to further understand the complex mechanisms in place. Avdoshina and co-workers have demonstrated that the TLR4 activator endotoxin LPS increases CCL5 release in primary cultures of microglia (Avdoshina et al. 2010). This knowledge taken together with the findings from a study conducted by Roscic-Mrkic, which suggest that morphine's ability to induce proliferation is via the MAPK pathway, which has also been activated by CCL5 (Roscic-Mrkic et al. 2003), highlights the complex signalling pathways involved. In order to fully understand these elaborate mechanisms, hence the relationship between the opioid, chemokine and TLRs, future studies should be designed with the aim of testing the hypothesis that that heteromerisation these receptors may play a crucial role in modulating analgesia and addiction.

9   Immediate Clinical Implications of Opioid-Induced Cytokine Signalling

The rank-order analgesic potency for opioids commonly utilised in medical settings has been determined both experimentally and through decades of clinical experience (Analgesic Expert Group 2007). To avoid complications associated with high-potency opioids, lower-potency opioids, such as codeine, are employed preferentially in the community, resulting in widespread use. In fact, guidelines, including the World Health Organization's 'pain ladder', routinely recommend the use of 'mild opioids' before stepping up to high-potency opioids like morphine (World Health Organization 1996). The general perception of greater safety and reduced abuse potential with lower-potency opioids has lead not only to the frequent prescribing of codeine but also to prevalent codeine self-medication in countries where the drug is available over the counter (Abbott and Fraser 1998; Harrison et al. 2012). Despite the well-understood differences in acute analgesic efficacy, the variability in central neuroimmune signalling between opioids of differing analgesic potency remains to be elucidated.

In the clinical setting, a condition in which differences in the ability of opioids to initiate neuroimmune signalling may be of particular importance is opioid-overuse headache. Opioid-overuse headache is a particularly onerous subtype of medication-overuse headache, wherein frequent analgesic intake results in exacerbation of a headache disorder (Headache Classification Committee of the International Headache Society 2006). As practice guidelines recommend against the use of potent opioids in the management of headache (Kennis et al. 2012) and many patients with headache disorders elect to self-medicate, a significant proportion of opioid-overuse headache patients develop the condition following the use of over-the-counter codeine products (Ravishankar 2008); thus, the propensity for codeine to induce opioid-overuse headache relative to other opioids is of importance.

Notably, opioid-overuse headache is confined to patients who already suffer from a pre-existing primary headache disorder such as migraine (Lance et al. 1988). Thus, it cannot be considered simply an adverse effect of opioid therapy and should instead be understood as an interaction between the headache disorder and opioid exposure. Although the pathophysiology behind opioid-overuse headache has not yet been confirmed, it has been hypothesised that the selective tendency of headache patients to develop opioid-overuse headache may arise due to alterations in neuroimmune signalling (Johnson et al. 2012), and a number of clinical observations and experimental findings support involvement of the neuroimmune system in this disorder (Meng and Cao 2007).

In headache patients, glial activation is thought to contribute to neuronal hypersensitivity, even in the absence of opioid use (Thalakoti et al. 2007). It is well established that calcitonin-gene-related peptide (CGRP) is released during migraine attacks, and when exposed to CGRP, glial cells release a variety of proinflammatory cytokines including IL-1β and IL-6 which could facilitate headache pain (Thalakoti et al. 2007; Capuano et al. 2009). The cumulative glial activation resulting from CGRP release and opioid exposure is likely to be greater than that caused by CGRP alone, potentially explaining the increase in headache observed following regular opioid treatment of migraine.

Preclinically the role of glial activation in headache following opioid exposure has been clearly demonstrated (Wieseler et al. 2011). Using a rodent model in which headache pain was assessed via the surrogate marker of facial allodynia, it was found that pre-exposure to morphine leads to allodynia when inflammatory 'soup' is applied to the dura in doses that do not cause allodynia in opioid-naïve rats (Wieseler et al. 2011). The involvement of glial activation in the facilitation of head pain was ascertained through administration of the glial-attenuating drug ibudilast concurrently with morphine, which prevented the presentation of facial allodynia (Wieseler et al. 2010).

Evidence also exists suggesting a role specifically for the TLR signalling pathway in medication-overuse headache. In a clinical study the TLR signalling pathway was identified using gene ontology in an analysis of the genomic expression patterns in medication-overuse headache patients that respond to medication withdrawal, alluding again to altered immunity in this condition (Hershey et al. 2011). Moreover, in silico docking simulations indicate codeine may dock to MD2 (Johnson et al. 2012) as morphine does (Eidson and Murphy 2013), suggesting codeine has potential to induce TLR4-dependent pain enhancement, independent of metabolic conversion to morphine. If codeine is able to directly activate the TLR4–MD2 complex, it may lead to far greater increases in pain sensitivity as compared to equianalgesic doses of morphine, as much larger doses of codeine must be administered to provide the same degree of pain relief.

The vast majority of studies investigating the neuroimmune consequences of opioids have focused upon glial activation within the spinal cord; however, to fully appreciate the potential risks, it must be determined if these actions can be generalised to other regions, for example, the trigeminal ganglion, a region of importance in headache pathology. Further research evaluating the neuroimmune actions of different opioids must be conducted to allow the risks to be weighed against the benefits of each treatment option, allowing appropriate drug selection and safe and effective clinical use.

10 Sex Differences in Analgesics

Sex differences in pain and analgesia are now well documented within both the experimental and clinical pain literature. Considerable evidence indicates diverse effectiveness of opioid analgesics in females versus males. Animal studies demonstrate greater analgesia in males; however, human studies reveal the opposite, with robust analgesic responses to opioids in females compared to males (Fillingim and Ness 2000). The existence of developmental and cycling hormone pain and analgesia profiles strongly suggest gonadal steroid hormone manipulation of nociception (Stoffel et al. 2003).

The hormones produced by the ovaries and testes are collectively referred to as the gonadal steroid hormones. The testes are responsible for the production of androgens: testosterone and dihydrotesterone. The ovaries produce both oestrogens (e.g. oestradiol, oestrone and oestriol) and progestins (e.g. progesterone). Although the majority of gonadal steroid hormones are produced in each sex from their respective sex organs, both oestrogens and androgens are present in both sexes; the adrenal cortex is responsible for the production of androgens, the testes known to produce oestrogens and the ovaries in turn producing testosterone (Craft et al. 2004).

The precise mechanisms underlying the role of gonadal steroid hormone manipulation of opioid analgesia are not completely understood. However, their ability to influence nociceptive sensitivity has been recognised both during development (organisational effects) and throughout adulthood (activational effects) (Craft et al. 2004). In addition to altering reproductive physiology and behaviour, the addition of testosterone or oestrogen, in addition to the surgical removal of the ovaries and testes, has been demonstrated to alter opioid analgesia. Androgenisation of neonatal females has been shown to produce more robust morphine analgesia comparable to intact adult males. Moreover, gonadectomy has desensitised morphine analgesia in males, reporting analogous results to adult intact females (Krzanowska and Bodnar 1999; Cicero et al. 2002). Evidently, the manipulation of neonatal gonadal hormones alters and removes the physiological sex differences in opioid analgesia. Furthermore, as (Fillingim and Ness 2000), these results suggest the pathways involved in opioid analgesia are sensitive to gonadal steroid hormones during development.

Studies that have investigated the activational role of gonadal steroid hormones have further demonstrated their contribution to opioid analgesia. Preclinical investigations have revealed this steroid hormone influence, investigating opioid analgesia in (a) gonadally intact and gonadectomised rodents, (b) gonadectomised rodents with and without steroid hormone replacement and (c) across the female menstrual and rodent oestrous cycles (Craft et al. 2004). In a significant number of studies, opioid analgesia was found significantly more potent in intact male rodents compared to gonadectomised subjects. The addition of testosterone in gonadectomised males has also confirmed an association between gonadal steroid hormones and nociceptive pathways, with greater opioid analgesia following testosterone replacement in several preclinical investigations (Ratka 1984; Rao and

Saifi 1985; Stoffel et al. 2003). Despite these findings, it must also be noted that some have in fact demonstrated the opposite effect or the failure to influence opioid analgesia utilising testosterone (Kepler et al. 1989, 1991; Candido et al. 1992).

A gonadal hormone contribution to opioid analgesia has further been established in cycling female rodents. Largely, these studies demonstrate reduced sensitivity to opioids throughout oestrus, the oestrus cycle phase characterised by low levels of 17β oestradiols (Banerjee et al. 1983; Stoffel et al. 2003). Preclinical investigations which have investigated gonadal steroid hormone replacement in gonadectomised females have further implicated an association between 17beta oestradiol and opioid analgesia, with reduced analgesia in 17β oestradiol-supplemented females compared to gonadectomised female subjects (Ryan and Maier 1988; Berglund and Simpkins 1988; Ratka and Simpkins 1990, 1991; Ali et al. 1995). Consequently, despite variability across studies, the majority of findings imply that oestrogen may be responsible for the changes in opioid analgesia across the rodent oestrus cycle. Notably, these findings further suggest that the reduced effectiveness of opioid analgesia in females may result from an association between the 17beta oestradiol and the nociceptive pathways.

As previously discussed, opioids, including morphine, activate not only classical opioid receptors but also TLRs, specifically TLR4, resulting in the production of pain-enhancing proinflammatory cytokines (Hutchinson et al. 2010; Wang et al. 2012). This exacerbated release of proinflammatory cytokines has been demonstrated to counteract the analgesic efficacy of opioids (Hutchinson et al. 2008a, 2010, 2012). Considering that TLR4-mediated responses are more robust in the female sex (Berglund and Simpkins 1988; Kahlke et al. 2000; Marriott et al. 2006; Rettew et al. 2009) and the predominant female sex hormone 17beta oestradiol has been found to stimulate the activation of TLR4 signalling components such as NFκB, resulting in the release of proinflammatory mediators known to play a role in nociception (Soucy et al. 2005; Rettew et al. 2009; Calippe et al. 2010), researchers have hypothesised whether the reduced potency of opioids in females lies in part to an association between oestrogens and opioids on TLR4. Current unpublished data from the University of Colorado investigating the links between the analgesic efficiency of opioids and TLR4 have in fact revealed a correlation between elevated oestrogen and opioid-induced hyperalgesia, resulting in the reduced efficacy of opioids in females. Although unpublished, these results highlight gonadal steroid hormone manipulation of nociception and the need to further investigate the mechanistic link between gonadal steroid hormones and TLRs in the aim to better treat female pain.

11 Conclusion

It is now widely accepted that exclusively considering neuronal opioid activity provides an incomplete understanding of the initiation and maintenance of opioid tolerance, hyperalgesia and dependence. In this chapter, we have highlighted evidence supporting the hypothesis that the release of proinflammatory mediators is initiated by activation of a non-stereoselective receptor such as the innate-immune toll-like receptor 4. The activation of non-neuronal cells within the CNS can profoundly affect the neuronal homeostatic environment leading to significant alterations in neuronal firing. There is now accumulating evidence to suggest that the initiation and maintenance of tolerance and enhanced pain states are not likely to be attributed to one mediator but rather a combination of many. Of note, the consequence of such central immune signalling substantially modifies not only the development of opioid tolerance, hyperalgesia and dependence but also acute opioid analgesia.

The current work on central immune signalling complements the existing body of published findings on neuronal-mediated opioid side effects, such as receptor internalisation and recycling. Of note, the potential of separating the negative side effects from the beneficial actions by targeting opioid-induced glial activation using blood–brain barrier permeable pharmacotherapies such as minocycline, ibudilast or (+)-opioid antagonists has immense clinical utility. For example, development of non-opioid treatments for chronic pain and opioid dependence would enable co-administration with opioids with the possibility of greater efficacy and decreased side effects. The development of pharmaceuticals which selectively regulate signalling pathways implicated in the desired or undesired effects in a tissue- or cell-specific manner promises greater success for opioid use in the future.

References

Abbott FV, Fraser MI (1998) Use and abuse of over-the-counter analgesic agents. J Psychiatry Neurosci 23:13–34

Ali BH, Sharif SI, Elkadi A (1995) Sex differences and the effect of gonadectomy on morphine-induced antinociception and dependence in rats and mice. Clin Exp Pharmacol Physiol 22:342–344

Analgesic Expert Group (2007) Getting to know your analgesics and adjuvants. In: Therapeutic guidelines analgesic. Version 5. Therapeutic guidelines limited, Melbourne, p 44

Araque A, Navarrete M (2010) Glial cells in neuronal network function. Philos Trans R Soc Lond B Biol Sci 365:2375–2381. doi:10.1098/rstb.2009.0313

Asensio VC, Campbell IL (1999) Chemokines in the CNS: plurifunctional mediators in diverse states. Trends Neurosci 22:504–512

Avdoshina V, Biggio F, Palchik G et al (2010) Morphine induces the release of CCL5 from astrocytes: potential neuroprotective mechanism against the HIV protein gp120. Glia 58:1630–1639. doi:10.1002/glia.21035

Baeuerle PA, Henkel T (1994) Function and activation of NF-kappa B in the immune system. Annu Rev Immunol 12:141–179

Banerjee P, Chatterjee TK, Ghosh JJ (1983) Ovarian steroids and modulation of morphine-induced analgesia and catalepsy in female rats. Eur J Pharmacol 96:291–294

Bederson JB, Fields HL, Barbaro NM (1990) Hyperalgesia during naloxone-precipitated withdrawal from morphine is associated with increased on-cell activity in the rostral ventromedial medulla. Somatosens Mot Res 7:185–203

Belkowski SM, Alicea C, Eisenstein TK et al (1995) Inhibition of interleukin-1 and tumor necrosis factor-alpha synthesis following treatment of macrophages with the kappa opioid agonist U50, 488H. J Pharmacol Exp Ther 273:1491–1496

Ben Achour S, Pascual O (2010) Glia: the many ways to modulate synaptic plasticity. Neurochem Int 57:440–445. doi:10.1016/j.neuint.2010.02.013

Berglund LA, Simpkins JW (1988) Alterations in brain opiate receptor mechanisms on proestrous afternoon. Neuroendocrinology 48:394–400

Bland ST, Hutchinson MR, Maier SF et al (2009) The glial activation inhibitor AV411 reduces morphine-induced nucleus accumbens dopamine release. Brain Behav Immun 23:492–497. doi:10.1016/j.bbi.2009.01.014

Broom DC, Samad TA, Kohno T et al (2004) Cyclooxygenase 2 expression in the spared nerve injury model of neuropathic pain. Neuroscience 124:891–900. doi:10.1016/j.neuroscience.2004.01.003

Bryan L, Kordula T, Spiegel S, Milstien S (2008) Regulation and functions of sphingosine kinases in the brain. Biochim Biophys Acta 1781:459–466. doi:10.1016/j.bbalip.2008.04.008

Buchanan MM, Hutchinson M, Watkins LR, Yin H (2010) Toll-like receptor 4 in CNS pathologies. J Neurochem 114:13–27. doi:10.1111/j.1471-4159.2010.06736.x

Calippe B, Douin-Echinard V, Delpy L et al (2010) 17Beta-estradiol promotes TLR4-triggered proinflammatory mediator production through direct estrogen receptor alpha signaling in macrophages in vivo. J Immunol 185:1169–1176. doi:10.4049/jimmunol.0902383

Campbell LA, Avdoshina V, Rozzi S, Mocchetti I (2013) CCL5 and cytokine expression in the rat brain: differential modulation by chronic morphine and morphine withdrawal. Brain Behav Immun 34:130–140. doi:10.1016/j.bbi.2013.08.006

Candido J, Lutfy K, Billings B et al (1992) Effect of adrenal and sex hormones on opioid analgesia and opioid receptor regulation. Pharmacol Biochem Behav 42:685–692

Capuano A, De Corato A, Lisi L et al (2009) Proinflammatory-activated trigeminal satellite cells promote neuronal sensitization: relevance for migraine pathology. Mol Pain 5:43. doi:10.1186/1744-8069-5-43

Célérier E, González JR, Maldonado R et al (2006) Opioid-induced hyperalgesia in a murine model of postoperative pain: role of nitric oxide generated from the inducible nitric oxide synthase. Anesthesiology 104:546

Chao CC, Molitor TW, Close K et al (1993) Morphine inhibits the release of tumor necrosis factor in human peripheral blood mononuclear cell cultures. Int J Immunopharmacol 15:447–453

Chapman GA, Moores K, Harrison D et al (2000) Fractalkine cleavage from neuronal membranes represents an acute event in the inflammatory response to excitotoxic brain damage. J Neurosci 20:RC87

Chapman V, Honoré P, Buritova J, Besson JM (1995) Cholecystokinin B receptor antagonism enhances the ability of a low dose of morphine to reduce c-Fos expression in the spinal cord of the rat. Neuroscience 67:731–739

Chen C, Li J, Bot G et al (2004) Heterodimerization and cross-desensitization between the μ-opioid receptor and the chemokine CCR5 receptor. Eur J Pharmacol 483:175–186. doi:10.1016/j.ejphar.2003.10.033

Chen X, Geller EB, Rogers TJ, Adler MW (2007a) Rapid heterologous desensitization of antinociceptive activity between mu or delta opioid receptors and chemokine receptors in rats. Drug Alcohol Depend 88:36–41. doi:10.1016/j.drugalcdep.2006.09.010

Chen X, Geller EB, Rogers TJ, Adler MW (2007b) The chemokine CX3CL1/fractalkine interferes with the antinociceptive effect induced by opioid agonists in the periaqueductal grey of rats. Brain Res 1153:52–57. doi:10.1016/j.brainres.2007.03.066

Chen Y, Sommer C (2009) The role of mitogen-activated protein kinase (MAPK) in morphine tolerance and dependence. Mol Neurobiol 40:101–107. doi:10.1007/s12035-009-8074-z

Cicero TJ, Nock B, O'Connor L, Meyer ER (2002) Role of steroids in sex differences in morphine-induced analgesia: activational and organizational effects. J Pharmacol Exp Ther 300:695–701

Craft RM, Mogil JS, Aloisi AM (2004) Sex differences in pain and analgesia: the role of gonadal hormones. Eur J Pain 8:397–411. doi:10.1016/j.ejpain.2004.01.003

Cui Y, Chen Y, Zhi J-L et al (2006) Activation of p38 mitogen-activated protein kinase in spinal microglia mediates morphine antinociceptive tolerance. Brain Res 1069:235–243. doi:10.1016/j.brainres.2005.11.066

Cui Y, Liao X-X, Liu W et al (2008) A novel role of minocycline: attenuating morphine antinociceptive tolerance by inhibition of p38 MAPK in the activated spinal microglia. Brain Behav Immun 22:114–123. doi:10.1016/j.bbi.2007.07.014

Dame JB, Juul SE (2000) The distribution of receptors for the pro-inflammatory cytokines interleukin (IL)-6 and IL-8 in the developing human fetus. Early Hum Dev 58:25–39

De A, Krueger JM, Simasko SM (2003) Tumor necrosis factor alpha increases cytosolic calcium responses to AMPA and KCl in primary cultures of rat hippocampal neurons. Brain Res 981:133–142

de Araujo Lucas G, Alster P, Brodin E, Wiesenfeld-Hallin Z (1998) Differential release of cholecystokinin by morphine in rat spinal cord. Neurosci Lett 245:13–16

De Leo JA, Tawfik VL, LaCroix-Fralish ML (2006) The tetrapartite synapse: path to CNS sensitization and chronic pain. Pain 122:17–21. doi:10.1016/j.pain.2006.02.034

Ding XZ, Bayer BM (1993) Increases of CCK mRNA and peptide in different brain areas following acute and chronic administration of morphine. Brain Res 625:139–144

Dourish CT, O'Neill MF, Coughlan J et al (1990) The selective CCK-B receptor antagonist L-365,260 enhances morphine analgesia and prevents morphine tolerance in the rat. Eur J Pharmacol 176:35–44

Eidson LN, Murphy AZ (2013) Blockade of toll-like receptor 4 attenuates morphine tolerance and facilitates the pain relieving properties of morphine. J Neurosci 33:15952–15963. doi:10.1523/JNEUROSCI.1609-13.2013

El-Hage N, Bruce-Keller AJ, Knapp PE, Hauser KF (2008) CCL5/RANTES gene deletion attenuates opioid-induced increases in glial CCL2/MCP-1 immunoreactivity and activation in HIV-1 Tat-exposed mice. J Neuroimmune Pharmacol 3:275–285. doi:10.1007/s11481-008-9127-1

Emch GS, Hermann GE, Rogers RC (2001) TNF-alpha-induced c-Fos generation in the nucleus of the solitary tract is blocked by NBQX and MK-801. Am J Physiol Regul Integr Comp Physiol 281:R1394–R1400

Fairbanks CA, Wilcox GL (2000) Spinal plasticity of acute opioid tolerance. J Biomed Sci 7:200–212

Faris PL, Komisaruk BR, Watkins LR, Mayer DJ (1983) Evidence for the neuropeptide cholecystokinin as an antagonist of opiate analgesia. Science 219:310–312

Ferré S, Baler R, Bouvier M et al (2009) Building a new conceptual framework for receptor heteromers. Nat Chem Biol 5:131–134. doi:10.1038/nchembio0309-131

Fields H (1992) Is there a facilitating component to central pain modulation? APS J 1:139–141

Fields HL, Vanegas H, Hentall ID, Zorman G (1983) Evidence that disinhibition of brain stem neurones contributes to morphine analgesia. Nature 306:684–686

Fillingim RB, Ness TJ (2000) Sex-related hormonal influences on pain and analgesic responses. Neurosci Biobehav Rev 24:485–501

Freedman NJ, Lefkowitz RJ (1996) Desensitization of G protein-coupled receptors. Recent Prog Horm Res 51:319–351; discussion 352–3

Freeman SE, Patil VV, Durham PL (2008) Nitric oxide-proton stimulation of trigeminal ganglion neurons increases mitogen-activated protein kinase and phosphatase expression in neurons and satellite glial cells. Neuroscience 157:542–555. doi:10.1016/j.neuroscience.2008.09.035

Gainetdinov RR, Premont RT, Bohn LM et al (2004) Desensitization of G protein-coupled receptors and neuronal functions. Annu Rev Neurosci 27:107–144. doi:10.1146/annurev.neuro.27.070203.144206

Goldstein A, Lowney LI, Pal BK (1971) Stereospecific and nonspecific interactions of the morphine congener levorphanol in subcellular fractions of mouse brain. Proc Natl Acad Sci U S A 68:1742–1747

González-Hernández T, Rustioni A (1999) Expression of three forms of nitric oxide synthase in peripheral nerve regeneration. J Neurosci Res 55:198–207

Grace PM, Hutchinson MR, Bishop A et al (2011) Adoptive transfer of peripheral immune cells potentiates allodynia in a graded chronic constriction injury model of neuropathic pain. Brain Behav Immun 25:503–513. doi:10.1016/j.bbi.2010.11.018

Grimm MC, Ben-Baruch A, Taub DD et al (1998) Opiates transdeactivate chemokine receptors: delta and mu opiate receptor-mediated heterologous desensitization. J Exp Med 188:317–325

Guo R-X, Zhang M, Liu W et al (2009) NMDA receptors are involved in upstream of the spinal JNK activation in morphine antinociceptive tolerance. Neurosci Lett 467:95–99. doi:10.1016/j.neulet.2009.10.013

Harrison CM, Charles J, Henderson J, Britt H (2012) Opioid prescribing in Australian general practice. Med J Aust 196:380–381. doi:10.5694/mja12.10168

Headache Classification Committee of the International Headache Society (2006) New appendix criteria open for a broader concept of chronic migraine. Cephalalgia 26(6):742–746

Heinisch S, Palma J, Kirby LG (2011) Interactions between chemokine and mu-opioid receptors: anatomical findings and electrophysiological studies in the rat periaqueductal grey. Brain Behav Immun 25:360–372. doi:10.1016/j.bbi.2010.10.020.Interactions

Heinricher M (2004) Neural basis for the hyperalgesic action of cholecystokinin in the rostral ventromedial medulla. J Neurophysiol 92:1982–1989

Heinricher M, McGaraughty S, Tortorici V (2001) Circuitry underlying antiopioid actions of cholecystokinin within the rostral ventromedial medulla. J Neurophysiol 85:280–286

Heinricher M, Morgan M, Fields HL (1992) Direct and indirect actions of morphine on medullary neurons that modulate nociception. Neuroscience 48(3):533–543

Heinricher MM, Morgan MM, Tortorici V, Fields HL (1994) Disinhibition of off-cells and antinociception produced by an opioid action within the rostral ventromedial medulla. Neuroscience 63:279–288

Heinricher MM, Tortorici V (1994) Interference with GABA transmission in the rostral ventromedial medulla: disinhibition of off-cells as a central mechanism in nociceptive modulation. Neuroscience 63:533–546

Hershey AD, Burdine D, Kabbouche MA, Powers SW (2011) Genomic expression patterns in medication overuse headaches. Cephalalgia 31:161–171. doi:10.1177/0333102410373155

Holguin A, O'Connor KA, Biedenkapp J et al (2004) HIV-1 gp120 stimulates proinflammatory cytokine-mediated pain facilitation via activation of nitric oxide synthase-I (nNOS). Pain 110:517–530. doi:10.1016/j.pain.2004.02.018

Holmes GM, Hebert SL, Rogers RC, Hermann GE (2004) Immunocytochemical localization of TNF type 1 and type 2 receptors in the rat spinal cord. Brain Res 1025:210–219. doi:10.1016/j.brainres.2004.08.020

Horvath RJ, DeLeo JA (2009) Morphine enhances microglial migration through modulation of P2X4 receptor signaling. J Neurosci 29:998–1005. doi:10.1523/JNEUROSCI.4595-08.2009

Horvath RJ, Romero-Sandoval EA, De Leo JA (2010) Inhibition of microglial P2X4 receptors attenuates morphine tolerance, Iba1, GFAP and mu opioid receptor protein expression while enhancing perivascular microglial ED2. Pain 150:401–413. doi:10.1016/j.pain.2010.02.042

Hutchinson MR, Coats BD, Lewis SS et al (2008a) Proinflammatory cytokines oppose opioid-induced acute and chronic analgesia. Brain Behav Immun 22:1178–1189. doi:10.1016/j.bbi.2008.05.004

Hutchinson MR, Lewis SS, Coats BD et al (2009) Reduction of opioid withdrawal and potentiation of acute opioid analgesia by systemic AV411 (ibudilast). Brain Behav Immun 23:240–250. doi:10.1016/j.bbi.2008.09.012

Hutchinson MR, Northcutt AL, Chao LW et al (2008b) Minocycline suppresses morphine-induced respiratory depression, suppresses morphine-induced reward, and enhances systemic morphine-induced analgesia. Brain Behav Immun 22:1248–1256. doi:10.1016/j.bbi.2008.07.008

Hutchinson MR, Northcutt AL, Hiranita T et al (2012) Opioid activation of toll-like receptor 4 contributes to drug reinforcement. J Neurosci 32:11187–11200. doi:10.1523/JNEUROSCI. 0684-12.2012

Hutchinson MR, Shavit Y, Grace PM et al (2011) Exploring the neuroimmunopharmacology of opioids: an integrative review of mechanisms of central immune signaling and their implications for opioid analgesia. Pharmacol Rev 63:772–810. doi:10.1124/pr.110.004135

Hutchinson MR, Zhang Y, Brown K et al (2008c) Non-stereoselective reversal of neuropathic pain by naloxone and naltrexone: involvement of toll-like receptor 4 (TLR4). Eur J Neurosci 28:20–29. doi:10.1111/j.1460-9568.2008.06321.x

Hutchinson MR, Zhang Y, Shridhar M et al (2010) Evidence that opioids may have toll-like receptor 4 and MD-2 effects. Brain Behav Immun 24:83–95. doi:10.1016/j.bbi.2009.08.004

Inoue K (2006) The function of microglia through purinergic receptors: neuropathic pain and cytokine release. Pharmacol Ther 109:210–226. doi:10.1016/j.pharmthera.2005.07.001

Itoh S, Katsuura G, Maeda Y (1982) Caerulein and cholecystokinin suppress beta-endorphin-induced analgesia in the rat. Eur J Pharmacol 80:421–425

Ji R-R, Gereau RW, Malcangio M, Strichartz GR (2009) MAP kinase and pain. Brain Res Rev 60:135–148. doi:10.1016/j.brainresrev.2008.12.011

Johnson JL, Hutchinson MR, Williams DB, Rolan P (2012) Medication-overuse headache and opioid-induced hyperalgesia: a review of mechanisms, a neuroimmune hypothesis and a novel approach to treatment. Cephalalgia 33:52–64. doi:10.1177/0333102412467512

Johnston IN, Milligan ED, Wieseler-Frank J et al (2004) A role for proinflammatory cytokines and fractalkine in analgesia, tolerance, and subsequent pain facilitation induced by chronic intrathecal morphine. J Neurosci 24:7353–7365. doi:10.1523/JNEUROSCI. 1850-04.2004

Juni A, Klein G, Pintar JE, Kest B (2007) Nociception increases during opioid infusion in opioid receptor triple knock-out mice. Neuroscience 147:439–444. doi:10.1016/j.neuroscience.2007.04.030

Kahlke V, Angele MK, Ayala A et al (2000) Immune dysfunction following trauma-haemorrhage: influence of gender and age. Cytokine 12:69–77. doi:10.1006/cyto.1999.0511

Kao S-C, Zhao X, Lee C-Y et al (2012) Absence of μ opioid receptor mRNA expression in astrocytes and microglia of rat spinal cord. Neuroreport 23:378–384. doi:10.1097/WNR.0b013e3283522e1b

Kawasaki Y, Zhang L, Cheng J-K, Ji R-R (2008) Cytokine mechanisms of central sensitization: distinct and overlapping role of interleukin-1beta, interleukin-6, and tumor necrosis factor-alpha in regulating synaptic and neuronal activity in the superficial spinal cord. J Neurosci 28:5189–5194. doi:10.1523/JNEUROSCI. 3338-07.2008

Kennis K, Kernick D, O'Flynn N (2012) Diagnosis and management of headaches in young people and adults: NICE guideline. Br J Gen Pract 63(613):443–445. doi:10.3399/bjgp13X670895

Kepler KL, Kest B, Kiefel JM et al (1989) Roles of gender, gonadectomy and estrous phase in the analgesic effects of intracerebroventricular morphine in rats. Pharmacol Biochem Behav 34:119–127

Kepler KL, Standifer KM, Paul D et al (1991) Gender effects and central opioid analgesia. Pain 45:87–94

Kolesnikov YA, Pick CG, Ciszewska G, Pasternak GW (1993) Blockade of tolerance to morphine but not to kappa opioids by a nitric oxide synthase inhibitor. Proc Natl Acad Sci U S A 90:5162–5166

Krzanowska EK, Bodnar RJ (1999) Morphine antinociception elicited from the ventrolateral periaqueductal gray is sensitive to sex and gonadectomy differences in rats. Brain Res 821:224–230

Lance F, Parkes C, Wilkinson M (1988) Does analgesic abuse cause headaches de novo? Headache 28:61–62

Li Y, Han JS (1989) Cholecystokinin-octapeptide antagonizes morphine analgesia in periaqueductal gray of the rat. Brain Res 480:105–110

Liu Y, Blackbourn DJ, Chuang LF et al (1992) Effects of in vivo and in vitro administration of morphine sulfate upon rhesus macaque polymorphonuclear cell phagocytosis and chemotaxis. J Pharmacol Exp Ther 263:533–539

Loredo GA, Benton HP (1998) ATP and UTP activate calcium-mobilizing P2U-like receptors and act synergistically with interleukin-1 to stimulate prostaglandin E2 release from human rheumatoid synovial cells. Arthritis Rheum 41:246–255. doi:10.1002/1529-0131(199802) 41:2<246::AID-ART8>3.0.CO;2-I

Machelska H, Ziólkowska B, Mika J et al (1997) Chronic morphine increases biosynthesis of nitric oxide synthase in the rat spinal cord. Neuroreport 8:2743–2747

Mantyh PW, Hunt SP (1984) Evidence for cholecystokinin-like immunoreactive neurons in the rat medulla oblongata which project to the spinal cord. Brain Res 291:49–54

Marriott I, Bost KL, Huet-Hudson YM (2006) Sexual dimorphism in expression of receptors for bacterial lipopolysaccharides in murine macrophages: a possible mechanism for gender-based differences in endotoxic shock susceptibility. J Reprod Immunol 71:12–27. doi:10.1016/j.jri. 2006.01.004

McMahon SB, Malcangio M (2009) Current challenges in glia-pain biology. Neuron 64:46–54. doi:10.1016/j.neuron.2009.09.033

Meller S, Pechman PS, Gebhart GF, Maves TJ (1992) Nitric oxide mediates the thermal hyperalgesia produced in a model of neuropathic pain in the rat. Neuroscience 50(1):7–10

Meller ST, Gebhart GF (1993) Nitric oxide (NO) and nociceptive processing in the spinal cord. Pain 52:127–136

Meng ID, Cao L (2007) From migraine to chronic daily headache: the biological basis of headache transformation. Headache 47(8):1251–1258

Milligan E, Zapata V, Schoeniger D et al (2005) An initial investigation of spinal mechanisms underlying pain enhancement induced by fractalkine, a neuronally released chemokine. Eur J Neurosci 22:2775–2782. doi:10.1111/j.1460-9568.2005.04470.x

Milligan ED, Twining C, Chacur M et al (2003) Spinal glia and proinflammatory cytokines mediate mirror-image neuropathic pain in rats. J Neurosci 23:1026–1040

Milligan ED, Watkins LR (2009) Pathological and protective roles of glia in chronic pain. Nat Rev Neurosci 10:23–36. doi:10.1038/nrn2533

Milligan ED, Zapata V, Chacur M et al (2004) Evidence that exogenous and endogenous fractalkine can induce spinal nociceptive facilitation in rats. Eur J Neurosci 20:2294–2302. doi:10.1111/j.1460-9568.2004.03709.x

Morioka N, Inoue A, Hanada T et al (2002) Nitric oxide synergistically potentiates interleukin-1 beta-induced increase of cyclooxygenase-2 mRNA levels, resulting in the facilitation of substance P release from primary afferent neurons: involvement of cGMP-independent mechanisms. Neuropharmacology 43:868–876

Morioka T, Kalehua AN, Streit WJ (1991) The microglial reaction in the rat dorsal hippocampus following transient forebrain ischemia. J Cereb Blood Flow Metab 11:966–973. doi:10.1038/jcbfm.1991.162

Murphy PM, Baggiolini M, Charo IF et al (2000) International union of pharmacology. XXII. Nomenclature for chemokine receptors. Pharmacol Rev 52:145–176

Muscoli C, Cuzzocrea S, Ndengele MM et al (2007) Therapeutic manipulation of peroxynitrite attenuates the development of opiate-induced antinociceptive tolerance in mice. J Clin Invest 117:3530–3539. doi:10.1172/JCI32420

Muscoli C, Doyle T, Dagostino C et al (2010) Counter-regulation of opioid analgesia by glial-derived bioactive sphingolipids. J Neurosci 30:15400–15408. doi:10.1523/JNEUROSCI. 2391-10.2010

Mustafa S, Ayoub MA, Pfleger KDG (2010) Uncovering GPCR heteromer-biased ligands. Drug Discov Today Technol 7:e77–e85. doi:10.1016/j.ddtec.2010.06.003

Mustafa S, Pfleger KDG (2011) G protein-coupled receptor heteromer identification technology: identification and profiling of GPCR heteromers. J Lab Autom 16:285–291. doi:10.1016/j.jala. 2011.03.002

Mustafa S, See HB, Seeber RM et al (2012) Identification and profiling of novel α1A-adrenoceptor-CXC chemokine receptor 2 heteromer. J Biol Chem 287:12952–12965. doi:10.1074/jbc.M111.322834

Nayak D, Huo Y, Kwang WXT et al (2010) Sphingosine kinase 1 regulates the expression of proinflammatory cytokines and nitric oxide in activated microglia. Neuroscience 166:132–144. doi:10.1016/j.neuroscience.2009.12.020

Oka T, Aou S, Hori T (1994) Intracerebroventricular injection of interleukin-1 beta enhances nociceptive neuronal responses of the trigeminal nucleus caudalis in rats. Brain Res 656:236–244

Okada T, Kajimoto T, Jahangeer S, Nakamura S-I (2009) Sphingosine kinase/sphingosine 1-phosphate signalling in central nervous system. Cell Signal 21:7–13. doi:10.1016/j.cellsig.2008.07.011

Ossipov MH, Lai J, Vanderah TW, Porreca F (2003) Induction of pain facilitation by sustained opioid exposure: relationship to opioid antinociceptive tolerance. Life Sci 73:783–800

Pan ZZ, Williams JT, Osborne PB (1990) Opioid actions on single nucleus raphe magnus neurons from rat and guinea-pig in vitro. J Physiol Lond 427:519–532

Pautz A, Franzen R, Dorsch S et al (2002) Cross-talk between nitric oxide and superoxide determines ceramide formation and apoptosis in glomerular cells. Kidney Int 61:790–796. doi:10.1046/j.1523-1755.2002.00222.x

Pellis NR, Harper C, Dafny N (1986) Suppression of the induction of delayed hypersensitivity in rats by repetitive morphine treatments. Exp Neurol 93:92–97

Perry VH, Hume DA, Gordon S (1985) Immunohistochemical localization of macrophages and microglia in the adult and developing mouse brain. Neuroscience 15:313–326. doi:10.1016/0306-4522(85)90215-5

Peterson PK, Sharp B, Gekker G et al (1987) Opioid-mediated suppression of interferon-gamma production by cultured peripheral blood mononuclear cells. J Clin Invest 80:824–831. doi:10.1172/JCI113140

Pyne S, Lee SC, Long J, Pyne NJ (2009) Role of sphingosine kinases and lipid phosphate phosphatases in regulating spatial sphingosine 1-phosphate signalling in health and disease. Cell Signal 21:14–21. doi:10.1016/j.cellsig.2008.08.008

Quan N, He L, Lai W (2003) Endothelial activation is an intermediate step for peripheral lipopolysaccharide induced activation of paraventricular nucleus. Brain Res Bull 59:447–452

Raghavendra V, Rutkowski MD, DeLeo JA (2002) The role of spinal neuroimmune activation in morphine tolerance/hyperalgesia in neuropathic and sham-operated rats. J Neurosci 22:9980–9989

Raghavendra V, Tanga FY, DeLeo JA (2004) Attenuation of morphine tolerance, withdrawal-induced hyperalgesia, and associated spinal inflammatory immune responses by propentofylline in rats. Neuropsychopharmacology 29:327–334. doi:10.1038/sj.npp.1300315

Raiteri M, Paudice P (1993) Release of cholecystokinin in the central nervous system. Neurochem Int 26:519–527

Raivich G (2005) Like cops on the beat: the active role of resting microglia. Trends Neurosci 28:571–573. doi:10.1016/j.tins.2005.09.001

Rao SS, Saifi AQ (1985) Influence of testosterone on morphine analgesia in albino rats. Indian J Physiol Pharmacol 29:103–106

Ratka A (1984) Interaction of morphine and steroid hormones in the postirradiation disease in rats. Pol J Pharmacol Pharm 36(1):41–49

Ratka A, Simpkins JW (1990) A modulatory role for luteinizing hormone-releasing hormone in nociceptive responses of female rats. Endocrinology 127:667–673. doi:10.1210/endo-127-2-667

Ratka A, Simpkins JW (1991) Effects of estradiol and progesterone on the sensitivity to pain and on morphine-induced antinociception in female rats. Horm Behav 25:217–228

Ravishankar K (2008) Medication overuse headache in India. Cephalalgia 28:1223–1226. doi:10.1111/j.1468-2982.2008.01731.x

Rettew JA, Huet YM, Marriott I (2009) Estrogens augment cell surface TLR4 expression on murine macrophages and regulate sepsis susceptibility in vivo. Endocrinology 150:3877–3884. doi:10.1210/en.2009-0098

Rezayat M, Nikfar S, Zarrindast MR (1994) CCK receptor activation may prevent tolerance to morphine in mice. Eur J Pharmacol 254:21–26

Rock RB, Hu S, Sheng WS, Peterson PK (2006) Morphine stimulates CCL2 production by human neurons. J Neuroinflammation 3:32. doi:10.1186/1742-2094-3-32

Roscic-Mrkic B, Fischer M, Leemann C et al (2003) RANTES (CCL5) uses the proteoglycan CD44 as an auxiliary receptor to mediate cellular activation signals and HIV-1 enhancement. Blood 102:1169–1177. doi:10.1182/blood-2003-02-0488

Ryan SM, Maier SF (1988) The estrous cycle and estrogen modulate stress-induced analgesia. Behav Neurosci 102:371–380

Samad TA, Moore KA, Sapirstein A et al (2001) Interleukin-1beta-mediated induction of Cox-2 in the CNS contributes to inflammatory pain hypersensitivity. Nature 410:471–475. doi:10.1038/35068566

Schildknecht S, Pape R, Müller N et al (2011) Neuroprotection by minocycline caused by direct and specific scavenging of peroxynitrite. J Biol Chem 286:4991–5002. doi:10.1074/jbc.M110.169565

Shavit Y, Wolf G, Goshen I et al (2005) Interleukin-1 antagonizes morphine analgesia and underlies morphine tolerance. Pain 115:50–59. doi:10.1016/j.pain.2005.02.003

Simpkins CO, Dickey CA, Fink MP (1984) Human neutrophil migration is enhanced by beta-endorphin. Life Sci 34:2251–2255

Smith K (2010) Neuroscience: settling the great glia debate. Nature 468:160–162. doi:10.1038/468160a

Snyder SH (1992) Nitric oxide: first in a new class of neurotransmitters? Science 257:494–496. doi:10.1126/science.1353273

Song P, Zhao ZQ (2001) The involvement of glial cells in the development of morphine tolerance. Neurosci Res 39:281–286

Soucy G, Boivin G, Labrie F, Rivest S (2005) Estradiol is required for a proper immune response to bacterial and viral pathogens in the female brain. J Immunol 174:6391–6398

Sperlágh B, Baranyi M, Haskó G, Vizi ES (2004) Potent effect of interleukin-1 beta to evoke ATP and adenosine release from rat hippocampal slices. J Neuroimmunol 151:33–39. doi:10.1016/j.jneuroim.2004.02.004

Stellwagen D, Beattie EC, Seo JY, Malenka RC (2005) Differential regulation of AMPA receptor and GABA receptor trafficking by tumor necrosis factor-alpha. J Neurosci 25:3219–3228. doi:10.1523/JNEUROSCI.4486-04.2005

Stievano L, Piovan E, Amadori A (2004) C and CX3C chemokines: cell sources and physiopathological implications. Crit Rev Immunol 24:205–228

Stoffel EC, Ulibarri CM, Craft RM (2003) Gonadal steroid hormone modulation of nociception, morphine antinociception and reproductive indices in male and female rats. Pain 103:285–302. doi:10.1016/s0304-3959(02)00457-8

Suzuki S, Chuang LF, Yau P et al (2002) Interactions of opioid and chemokine receptors: oligomerization of mu, kappa, and delta with CCR5 on immune cells. Exp Cell Res 280(2):192–200. doi:10.1006/excr.2002.5638

Szabo I, Chen X-H, Xin L et al (2002) Heterologous desensitization of opioid receptors by chemokines inhibits chemotaxis and enhances the perception of pain. Proc Natl Acad Sci U S A 99:10276–10281. doi:10.1073/pnas.102327699

Szabo I, Wetzel MA, Zhang N et al (2003) Selective inactivation of CCR5 and decreased infectivity of R5 HIV-1 strains mediated by opioid-induced heterologous desensitization. J Leukoc Biol 74(6):1074–1082. doi:10.1189/jlb.0203067.1

Takagi K, Fukuda H, Watanabe M (1960) Studies on antitussives. III. (+)-morphine. Yakugaku Zasshi 80:1506–1509

Takayama N, Ueda H (2005) Morphine-induced chemotaxis and brain-derived neurotrophic factor expression in microglia. J Neurosci 25:430–435. doi:10.1523/JNEUROSCI. 3170-04.2005

Tanga FY, Nutile-McMenemy N, DeLeo JA (2005) The CNS role of Toll-like receptor 4 in innate neuroimmunity and painful neuropathy. Proc Natl Acad Sci U S A 102:5856–5861. doi:10.1073/pnas.0501634102

Tao F, Tao Y-X, Mao P et al (2003) Intact carrageenan-induced thermal hyperalgesia in mice lacking inducible nitric oxide synthase. Neuroscience 120:847–854

Taub DD, Eisenstein TK, Geller EB et al (1991) Immunomodulatory activity of mu- and kappa-selective opioid agonists. Proc Natl Acad Sci U S A 88:360–364

Tawfik VL, LaCroix-Fralish ML, Bercury KK et al (2006) Induction of astrocyte differentiation by propentofylline increases glutamate transporter expression in vitro: heterogeneity of the quiescent phenotype. Glia 54:193–203. doi:10.1002/glia.20365

Thalakoti S, Patil VV, Damodaram S et al (2007) Neuron-glia signaling in trigeminal ganglion: implications for migraine pathology. Headache 47:1008–1023. doi:10.1111/j.1526-4610.2007.00854.x; discussion 24–5

Thomas J, Hutchinson MR (2012) Exploring neuroinflammation as a potential avenue to improve the clinical efficacy of opioids. Expert Rev Neurother 12:1311–1324. doi:10.1586/ern.12.125

Trang T, Beggs S, Wan X, Salter MW (2009) P2X4-receptor-mediated synthesis and release of brain-derived neurotrophic factor in microglia is dependent on calcium and p38-mitogen-activated protein kinase activation. J Neurosci 29:3518–3528. doi:10.1523/JNEUROSCI. 5714-08.2009

Triantafilou M, Lepper PM, Briault CD et al (2008) Chemokine receptor 4 (CXCR4) is part of the lipopolysaccharide "sensing apparatus". Eur J Immunol 38:192–203. doi:10.1002/eji.200636821

van Epps DE, Saland L (1984) Beta-endorphin and met-enkephalin stimulate human peripheral blood mononuclear cell chemotaxis. J Immunol 132:3046–3053

Vanderah TW, Lai J, Yamamura HI, Porreca F (1994) Antisense oligodeoxynucleotide to the CCKB receptor produces naltrindole- and [Leu5]enkephalin antiserum-sensitive enhancement of morphine antinociception. Neuroreport 5:2601

Vanderhaeghen JJ, Signeau JC, Gepts W (1975) New peptide in the vertebrate CNS reacting with antigastrin antibodies. Nature 257:604–605

Verge GM, Milligan ED, Maier SF et al (2004) Fractalkine (CX3CL1) and fractalkine receptor (CX3CR1) distribution in spinal cord and dorsal root ganglia under basal and neuropathic pain conditions. Eur J Neurosci 20:1150–1160. doi:10.1111/j.1460-9568.2004.03593.x

Viviani B, Bartesaghi S, Gardoni F et al (2003) Interleukin-1beta enhances NMDA receptor-mediated intracellular calcium increase through activation of the Src family of kinases. J Neurosci 23:8692–8700

Wang X, Loram LC, Ramos K et al (2012) Morphine activates neuroinflammation in a manner parallel to endotoxin. Proc Natl Acad Sci U S A 109:6325–6330. doi:10.1073/pnas.1200130109

Wang Z, Ma W, Chabot J-G, Quirion R (2009) Cell-type specific activation of p38 and ERK mediates calcitonin gene-related peptide involvement in tolerance to morphine-induced analgesia. FASEB J 23:2576–2586. doi:10.1096/fj.08-128348

Wang Z, Ma W, Chabot J-G, Quirion R (2010) Morphological evidence for the involvement of microglial p38 activation in CGRP-associated development of morphine antinociceptive tolerance. Peptides 31:2179–2184. doi:10.1016/j.peptides.2010.08.020

Watkins L, Kinscheck I, Mayer D (1985a) Potentiation of morphine analgesia by the cholecystokinin antagonist proglumide. Brain Res 327:169–180

Watkins LR, Hansen MK, Nguyen KT et al (1999) Dynamic regulation of the proinflammatory cytokine, interleukin-1beta: molecular biology for non-molecular biologists. Life Sci 65:449–481

Watkins LR, Hutchinson MR, Johnston IN, Maier SF (2005) Glia: novel counter-regulators of opioid analgesia. Trends Neurosci 28:661–669. doi:10.1016/j.tins.2005.10.001

Watkins LR, Hutchinson MR, Ledeboer A et al (2007) Glia as the "bad guys": implications for improving clinical pain control and the clinical utility of opioids. Brain Behav Immun 21:131–146. doi:10.1016/j.bbi.2006.10.011

Watkins LR, Hutchinson MR, Rice KC, Maier SF (2009) The "toll" of opioid-induced glial activation: improving the clinical efficacy of opioids by targeting glia. Trends Pharmacol Sci 30:581–591. doi:10.1016/j.tips.2009.08.002

Watkins LR, Kinscheck I, Kaufman E, Miller J (1985b) Cholecystokinin antagonists selectively potentiate analgesia induced by endogenous opiates. Brain Res 327:181–190

Waxman AR, Arout C, Caldwell M et al (2009) Acute and chronic fentanyl administration causes hyperalgesia independently of opioid receptor activity in mice. Neurosci Lett 462:68–72. doi:10.1016/j.neulet.2009.06.061

Wetzel MA, Steele AD, Eisenstein TK et al (2000) μ-opioid induction of monocyte chemoattractant protein-1, RANTES, and IFN-γ-inducible protein-10 expression in human peripheral blood mononuclear cells. J Immunol 165(11):6519–6524

Wieseler J et al. (2010) Facial allodynia: involvement of glia and potentiation by prior morphine. Program No. 780.4/MM4. Neuroscience Meeting Planner. Society for Neuroscience, San Diego. Online

Wieseler J et al. (2011) Facial allodynia potentiation by supradural inflammatory mediators and morphine: a model of medication overuse headache. Program no. 178.09/NN19. Neuroscience Meeting Planner. Society for Neuroscience, Washington, DC. Online

Wiesenfeld-Hallin Z, Xu XJ (1996) The role of cholecystokinin in nociception, neuropathic pain and opiate tolerance. Regul Pept 65:23–28

World Health Organization (1996) Cancer pain relief. World Health Organization, Geneva

Wu H-E, Hong J-S, Tseng LF (2007) Stereoselective action of (+)-morphine over (-)-morphine in attenuating the (-)-morphine-produced antinociception via the naloxone-sensitive sigma receptor in the mouse. Eur J Pharmacol 571:145–151. doi:10.1016/j.ejphar.2007.06.012

Wu H-E, Sun H-S, Terashivili M et al (2006) dextro- and levo-morphine attenuate opioid delta and kappa receptor agonist produced analgesia in mu-opioid receptor knockout mice. Eur J Pharmacol 531:103–107. doi:10.1016/j.ejphar.2005.12.012

Wu H-E, Thompson J, Sun H-S et al (2005) Antianalgesia: stereoselective action of dextro-morphine over levo-morphine on glia in the mouse spinal cord. J Pharmacol Exp Ther 314:1101–1108. doi:10.1124/jpet.105.087130

Xie JY, Herman DS, Stiller C-O et al (2005) Cholecystokinin in the rostral ventromedial medulla mediates opioid-induced hyperalgesia and antinociceptive tolerance. J Neurosci 25:409–416. doi:10.1523/JNEUROSCI. 4054-04.2005

Xie N, Li H, Wei D et al (2010) Glycogen synthase kinase-3 and p38 MAPK are required for opioid-induced microglia apoptosis. Neuropharmacology 59:444–451. doi:10.1016/j.neuropharm.2010.06.006

Yuan Y, Arnatt CK, El-Hage N, Dever SM, Jacob JC, Selley DE, Hauser KF, Zhang Y (2013) A bivalent ligand targeting the putative mu opioid receptor and chemokine receptor CCR5 heterodimers: binding affinity versus functional activities. Medchemcomm 4(5):847–851. doi:10.1039/c3md00080j

Zhang J, Shi XQ, Echeverry S et al (2007) Expression of CCR2 in both resident and bone marrow-derived microglia plays a critical role in neuropathic pain. J Neurosci 27:12396–12406. doi:10.1523/JNEUROSCI. 3016-07.2007

Zhang N, Rogers TJ, Caterina M, Oppenheim JJ (2004) Proinflammatory chemokines, such as C-C chemokine ligand 3, desensitize mu-opioid receptors on dorsal root ganglia neurons. J Immunol 173:594–599

Zhou Y, Sun YH, Zhang ZW, Han JS (1992) Accelerated expression of cholecystokinin gene in the brain of rats rendered tolerant to morphine. Neuroreport 3:1121–1123

The Role of Proteases in Pain

Jason J. McDougall and Milind M. Muley

Contents

1 Introduction 240
   1.1 Proteinase-Activated Receptor: Activation, Signal
       Transduction and Desensitisation 240
   1.2 Proteinase-Activated Receptor: Role in Physiology and Disease 243
2 Proteinase-Activated Receptor: Role in Pain 247
   2.1 PARs and Inflammatory Pain 247
   2.2 PARs and Neuropathic Pain 250
3 PARs as a Drug Target for Pain 253
4 Conclusion 255
References 255

Abstract

Proteinase-activated receptors (PARs) are a family of G protein-coupled receptor that are activated by extracellular cleavage of the receptor in the N-terminal domain. This slicing of the receptor exposes a tethered ligand which binds to a specific docking point on the receptor surface to initiate intracellular signalling. PARs are expressed by numerous tissues in the body, and they are involved in various physiological and pathological processes such as food digestion, tissue remodelling and blood coagulation. This chapter will summarise how serine proteinases activate PARs leading to the development of pain in several chronic pain conditions. The potential of PARs as a drug target for pain relief is also discussed.

J.J. McDougall (✉) • M.M. Muley
Departments of Pharmacology and Anaesthesia, Pain Management and Perioperative Medicine, Dalhousie University, 5850 College Street, Halifax, NS, Canada, B3H 4R2
e-mail: jason.mcdougall@dal.ca

© Springer-Verlag Berlin Heidelberg 2015
H.-G. Schaible (ed.), *Pain Control*, Handbook of Experimental Pharmacology 227,
DOI 10.1007/978-3-662-46450-2_12

Keywords

Proteinase-activated receptor • Pain • Inflammation

1 Introduction

Proteases are group of enzymes that participate in the breakdown of long-chain proteins by hydrolyzing peptide bonds. Alternative nomenclature includes proteolytic enzymes or proteinases (Barrett et al. 2004). Proteases are involved in a multitude of biological activities from the simple digestion of dietary proteins to highly regulated physiological processes (e.g. the blood clotting cascade, the complement system and cell apoptosis). Proteases are subdivided into six general categories (serine, threonine, cysteine, aspartate, glutamic acid, metalloproteases) based on the presence of their catalytic residues. Most of the proteinases are either exoproteinases which cleave one or a few amino acids from the N- or C-terminus of the polypeptide chain or endopeptidases that cleave the protein internally (Turk 2006). When an exopeptidase acts on the N-terminal, it releases single amino acid residues (amino-peptidases) or a dipeptide (dipeptidyl-peptidases) or a tripeptide (tripeptidyl-peptidases). Action on the free C-terminus releases a single residue (carboxypeptidases) or a dipeptide (peptidyl-dipeptidases). The endopeptidases are categorised into different types on the basis of catalytic mechanism, viz. aspartic endopeptidases, glutamic endopeptidases, cysteine endopeptidases, metalloendopeptidases, serine endopeptidases and threonine endopeptidases (Barrett 2001).

In addition to their proteolytic properties, proteases can also act as signalling molecules by cleaving the extracellular domain of a family of G protein-coupled receptors called proteinase-activated receptors (PARs). These receptors are highly expressed in platelets but also on endothelial cells, myocytes and neurones (Macfarlane et al. 2001). Four PARs have been identified so far (PAR1–4) which are expressed in a plethora of tissues throughout the body. PARs are activated by the action of serine proteases such as thrombin (acts on PAR1, 3 and 4) and trypsin (PAR2). Table 1 highlights some of the known major PAR-cleaving enzymes, activating peptide and antagonists.

1.1 Proteinase-Activated Receptor: Activation, Signal Transduction and Desensitisation

1.1.1 Activation

Unlike classical ligand-binding receptors, PARs have a unique mechanism of receptor activation. All PARs contain a discrete serine proteinase cleavage site within their extracellular N-terminus which when cleaved unmasks a new amino acid terminal known as a tethered ligand. The exposed tethered ligand then binds to the second extracellular loop of the receptor to initiate the recruitment of G proteins

Table 1 Common PAR-cleaving enzymes, activating peptides and antagonists

Proteinase-activated receptor (PAR)	Activating protease	Activating peptide	PAR antagonist
PAR1	Thrombin	TFLLR	FR-171113
	FXa		RWJ-56110
	Trypsin		RWJ-58259
	Granzyme A		BMS-200261
	APC		SCH-79797
	Gingipains-R		
PAR2	Trypsin	SLIGKV	FSLLRY-NH2
	Tryptase	SLIGRL	GB-83
	Factor VIIa	FLIGRL	GB-88
	Factor Xa		ENMD-1068
	Neutrophil elastase		
	Granzyme A		
	Matriptase		
	Acrosien		
	Gingipains-R		
	Proteinase-3		
PAR3	Thrombin	None	None
PAR4	Thrombin	GYPGQV	Pepducin P4pal-10
	Trypsin	AYPGKF	tcY-NH$_2$
	Plasmin		YD-3
	Bacterial gingipains		
	Cathepsin G		

and other signalling molecules to the intracellular domains of the receptors. PAR3 is a distinct receptor in that it acts as a cofactor for thrombin-mediated activation of the PAR4 receptor (Nakanishi-Matsui et al. 2000). For practical purposes, short synthetic peptides derived from the sequence of the unmasked tethered ligand have been developed which can also activate PARs in the absence of receptor proteolysis (Fig. 1). These experimental tools allow for the pharmacological study of PAR processes in vitro and in vivo without the confounding multifactorial effects of proteolytic activity. Hence, these short activating peptides have been used in studying the role of PARs in different physiological and pathophysiological situations (Russell and McDougall 2009). In addition to protease activation, some proteases can also disarm PAR signalling by cleaving the N-terminus downstream of the receptor-activating site, thereby detaching the tethered ligand making it unavailable for receptor activation (Fig. 1). A recent report confirmed that serine proteinase enzymes such as neutrophil elastase can disarm trypsin-mediated PAR2 signalling and at the same time activate PAR2 signalling selectively via a mitogen-

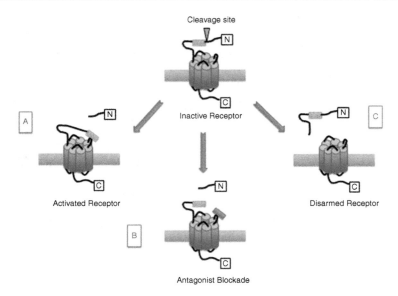

Fig. 1 Schematic representation of proteinase-activated receptor (PAR) modulation. (**a**) Inactive PARs can be cleaved by a protease revealing a tethered ligand (*orange box*) which binds to the active domain of the same receptor leading to cell signalling. (**b**) The PAR binding site can be blocked with a small molecule antagonist (*red box*). (**c**) Alternative proteases can remove a segment of the N-terminal including the binding domain leading to disarmament of the receptor

activated protein kinase (MAPK) pathway, without triggering an elevation in intracellular calcium levels (Ramachandran et al. 2011).

1.1.2 Signalling

PAR1

PARs couple with G proteins and activate multiple pathways and hence they can regulate various cellular functions. PAR1 interacts with several α-subunits particularly $G_{1\alpha}$, $G_{12/13\alpha}$ and $G_{q11\alpha}$. $G_{1\alpha}$ leading to inhibition of adenylyl cyclase (AC) to reduce cyclic adenosine monophosphate (cAMP) (Benka et al. 1995). $G_{q11\alpha}$ activates phospholipase Cβ (PLCβ) to generate inositol trisphosphate (InsP3), which mobilises Ca^{2+}, and diacylglycerol (DAG), which activates protein kinase C (PKC). $G_{12/13\alpha}$ couples to guanine nucleotide exchange factors (GEF), resulting in activation of Rho, Rho-kinase (ROK) and serum response elements (SRE) (Ossovskaya and Bunnett 2004). Recent studies suggest that the cleaved N-terminal domain of PAR1 is released and exhibits biological activity in certain settings and has been termed 'parstatin' (Duncan and Kalluri 2009). PAR1 can activate the MAPK cascade by transactivation of the EGF receptor, through activation of PKC, Phosphoinositide 3-kinase (PI3K), Pyk2 and other mechanisms (Coughlin 2000).

PAR2

PAR2 couples to $G_{q\alpha}$ and stimulates the generation of InsP3 and mobilisation of Ca^{2+} in PAR2-transfected cell lines (Bohm et al. 1996). In the case of enterocytes and transfected epithelial cells, activation of PAR2 leads to arachidonic acid release and the generation of prostaglandins E_2 and $F_{1\alpha}$ (Kong et al. 1997). This suggests that PAR2 cleavage involves the activation of phospholipase A_2 and cyclooxygenase-1 (Kong et al. 1997). Also, it has been reported that PAR2 activates MAP kinases ERK1/2 and weakly stimulates MAP kinase p38, although c-Jun amino-terminal kinase is not activated (Belham et al. 1996; DeFea et al. 2000).

PAR3 and PAR4

PAR3 does not signal autonomously and is only considered as a cofactor for PAR4 activation by thrombin (Nakanishi-Matsui et al. 2000). However, in a recent report, it was observed that PAR3 can elicit Rho- and Ca^{2+}-dependent release of ATP from lung epithelial A549 cells and PAR4 couples to G_q and $G_{12/13}$ signalling effectors in order to signal through G proteins (Seminario-Vidal et al. 2009).

1.1.3 Desensitisation

The desensitisation mechanisms for the PARs are distinct but poorly understood. Phosphorylation of activated PAR1 takes place to uncouple it from G proteins and G protein-coupled receptor kinase 3 or 5 (GRK 3 or 5) which enhances PAR1 phosphorylation. Moreover, the β-arrestin 1 mechanism also contributes to desensitisation of PAR1. In the case of PAR2, the main mechanism of uncoupling is phosphorylation by PKC and other kinases, by binding both β-arrestin 1 and β-arrestin 2 leading to a rapid uncoupling from G protein signalling at the cell surface. The mechanisms contributing to the desensitisation of PAR3 and PAR4 are unknown; however, it has been suggested that receptor internalisation may contribute to the termination of PAR4 signalling (Soh et al. 2010).

1.2 Proteinase-Activated Receptor: Role in Physiology and Disease

1.2.1 Cardiovascular System

Protease signalling through PARs contributes to both normal homeostasis and various cardiovascular disease states such as thrombosis and atherosclerosis. Evidence that PARs are involved in cardiovascular homeostasis comes from the observation that the receptors are expressed on platelets, endothelial cells and smooth muscle cells (Nelken et al. 1992; Takada et al. 1998). PAR1, but not PAR2, is expressed on rat cardiac fibroblast (Steinberg 2005). A study, however, showed the presence of PAR2 in rat cardiac fibroblast (Murray et al. 2012). PAR1 and PAR4 mainly coordinate thrombin-mediated platelet aggregation. After activation, PAR1 rapidly transmits a signal across the plasma membrane to G proteins, which results in the formation of platelet–platelet aggregates. It also causes stimulation of G_q proteins which culminates in a rapid rise in intracellular calcium and activation of the GP IIb/IIIa ($\alpha_{IIb}\beta_3$) fibrinogen receptor. PAR4 is cleaved and signals more slowly but, despite its slower response, generates the majority of the intracellular calcium flux and does not require additional input from ADP receptor to form stable platelet clumping. Blockade of thrombin binding to mouse PAR3 or knockout of the PAR3 gene inhibited mouse platelet aggregation indicating the importance of PAR3 for thrombin signalling in mouse platelets. However, when mouse PAR3 was overexpressed in mouse platelets, it did not trigger thrombin signalling. Mouse platelets express both PAR3 and PAR4, so Matsui et al. carried out a series of experiments and showed that PAR3 and PAR4 interact with each other and PAR3 functions as a cofactor in cleavage and activation of PAR4 by thrombin (Nakanishi-Matsui et al. 2000). PAR2 is expressed by numerous cell types within the cardiovascular system. Functional PAR2 expression has been demonstrated on vascular endothelium, smooth muscle cells and cardiomyocytes (Steinberg 2005; Sabri et al. 2000). Armed with this information, a group of researchers carried out an investigation which involved evaluation of the role of PAR2 in a cardiac ischaemia/reperfusion injury model (Antoniak et al. 2010). It was demonstrated that PAR2 deficiency reduced myocardial infarction and heart remodelling after ischaemia/reperfusion injury. In another study, it has been reported that PAR2 contributes to the pathogenesis of heart hypertrophy and failure. In a similar study, it was demonstrated that cardiomyocyte-specific overexpression of PAR2 led to pathological heart hypertrophy associated with cardiac fibrosis (Antoniak et al. 2013). Pathological remodelling of the heart in αMHC-PAR2 mice was accompanied by increased ANP: Atrial natriuretic peptide, BNP: B-type natriuretic peptide and βMHC expression and decreased MHC expression (Antoniak et al. 2013).

1.2.2 Nervous System

Various reports available in the literature show that thrombin changes the morphology of neurones and astrocytes, induces glial cell proliferation and even exerts, depending on the concentration applied, either cytoprotective or cytotoxic effects on neurones (Wang and Reiser 2003). Thrombin induces various neuronal changes such as neurite retraction, cell rounding, NMDA receptor potentiation and protection from cell death which are all mediated by PAR1 (Jalink and Moolenaar 1992; Turnell et al. 1995; Gingrich et al. 2000). PAR1 agonists also stimulate proliferation and shape changes in astrocytes, which results in the release of endothelin-1 and nerve growth factor and to inhibit the expression of glutamate receptors (Beecher et al. 1994). When thrombin was infused into the brain, it has been shown to reproduce inflammatory signs observed after injury in the CNS (Suidan et al. 1996). Although expression of the other two thrombin receptors (PAR3 and PAR4) in the brain has also been detected by several studies, the physiological role of PAR3 and PAR4 in neuronal differentiation is presently unknown (Wang and Reiser 2003). In a recent study, animals subjected to transient middle cerebral artery occlusion followed by reperfusion showed an increase in PAR2 expression. Also, there was significant decrease in the neuronal expression of phosphorylated extracellular signal-regulated kinase (p-ERK) in PAR2 KO mice (Jin et al. 2005).

The p-ERK is mainly responsible for regulating cell survival under normal and pathological conditions (Bonni et al. 1999). In the case of PAR2 KO mice, the infarct volume was increased significantly. Also, astrocyte activation was reduced in PAR2 KO mice. Astrocytes are important cells as they provide structural, trophic and metabolic support to neurones and thereby regulate synaptic activity (Stoll et al. 1998). So, it is clear that PAR2 gene deficiency increases brain injury. The role of PAR2 in neurodegenerative disorders is uncertain; however, PAR2 exerted protective effects in neurones, but its activation in glia was pathogenic with secretion of neurotoxic factors and suppression of astrocytic anti-inflammatory mechanisms (Afkhami-Goli et al. 2007).

1.2.3 Gastrointestinal System

PARs have an important role in controlling gastrointestinal function since the digestive system produces, secretes and therefore is exposed to many different proteinases Vergnolle 2000. In addition to their digestive effects, gastrointestinal proteinases are involved in local tissue remodelling, blood coagulation, nutrient absorption and gut motility (Kawabata et al. 2001). During inflammation, infiltrating immune cells release serine proteinases such as thrombin, trypsin and mast cell tryptase which can subsequently cleave PARs. PAR2 agonists and trypsin present in the intestinal lumen stimulate the generation of inositol 1,4,5-triphosphate, arachidonic acid release and secretion of prostaglandin E2 (PGE2) and F1α from enterocytes (Kong et al. 1997). Since prostaglandins are known to regulate gastric secretion, intestinal transport and motility, this PAR2 pathway is important for the management of intestinal function (Eberhart and Dubois 1995). PGE2 is involved in the protection of cells in the upper intestine against digestion by pancreatic trypsin, and this is mediated by PARs in the epithelium (Kong et al. 1997). In a study by Cocks et al., it was reported that activation of PAR2, which co-localises with trypsin in airway epithelium, induces the relaxation of airway preparations by the release of PGE2, and it offers bronchoprotection. The bronchial epithelium resembles intestinal epithelium in terms of its morphology. So, it is possible that PAR2 may exert protective effect on the intestinal epithelium (Cocks et al. 1999a, b).

PAR2 present on enterocytes can be activated by trypsin, tryptase, while PAR2 effector cells like inflammatory cells, fibroblasts or neurones result in the secretion of eicosanoids (Kong et al. 1997). PAR1 and PAR2 activation can alter gastrointestinal motility since they are highly expressed by gastrointestinal smooth muscle cells (Corvera et al. 1997). When PAR1 and PAR2 were activated, they induced contractions in gastric smooth muscle (Saifeddine et al. 1996). Also, indomethacin blocked gastric contractions which indicates that the PAR1- and PAR2-induced contractions are prostaglandin mediated (Cocks et al. 1999b). In a study, the role for PARs in the modulation of motility of the rat oesophageal muscularis mucosae was observed (Kawabata et al. 2000). PAR1-activating peptides, but not the PAR1-inactive peptide, evoked a marked contraction in smooth muscle. However, PAR2 and PAR4 agonists caused negligible muscular contraction (Kawabata et al. 2000).

1.2.4 Musculoskeletal System

Individual PARs are expressed in a wide variety of musculoskeletal tissues such as bone, articular cartilage, menisci, synovium and muscle (Russell and McDougall 2009; McDougall and Linton 2012; Chinni et al. 2000). PAR1 and PAR2 are expressed in bone marrow stromal cells (Smith et al. 2004). Human osteoblasts express PAR1, 2 and 3, and mouse osteoblasts express PAR1, 2 and 4, whereas rat osteoblasts have been shown to express PAR1 and 2 only (Jenkins et al. 1993; Pagel et al. 2003; Bluteau et al. 2006). Activation of either PAR1 or PAR4 on osteoblasts using specific receptor-activating peptides causes a rapid mobilisation of intracellular calcium (Jenkins et al. 1993; Pagel et al. 2003). Thrombin is also involved in stimulating proliferation of bone marrow stromal cells in a PAR1-dependent manner (Song et al. 2005). Thrombin also decreases alkaline phosphatase activity in osteoblasts which is a marker of osteoblast differentiation (Abraham and Mackie 1999). Elsewhere, thrombin has been reported to be involved in stimulating bone resorption in an organ culture of neonatal mouse skull bones and foetal rat long bones (Gustafson and Lerner 1983). The resorption induced by thrombin can be inhibited by indomethacin (Gustafson and Lerner 1983; Hoffmann et al. 1986). This action could be related to thrombin stimulating the release of PGE2 and IL-6 by osteoblastic cells, and this effect is mediated by stimulation of osteoclast differentiation (Mackie et al. 2008). In chondrocytes which are obtained from osteoarthritis cartilage, the expression of PAR2 has been found to be higher when compared with normal cartilage (Xiang et al. 2006). PAR2 and PAR3 are expressed by cartilage in the embryonic mouse skeleton (Abraham et al. 1998). Elsewhere, it has been reported that thrombin triggers proliferation of chondrocytes isolated from human articular cartilage (Kirilak et al. 2006). Myoblasts express PAR1 and PAR2, whereas muscle fibres express PAR1, 2 and 4 (Chinni et al. 2000; Jenkins et al. 1993). When myoblasts were stimulated with thrombin or a PAR1-activating peptide, it resulted in mobilisation of calcium (Mackie et al. 2008). A study was carried out to explore the role of PAR2 activation in four different models of arthritis and in human arthritic synovium (Busso et al. 2007). In the adjuvant-induced arthritis model, arthritic symptoms were significantly decreased in PAR2-deficient mice and also in the presence of anti-mBSAIgG antibodies (Busso et al. 2007). No difference in arthritis severity was seen in mice with ZIA, K/BxN serum-induced arthritis or CFA-induced arthritis. Expression of PAR2 in rheumatoid arthritis synovium was significantly higher than in osteoarthritis synovium (Busso et al. 2007). In another study, when an antagonist of PAR2 (GB88) was tested for its efficacy, it attenuated PAR2 signalling, macrophage activation, mast cell degranulation and collagen-induced arthritis in rats (Lohman et al. 2012). McDougall et al. evaluated the role of PAR4 in synovial blood flow which is increased during inflammation. When kaolin/carrageenan was injected into the knees of the animals, they showed an increase in synovial blood flow. Treatment of these inflamed knees with the PAR4 antagonist pepducin P4pal-10 reduced the hyperaemia associated with acute synovitis (McDougall et al. 2009).

Persistent neck pain is a major cause of disability and the cervical facet joint is a common source of neck pain (Barnsley et al. 1995). Rats subjected to a painful joint distraction and receiving an injection of ketorolac either immediately or 1 day later showed an increase in spinal PAR1 and astrocytic PAR1 expression. The astrocytic PAR1 was returned to sham levels when ketorolac was administered on day 1 but not after the immediate administration. However, spinal PAR1 was significantly reduced by ketorolac independent of timing. This indicates that spinal astrocyte expression of PAR1 is involved in the maintenance of facet joint-mediated pain (Dong et al. 2013).

2 Proteinase-Activated Receptor: Role in Pain

2.1 PARs and Inflammatory Pain

Pain is a natural response to noxious environmental stimuli and warns the body of actual or impending damage. In addition to this physiologically appropriate acute pain response, long-lasting pain is maladaptive and serves no functional benefit to the organism. Chronic pain is typically a consequence of an underlying malady and is very difficult to treat across the lifespan. Inflammatory diseases, such as rheumatoid arthritis and Crohn's disease, can be extremely painful and difficult to manage. During nociceptive pain, afferent nerve fibres are activated directly by a noxious environmental stimulus, and the resulting pain response lasts for only a relatively short period. During inflammation, however, the afferent fibres are continuously bombarded by inflammatory mediators culminating in protracted pain. This continuous peripheral drive leads to plasticity changes in the central nervous system resulting in chronic pain. While the list of inflammatory mediators and their distinct receptors involved in inflammatory pain is escalating, recent reports suggest an important role for PARs in pain signalling pathways.

2.1.1 PAR1

It has been shown that PAR1 is expressed on sensory neurones, but their role in nociceptive signalling remains under investigation. Asfaha et al. (2002) studied the effect of PAR1 activation on nociceptive response by thermal and mechanical stimuli. When thrombin was injected into the paws of rodents, it increased the nociceptive threshold and withdrawal latency indicative of an anti-nociceptive effect. While intraplantar injection of carrageenan produces a classical inflammatory pain response, co-administration of the compound with thrombin resulted in a reduction in mechanical and thermal hyperalgesia. Similarly in another study, when thrombin and the PAR1 agonist TFLLR-NH$_2$ were injected intraplantarly, their effects significantly attenuated the hyperalgesia in rats treated with carrageenan (Kawabata et al. 2002). The mechanism by which PAR1 ameliorates inflammatory hyperalgesia was elucidated by Martin et al. (2009) who demonstrated that PAR1 agonism triggers the production of proenkephalin and the activation of opioid receptors.

2.1.2 PAR2

PAR2 is cleaved by trypsin, mast cell tryptase, chymase, neutrophil elastase but not by thrombin. Peptide agonists that trigger PAR2 can lead to acute inflammation, in part via a neurogenic mechanism (Steinhoff et al. 2000). It has been confirmed that PAR2 is expressed by primary afferent neurons, and PAR2 agonists trigger the peripheral release of inflammatory neuropeptides such as substance P and calcitonin gene-related peptide (Steinhoff et al. 2000). Intraplantar injection of sub-inflammatory doses of PAR2 agonists in rats and mice induced a prolonged thermal and mechanical hyperalgesia and elevated fos protein expression in the dorsal horn, indicating that peripheral PAR2 stimulation leads to increased electrochemical activity of spinal neurones (Vergnolle et al. 2001). Interestingly, this hyperalgesia was not present in mice lacking substance P-preferring NK-1 receptors or preprotachykinin-A or in rats treated with an intrathecal injection of a NK-1 antagonist. These observations further support the suggestion that substance P is involved in PAR2-mediated pain responses (Vergnolle et al. 2001).

In joints, PAR2 has emerged as a new therapeutic target for arthritis (Russell and McDougall 2009). PAR2 is expressed in several cell types where its cleavage by serine proteinases is involved in the pathogenesis of inflammatory arthritis by mechanisms that are as yet unclear (Russell et al. 2012). However, there have been a few studies attempting to explore neuronal and inflammatory changes in joints after PAR2 activation. Using the retrograde neuronal tracer Fluoro-Gold, Russell et al. (2012) identified the expression of PAR2 in rat knee joint L3–L5 DRG cells. Additionally, it was found that activation of PAR2 by the selective activating peptide 2-furoyl-LIGRLO-NH2 increased joint nociceptor fibre firing rate during normal and noxious rotation (Russell et al. 2012). Furthermore, intravital microscopy experiments showed significantly increased leukocyte rolling and adhesion in response to PAR2 stimulation (Russell et al. 2012). All these effects were blocked by pretreatment with a TRPV1 or NK-1 receptor-selective antagonist. In another study, it has been shown that intra-articular injections of the PAR2-activating peptide, SLIGRL-NH2, caused swelling, cytokine release and increased sensitivity to pain in the mouse knee joint (Helyes et al. 2010). The secondary mechanical allodynia and change in weight distribution induced by intra-articular SLIGRL-NH2 were also found to be TRPV1 dependent (Helyes et al. 2010). In other joints, PAR2 is expressed in the lining layer of the rat temporomandibular joint (TMJ) synovium, as well as in a high proportion of the trigeminal ganglion neurones that innervate this joint (Denadai-Souza et al. 2010). When PAR2 agonists were injected by the intra-articular route into the TMJ, they triggered a dose-dependent increase in plasma extravasation, neutrophil influx and induction of mechanical allodynia, and these effects were inhibited by a NK-1 receptor antagonist (Denadai-Souza et al. 2010).

There are a number of ion channels involved in modulating inflammatory pain, viz. TRPV1, TRPA1 and P2X3. Inflammatory mediators such as substance P and bradykinin potentiate currents through ATP receptor channels containing the P2X3 subunit (Paukert et al. 2001). PAR2-induced neurogenic inflammation causes an increase in P2X3 currents, evoked by α- and β-methylene ATP in DRG neurones

(Wang et al. 2012). Thus, it has been proven that the functional interaction of the PAR2 and P2X3 in primary sensory neurones could contribute to the generation of inflammatory pain.

It has also been demonstrated that PAR2 activation induces visceral pain (Kawao et al. 2004). A study was undertaken to evaluate the effect of activation of PAR2 on colonic motility. PAR2 agonists administered intraluminally induced contractions of the colon and produced hypersensitivity to colorectal distension (Suckow et al. 2012). Lesioning of TRPV1 neurones by capsaicin treatment eliminated this enhancement in contraction which indicates that TRPV1/PAR2 expressing primary afferent neurones mediate an extrinsic motor reflex pathway in the colon (Suckow et al. 2012). In another study, PAR2 agonist administration induced sustained, concentration-dependent contraction of oesophageal longitudinal smooth muscle strips. Capsaicin desensitisation, substance P desensitisation or application of the selective neurokinin-2 (NK-2) receptor antagonist MEN 10376 blocked these contractions (Paterson et al. 2007). This pathway is similar to the pathway involved in acid-induced longitudinal smooth muscle contraction and oesophageal shortening (Paterson et al. 2007). It could be possible that acid-induced longitudinal smooth muscle contraction may involve mast cell-derived mediators that activate capsaicin-sensitive neurones via PAR-2 and hence modulation of PAR2 might be useful in treatment of oesophageal pain and hiatus hernia (Liu et al. 2010). PAR2 is also involved in mediating inflammatory pain in acute pancreatitis (Ceppa et al. 2011). Exogenous trypsin injected at a sub-inflammatory dose caused increased c-fos immunoreactivity, which is an indicator of spinal nociceptor activation. There were no signs of inflammation at this dose as indicated by serum amylase and myeloperoxidase levels. Trypsin IV and P23 injected at similar doses resulted in an increase of some inflammatory end points and caused a more robust effect on nociception; these effects were blocked by the trypsin inhibitor melagatran (Ceppa et al. 2011). Trypsin IV and rat P23 activate PAR2 and are resistant to pancreatic trypsin inhibitors, and hence they contribute to pancreatic inflammation and pain (Ceppa et al. 2011). Elsewhere, caerulein (an oligopeptide that stimulates smooth muscle and increases digestive secretions) administered at a single dose increased abdominal sensitivity to stimulation by von Frey hairs, without causing pancreatitis in PAR2 KO mice. Multiple administrations increased the severity of abdominal allodynia/hyperalgesia in PAR2 KO as compared to WT mice. When a PAR2-AP was co-administered with caerulein, it abolished hyperalgesia/allodynia in WT mice but not in PAR2 KO mice. These results clearly indicate that PAR2 attenuates pancreatitis-related hyperalgesia/allodynia without affecting the disease (Kawabata et al. 2006).

Phosphoinositide 3-kinases (PI3Ks) have also been implicated in dermal mechanosensitivity where touch was evaluated. PI3Kγ gene deletion increased scratching behaviours in histamine-dependent and PAR2-dependent itch, whereas PI3Kg-deficient mice were not able to enhance scratching in chloroquine-induced itch (Lee et al. 2011). Furthermore, deletion of the PI3Kγ gene does not affect behavioural licking responses to intraplantar injections of formalin or mechanical allodynia in a chronic inflammatory pain model (Lee et al. 2011). These findings suggest that PI3Kγ contributes to behavioural itching induced by histamine and PAR2 agonist but not a chloroquine agonist (Lee et al. 2011).

2.1.3 PAR3

There have been no specific studies conducted evaluating the role of PAR3 in inflammatory pain. In an investigation carried out by Zhu et al. (2005), however, it was observed that PAR3 mRNA was expressed in 41 % of rat DRGs. It was also observed that 84 % of PAR3 positive cells co-localised with CGRP suggesting that PAR3 could be involved in peripheral nociceptive mechanisms.

2.1.4 PAR4

PAR4 is the most recently discovered member of the PAR family and is primarily cleaved by thrombin, trypsin and cathepsin G, as well as small synthetic peptides (Hollenberg and Compton 2002). Asfaha et al. (2007) first demonstrated the expression of PAR4 on sensory neurones isolated from rat DRG establishing the possibility that PAR4 is involved in modulating pain transmission. The researchers also showed that PAR4 co-localises with CGRP and SP. Initial studies found that intraplantar injection of PAR4-activating peptides increased the nociceptive threshold in response to thermal and mechanical noxious stimuli indicative of an analgesic role for PAR4 (Asfaha et al. 2007). Similarly, in a colorectal distension model of gastrointestinal pain, PAR4-activating peptides reduced nociception as measured by abdominal muscle contraction (Annahazi et al. 2009). Intra-colonic pretreatment of animals with the tight junction blocker 2,4,6-triaminopyrimidine inhibited PAR4 analgesia suggesting that PAR4-activating peptides need direct access to colonic nerve terminals in order to reduce pain (Annahazi et al. 2012).

In contrast to the gastrointestinal system, local activation of PAR4 receptors in knee joints results in an increase in joint blood flow, oedema and pain (McDougall et al. 2009). Using retrograde tracing techniques, it has been shown that PAR4 is expressed on approximately 60 % of knee joint primary afferents supporting the concept that PAR4 has the potential to modulate nociceptor activity (Russell et al. 2010). Indeed, intra-articular injection of the PAR4-activating peptide AYPGKF-NH$_2$ sensitises joint afferents leading to the generation of joint pain (Russell et al. 2010). These pro-nociceptive effects of PAR4 could be blocked by a bradykinin B$_2$ antagonist (HOE140) and following stabilisation of joint connective tissue mast cells (McDougall et al. 2009). Thus, in joints, PAR4 agonists can promote joint pain either by sensitising mechanosensory nerves directly within the joint or by causing the secondary release of bradykinin from synovial mast cells. It appears, therefore, that PAR4 can either promote or inhibit inflammatory pain in an organ-dependent manner.

2.2 PARs and Neuropathic Pain

Neuropathic pain is a complex, chronic pain state that is usually not accompanied with tissue injury. This type of pain could result from lesions or disorders of the peripheral and central nervous systems resulting in abnormal processing of sensory input. There are many factors which lead to the generation of neuropathic pain, viz. infectious agents, metabolic disease, neurodegenerative disease and physical trauma (Vecht 1989; Pasero 2004). PARs participate in the initiation and maintenance of neuropathic pain by mediating various actions such as an abnormal increase of algesic neurotransmitters such as substance P, calcitonin gene-related peptide, prostaglandins and kinins (Jin et al. 2009).

2.2.1 PAR1

PAR1 is not only expressed in platelets but also in the CNS where it is involved in various neurophysiological functions. Narita et al. (2005) investigated the role of PAR1 in the development of neuropathic pain after nerve injury. When sciatic nerve ligation was performed in animals, it induced thermal hyperalgesia and tactile allodynia which were suppressed by repeated intrathecal injection of hirudin, a specific and potent thrombin inhibitor (Narita et al. 2005). Since nerve ligation and thrombin both upregulate PAR1 expression in the dorsal horn of the spinal cord, it suggests that PAR1 is involved in neuropathic pain transmission.

2.2.2 PAR2

Chronic compression of dorsal root ganglia (CCD) is a neuropathic pain model which results in mechanical and thermal hyperalgesia in rats. The cAMP–protein kinase A (PKA) pathway is shown to be important for maintaining both DRG neuronal hyperexcitability and behaviourally expressed hyperalgesia (Song et al. 1999). Recently, it was revealed that PAR2 is involved in mediating increases of cAMP and PKA activity and also cAMP-dependent hyperexcitability and hyperalgesia in rats. When a PAR2 activator was administered into the intervertebral foramen in animals with CCD, it caused an increase in cAMP accumulation, mRNA and protein expression for PKA subunits and protein expression of PAR2 (Huang et al. 2012). In conjunction with these changes in the cAMP–PKA pathway, CCD caused neuronal hyperexcitability and thermal hyperalgesia which were prevented by pretreatment with a PAR2 antagonist (Huang et al. 2012). PAR2 has been reportedly involved in paclitaxel-induced neuropathic pain (Chen et al. 2011). When paclitaxel was administered in animals, it caused enhancement of mast cell tryptase levels in the spinal cord and DRG which led to neuronal activation of PAR2. These effects were blocked by administration of the selective PAR2 antagonist (FSLLRY-amide). Additionally, pain responses were blocked by antagonists of TRPV1, TRPV4 and TRPA1 implicating the involvement of these receptors in PAR2 sensitisation (Chen et al. 2011). Table 2 summarises the role of PARs in inflammatory and neuropathic pain.

2.2.3 PAR3 and PAR4

There are no prospective studies available exploring a role for PAR3 and PAR4 in neuropathic pain.

Table 2 Summary of role of PARs in inflammatory and neuropathic pain

PARs	Inflammatory pain	Neuropathic pain
PAR1	1. Increase in nociceptive threshold	1. Mediates long-lasting hyperalgesia and allodynia
	2. Decrease in mechanical and thermal hyperalgesia	2. Increase in spinal and astrocytic PAR1 expression after painful joint distraction injury
	3. Increase in production of proenkephalins	
	ANTI-NOCICEPTIVE	PRO-NOCICEPTIVE
PAR2	1. Increase in release of CGRP and SP	1. Causes increase in neuronal hyperexcitability and hyperalgesia in nerve injury model
	2. Prolonged thermal and mechanical hyperalgesia	2. Mediates paclitaxel-induced neuropathic pain by activating PKA or C pathway
	3. Joints	
	– Swelling, cytokine release, triggers pain	
	– Increase in joint afferent firing rate	
	– Activation of TRPV1, NK-1	
	4. Gastrointestinal system	
	– Increase in colonic motility	
	– Increase in contraction of oesophageal longitudinal smooth muscle	
	PRO-NOCICEPTIVE	PRO-NOCICEPTIVE
PAR3	1. Expressed in rat DRG	No studies evaluating role of PAR3 in neuropathic pain
PAR4	1. Gastrointestinal system	No studies evaluating role of PAR4 in neuropathic pain
	– Increase in nociceptive threshold in colorectal distension model	
	– Co-localised with CGRP and SP	
	ANTI-NOCICEPTIVE	
	2. Joints	
	– Causes sensitisation of joint afferents	
	– Causes joint pain mediated by release of bradykinin	
	– Pain is mast cell dependent	
	PRO-NOCICEPTIVE	

3 PARs as a Drug Target for Pain

Based on the pharmacological properties of PARs in models of disease, it is clear that these receptors participate in nociceptor modulation and pain transmission (Fig. 2) and could be a potential target for the control of pain.

Thrombin plays an important role in maintaining haemostatic balance, and it executes its action through PAR1 (Vu et al. 1991). This makes PAR1 an attractive target for the treatment of cardiovascular diseases. A PAR1 antagonist SCH 530348 discovered by Schering-Plough is currently being developed by Merck Sharp & Dohme Corp. for the treatment of non-emergent percutaneous coronary intervention. Another PAR1 antagonist, atopaxar (E5555), is in the late stages of clinical development for its safety and tolerability in patients with acute coronary syndromes or stable coronary artery disease on top of standard antiplatelet therapy (Ramachandran 2012). There are very few preclinical studies done in investigating the role of PAR1 in pain. However, from the findings of those studies, it is clear that

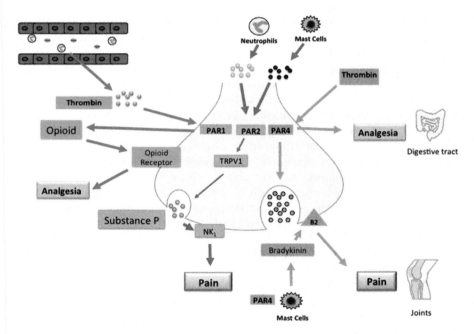

Fig. 2 Schematic representation of proteinase-activated receptor mechanisms of nociceptive modulation. PAR1 (*blue*) is activated by thrombin which then triggers the production of opioids which activate opioid receptors on sensory nerve terminals resulting in analgesia. PAR2 (*red*) present on nociceptors is cleaved by proteinases emanating from neutrophils and mast cells. PAR2 activation further stimulates TRPV1 which causes the release of substance P which upon binding to NK-1 receptors produces peripheral sensitisation and pain. PAR4 (*purple*) is activated by thrombin which in the gut results in analgesia. Conversely in joints, neuronal and mast cell-derived PAR4 activation causes bradykinin release which binds to neuronal B2 receptors producing peripheral sensitisation and pain

PAR1 holds anti-nociceptive properties in inflammatory pain and pro-nociceptive effects in neuropathic pain. Based on these studies, it is difficult to comment on the potential of PAR1 as a target for general pain, and clearly there is need for more studies to elucidate the role of PAR1 in pain signalling.

Among all the PARs, PAR2 has been studied in many inflammatory pain models and a few neuropathic pain models. It can be said that PAR2 plays a very important role in pain transmission, and it uniformly promotes pain sensation Vergnolle 2003. Different agents like trypsin inhibitors, antibodies for PAR2 and antagonists of PAR2 Barry et al. 2010 have been designed to see whether they block the actions of PAR2. These agents have been studied in different disease models. A trypsin inhibitor and PAR2 antibody prevented airway hyperresponsiveness and allergic airway inflammation induced by intranasal administration of cockroach extract (Arizmendi et al. 2011). PAR2 antagonists like ENMD-1068, GB 83 and GB 88 have been synthesised and tested in disease models, and they have shown good efficacy profile in these models (Kelso et al. 2006; Suen et al. 2012; Lohman et al. 2012; Barry et al. 2006). However, these agents have not been tested in chronic pain and neuropathic pain models. Recently, a report by Oliveira et al. (2013) showed promising efficacy of ENMD-1068 in preventing the development of post-operative nociception mediated by PAR2. It is known from the literature that PAR2 mediate pain by multiple mechanisms and blocking them could be an attractive strategy for the alleviating chronic pain symptoms.

PAR4 plays a dual role in pain processing depending on the specific organ in which it is expressed. In the gastrointestinal system, it increases the nociceptive threshold and exhibits anti-nociceptive property, whereas in joints, it increases blood flow, oedema and pain (McDougall et al. 2009). So, blocking PAR4 may be a good strategy for alleviating arthritic but not gastrointestinal pain. Covic et al. proposed a novel approach of designing the cell-penetrating peptides (pepducin) which modulate receptor activity either by activating or inhibiting receptors. These researchers developed PAR1- and PAR4-based pepducins antagonist and tested them for anti-haemostatic and anti-thrombotic effects (Covic et al. 2002). PAR1 pepducin antagonist Plpal-12 blocked 75–95 % of aggregation in response to the PAR1 extracellular ligand SFLLRN. PAR4 pepducin antagonist P4pal-10 blocked 50–80 % of aggregation in response to the PAR4 extracellular ligand AYPGKF (Covic et al. 2002). Russell et al. (2010) used pepducin P4pal-10 to inhibit the AYPGKF-NH2-induced increase in nociceptor firing rate in knee joints of rats indicating that this peptide antagonist may be useful to treat arthritis pain. Recently, a group of researchers characterised a non-peptide, selective, PAR4 receptor antagonist YD-3 for its efficacy in ex vivo platelet assays (Chen et al. 2008) and in an in vivo mouse model of angiogenesis (Lee et al. 2001). However, there is no evidence of its efficacy in inflammation and pain models.

4 Conclusion

Despite years of research in the area of pain, there is still no satisfactory and safe treatment for chronic pain conditions. It is clear that all PARs play some role in inflammation and pain signalling, and further development is required to advance this promising area of pain treatment.

References

Abraham AA, Jenkins AL, Stone SR, Mackie EJ (1998) Expression of the thrombin receptor in developing bone and associated tissues. J Bone Miner Res 13(5):818–827

Abraham LA, MacKie EJ (1999) Modulation of osteoblast-like cell behavior by activation of protease-activated receptor-1. J Bone Miner Res 14:1320–1329

Afkhami-Goli A, Noorbakhsh F, Keller AJ, Vergnolle N, Westaway D, Jhamandas JH, Andrade-Gordon P, Hollenberg MD, Arab H, Dyck RH, Power C (2007) Proteinase-activated receptor-2 exerts protective and pathogenic cell type-specific effects in Alzheimer's disease. J Immunol 179:5493–5503

Annahazi A, Dabek M, Gecse K, Salvador-Cartier C, Polizzi A, Rosztoczy A, Roka R, Theodorou V, Wittmann T, Bueno L, Eutamene H (2012) Proteinase-activated receptor-4 evoked colorectal analgesia in mice: an endogenously activated feed-back loop in visceral inflammatory pain. Neurogastroenterol Motil 24(1):76–85, e13

Annahazi A, Gecse K, Dabek M, Ait-Belgnaoui A, Rosztoczy A, Roka R, Molnar T, Theodorou V, Wittmann T, Bueno L, Eutamene H (2009) Fecal proteases from diarrheic-IBS and ulcerative colitis patients exert opposite effect on visceral sensitivity in mice. Pain 144:209–217

Antoniak S, Rojas M, Spring D, Bullard TA, Verrier ED, Blaxall BC, Mackman N, Pawlinski R (2010) Protease-activated receptor 2 deficiency reduces cardiac ischemia/reperfusion injury. Arterioscler Thromb Vasc Biol 30:2136–2142

Antoniak S, Sparkenbaugh EM, Tencati M, Rojas M, Mackman N, Pawlinski R (2013) Protease activated receptor-2 contributes to heart failure. PLoS One 8:e81733

Arizmendi NG, Abel M, Mihara K, Davidson C, Polley D, Nadeem A, El Mays T, Gilmore BF, Walker B, Gordon JR, Hollenberg MD, Vliagoftis H (2011) Mucosal allergic sensitization to cockroach allergens is dependent on proteinase activity and proteinase-activated receptor-2 activation. J Immunol 186:3164–3172

Asfaha S, Brussee V, Chapman K, Zochodne DW, Vergnolle N (2002) Proteinase-activated receptor-1 agonists attenuate nociception in response to noxious stimuli. Br J Pharmacol 135:1101–1106

Asfaha S, Cenac N, Houle S, Altier C, Papez MD, Nguyen C, Steinhoff M, Chapman K, Zamponi GW, Vergnolle N (2007) Protease-activated receptor-4: a novel mechanism of inflammatory pain modulation. Br J Pharmacol 150:176–185

Barnsley L, Lord SM, Wallis BJ, Bogduk N (1995) The prevalence of chronic cervical zygapophysial joint pain after whiplash. Spine (Phila Pa 1976) 20:20–25; discussion 26

Barrett AJ (2001) Proteolytic enzymes: nomenclature and classification. In: Beynon R, Bond JS (eds) Proteolytic enzymes. A practical approach, 2nd edn. Oxford University Press, Oxford, pp 1–21

Barrett AJ, Rawlings ND, Woessner JF Jr (eds) (2004) Handbook of proteolytic enzymes, 2nd edn. Academic, Amsterdam

Barry GD, Le GT, Fairlie DP (2006) Agonists and antagonists of protease activated receptors (PARs). Curr Med Chem 13(3):243–265

Barry GD, Suen JY, Le GT, Cotterell A, Reid RC, Fairlie DP (2010) Novel agonists and antagonists for human protease activated receptor 2. J Med Chem 53:7428–7440

Beecher KL, Andersen TT, Fenton JW 2nd, Festoff BW (1994) Thrombin receptor peptides induce shape change in neonatal murine astrocytes in culture. J Neurosci Res 37:108–115

Belham CM, Tate RJ, Scott PH, Pemberton AD, Miller HR, Wadsworth RM, Gould GW, Plevin R (1996) Trypsin stimulates proteinase-activated receptor-2-dependent and -independent activation of mitogen-activated protein kinases. Biochem J 320(Pt 3):939–946

Benka ML, Lee M, Wang GR, Buckman S, Burlacu A, Cole L, DePina A, Dias P, Granger A, Grant B et al (1995) The thrombin receptor in human platelets is coupled to a GTP binding protein of the G alpha q family. FEBS Lett 363:49–52

Bluteau G, Pilet P, Bourges X, Bilban M, Spaethe R, Daculsi G, Guicheux J (2006) The modulation of gene expression in osteoblasts by thrombin coated on biphasic calcium phosphate ceramic. Biomaterials 27:2934–2943

Bohm SK, Kong W, Bromme D, Smeekens SP, Anderson DC, Connolly A, Kahn M, Nelken NA, Coughlin SR, Payan DG, Bunnett NW (1996) Molecular cloning, expression and potential functions of the human proteinase-activated receptor-2. Biochem J 314(Pt 3):1009–1016

Bonni A, Brunet A, West AE, Datta SR, Takasu MA, Greenberg ME (1999) Cell survival promoted by the Ras-MAPK signaling pathway by transcription-dependent and -independent mechanisms. Science 286:1358–1362

Busso N, Frasnelli M, Feifel R, Cenni B, Steinhoff M, Hamilton J, So A (2007) Evaluation of protease-activated receptor 2 in murine models of arthritis. Arthritis Rheum 56:101–107

Ceppa EP, Lyo V, Grady EF, Knecht W, Grahn S, Peterson A, Bunnett NW, Kirkwood KS, Cattaruzza F (2011) Serine proteases mediate inflammatory pain in acute pancreatitis. Am J Physiol Gastrointest Liver Physiol 300:G1033–G1042

Chen HS, Kuo SC, Teng CM, Lee FY, Wang JP, Lee YC, Kuo CW, Huang CC, Wu CC, Huang LJ (2008) Synthesis and antiplatelet activity of ethyl 4-(1-benzyl-1H-indazol-3-yl)benzoate (YD-3) derivatives. Bioorg Med Chem 16:1262–1278

Chen Y, Yang C, Wang ZJ (2011) Proteinase-activated receptor 2 sensitizes transient receptor potential vanilloid 1, transient receptor potential vanilloid 4, and transient receptor potential ankyrin 1 in paclitaxel-induced neuropathic pain. Neuroscience 193:440–451

Chinni C, de Niese MR, Jenkins AL, Pike RN, Bottomley SP, Mackie EJ (2000) Protease-activated receptor-2 mediates proliferative responses in skeletal myoblasts. J Cell Sci 113 (Pt 24):4427–4433

Cocks TM, Fong B, Chow JM, Anderson GP, Frauman AG, Goldie RG, Henry PJ, Carr MJ, Hamilton JR, Moffatt JD (1999a) A protective role for protease-activated receptors in the airways. Nature 398:156–160

Cocks TM, Sozzi V, Moffatt JD, Selemidis S (1999b) Protease-activated receptors mediate apamin-sensitive relaxation of mouse and guinea pig gastrointestinal smooth muscle. Gastroenterology 116:586–592

Corvera CU, Dery O, McConalogue K, Bohm SK, Khitin LM, Caughey GH, Payan DG, Bunnett NW (1997) Mast cell tryptase regulates rat colonic myocytes through proteinase-activated receptor 2. J Clin Invest 100:1383–1393

Coughlin SR (2000) Thrombin signalling and protease-activated receptors. Nature 407:258–264

Covic L, Misra M, Badar J, Singh C, Kuliopulos A (2002) Pepducin-based intervention of thrombin-receptor signaling and systemic platelet activation. Nat Med 8:1161–1165

DeFea KA, Zalevsky J, Thoma MS, Dery O, Mullins RD, Bunnett NW (2000) beta-arrestin-dependent endocytosis of proteinase-activated receptor 2 is required for intracellular targeting of activated ERK1/2. J Cell Biol 148:1267–1281

Denadai-Souza A, Cenac N, Casatti CA, Camara PR, Yshii LM, Costa SK, Vergnolle N, Muscara MN (2010) PAR(2) and temporomandibular joint inflammation in the rat. J Dent Res 89:1123–1128

Dong L, Smith JR, Winkelstein BA (2013) Ketorolac reduces spinal astrocytic activation and PAR1 expression associated with attenuation of pain after facet joint injury. J Neurotrauma 30:818–825

Duncan MB, Kalluri R (2009) Parstatin, a novel protease-activated receptor 1-derived inhibitor of angiogenesis. Mol Interv 9:168–170

Eberhart CE, Dubois RN (1995) Eicosanoids and the gastrointestinal tract. Gastroenterology 109:285–301

Gingrich MB, Junge CE, Lyuboslavsky P, Traynelis SF (2000) Potentiation of NMDA receptor function by the serine protease thrombin. J Neurosci 20:4582–4595

Gustafson GT, Lerner U (1983) Thrombin, a stimulator of bone resorption. Biosci Rep 3:255–261

Helyes Z, Sandor K, Borbely E, Tekus V, Pinter E, Elekes K, Toth DM, Szolcsanyi J, McDougall JJ (2010) Involvement of transient receptor potential vanilloid 1 receptors in protease-activated receptor-2-induced joint inflammation and nociception. Eur J Pain 14:351–358

Hoffmann O, Klaushofer K, Koller K, Peterlik M, Mavreas T, Stern P (1986) Indomethacin inhibits thrombin-, but not thyroxin-stimulated resorption of fetal rat limb bones. Prostaglandins 31(4):601–608

Hollenberg MD, Compton SJ (2002) International Union of Pharmacology. XXVIII. Proteinase-activated receptors. Pharmacol Rev 54:203–217

Huang ZJ, Li HC, Cowan AA, Liu S, Zhang YK, Song XJ (2012) Chronic compression or acute dissociation of dorsal root ganglion induces cAMP-dependent neuronal hyperexcitability through activation of PAR2. Pain 153:1426–1437

Jalink K, Moolenaar WH (1992) Thrombin receptor activation causes rapid neural cell rounding and neurite retraction independent of classic second messengers. J Cell Biol 118:411–419

Jenkins AL, Bootman MD, Taylor CW, Mackie EJ, Stone SR (1993) Characterization of the receptor responsible for thrombin-induced intracellular calcium responses in osteoblast-like cells. J Biol Chem 268:21432–21437

Jin G, Hayashi T, Kawagoe J, Takizawa T, Nagata T, Nagano I, Syoji M, Abe K (2005) Deficiency of PAR-2 gene increases acute focal ischemic brain injury. J Cereb Blood Flow Metab 25:302–313

Jin C et al (2009) Protease-activated receptors in neuropathic pain: an important mediator between neuron and glia. J Med Coll PLA 24:244–249

Kawabata A, Kuroda R, Nagata N, Kawao N, Masuko T, Nishikawa H, Kawai K (2001) In vivo evidence that protease-activated receptors 1 and 2 modulate gastrointestinal transit in the mouse. Br J Pharmacol 133:1213–1218

Kawabata A, Kawao N, Kuroda R, Tanaka A, Shimada C (2002) The PAR-1-activating peptide attenuates carrageenan-induced hyperalgesia in rats. Peptides 23:1181–1183

Kawabata A, Kuroda R, Kuroki N, Nishikawa H, Kawai K (2000) Dual modulation by thrombin of the motility of rat oesophageal muscularis mucosae via two distinct protease-activated receptors (PARs): a novel role for PAR-4 as opposed to PAR-1. Br J Pharmacol 131:578–584

Kawabata A, Matsunami M, Tsutsumi M, Ishiki T, Fukushima O, Sekiguchi F, Kawao N, Minami T, Kanke T, Saito N (2006) Suppression of pancreatitis-related allodynia/hyperalgesia by proteinase-activated receptor-2 in mice. Br J Pharmacol 148:54–60

Kawao N, Ikeda H, Kitano T, Kuroda R, Sekiguchi F, Kataoka K, Kamanaka Y, Kawabata A (2004) Modulation of capsaicin-evoked visceral pain and referred hyperalgesia by protease-activated receptors 1 and 2. J Pharmacol Sci 94:277–285

Kelso EB, Lockhart JC, Hembrough T, Dunning L, Plevin R, Hollenberg MD, Sommerhoff CP, McLean JS, Ferrell WR (2006) Therapeutic promise of proteinase-activated receptor-2 antagonism in joint inflammation. J Pharmacol Exp Ther 316:1017–1024

Kirilak Y, Pavlos NJ, Willers CR, Han R, Feng H, Xu J, Asokananthan N, Stewart GA, Henry P, Wood D, Zheng MH (2006) Fibrin sealant promotes migration and proliferation of human articular chondrocytes: possible involvement of thrombin and protease-activated receptors. Int J Mol Med 17:551–558

Kong W, McConalogue K, Khitin LM, Hollenberg MD, Payan DG, Bohm SK, Bunnett NW (1997) Luminal trypsin may regulate enterocytes through proteinase-activated receptor 2. Proc Natl Acad Sci U S A 94:8884–8889

Lee B, Descalzi G, Baek J, Kim JI, Lee HR, Lee K, Kaang BK, Zhuo M (2011) Genetic enhancement of behavioral itch responses in mice lacking phosphoinositide 3-kinase-gamma (PI3Kgamma). Mol Pain 7:96

Lee FY, Lien JC, Huang LJ, Huang TM, Tsai SC, Teng CM, Wu CC, Cheng FC, Kuo SC (2001) Synthesis of 1-benzyl-3-(5′-hydroxymethyl-2′-furyl)indazole analogues as novel antiplatelet agents. J Med Chem 44:3746–3749

Liu H, Miller DV, Lourenssen S, Wells RW, Blennerhassett MG, Paterson WG (2010) Proteinase-activated receptor-2 activation evokes oesophageal longitudinal smooth muscle contraction via a capsaicin-sensitive and neurokinin-2 receptor-dependent pathway. Neurogastroenterol Motil 22(2):210–216, e67

Lohman RJ, Cotterell AJ, Barry GD, Liu L, Suen JY, Vesey DA, Fairlie DP (2012) An antagonist of human protease activated receptor-2 attenuates PAR2 signaling, macrophage activation, mast cell degranulation, and collagen-induced arthritis in rats. FASEB J 26:2877–2887

Macfarlane SR, Seatter MJ, Kanke T, Hunter GD, Plevin R (2001) Proteinase-activated receptors. Pharmacol Rev 53:245–282

Mackie EJ, Loh LH, Sivagurunathan S, Uaesoontrachoon K, Yoo HJ, Wong D, Georgy SR, Pagel CN (2008) Protease-activated receptors in the musculoskeletal system. Int J Biochem Cell Biol 40:1169–1184

Martin L et al (2009) Thrombin receptor: an endogenous inhibitor of inflammatory pain, activating opioid pathways. Pain 146:121–129

McDougall JJ, Linton P (2012) Neurophysiology of arthritis pain. Curr Pain Headache Rep 16:485–491

McDougall JJ, Zhang C, Cellars L, Joubert E, Dixon CM, Vergnolle N (2009) Triggering of proteinase-activated receptor 4 leads to joint pain and inflammation in mice. Arthritis Rheum 60:728–737

Murray DB, McLarty-Williams J, Nagalla KT, Janicki JS (2012) Tryptase activates isolated adult cardiac fibroblasts via protease activated receptor-2 (PAR-2). J Cell Commun Signal 6:45–51

Nakanishi-Matsui M, Zheng YW, Sulciner DJ, Weiss EJ, Ludeman MJ, Coughlin SR (2000) PAR3 is a cofactor for PAR4 activation by thrombin. Nature 404:609–613

Narita M, Usui A, Niikura K, Nozaki H, Khotib J, Nagumo Y, Yajima Y, Suzuki T (2005) Protease-activated receptor-1 and platelet-derived growth factor in spinal cord neurons are implicated in neuropathic pain after nerve injury. J Neurosci 25:10000–10009

Nelken NA, Soifer SJ, O'Keefe J, Vu TK, Charo IF, Coughlin SR (1992) Thrombin receptor expression in normal and atherosclerotic human arteries. J Clin Invest 90:1614–1621

Oliveira SM, Silva CR, Ferreira J (2013) Critical role of protease-activated receptor 2 activation by mast cell tryptase in the development of postoperative pain. Anesthesiology 118:679–690

Ossovskaya VS, Bunnett NW (2004) Protease-activated receptors: contribution to physiology and disease. Physiol Rev 84:579–621

Pagel CN, de Niese MR, Abraham LA, Chinni C, Song SJ, Pike RN, Mackie EJ (2003) Inhibition of osteoblast apoptosis by thrombin. Bone 33:733–743

Pasero C (2004) Pathophysiology of neuropathic pain. Pain Manag Nurs 5:3–8

Paterson WG, Miller DV, Dilworth N, Assini JB, Lourenssen S, Blennerhassett MG (2007) Intraluminal acid induces oesophageal shortening via capsaicin-sensitive neurokinin neurons. Gut 56:1347–1352

Paukert M, Osteroth R, Geisler HS, Brandle U, Glowatzki E, Ruppersberg JP, Grunder S (2001) Inflammatory mediators potentiate ATP-gated channels through the P2X(3) subunit. J Biol Chem 276:21077–21082

Ramachandran R (2012) Developing PAR1 antagonists: minding the endothelial gap. Discov Med 13(73):425–431

Ramachandran R, Mihara K, Chung H, Renaux B, Lau CS, Muruve DA, DeFea KA, Bouvier M, Hollenberg MD (2011) Neutrophil elastase acts as a biased agonist for proteinase-activated receptor-2 (PAR2). J Biol Chem 286:24638–24648

Russell FA, McDougall JJ (2009) Proteinase activated receptor (PAR) involvement in mediating arthritis pain and inflammation. Inflamm Res 58:119–126

Russell FA, Schuelert N, Veldhoen VE, Hollenberg MD, McDougall JJ (2012) Activation of PAR (2) receptors sensitizes primary afferents and causes leukocyte rolling and adherence in the rat knee joint. Br J Pharmacol 167:1665–1678

Russell FA, Veldhoen VE, Tchitchkan D, McDougall JJ (2010) Proteinase-activated receptor-4 (PAR4) activation leads to sensitization of rat joint primary afferents via a bradykinin B2 receptor-dependent mechanism. J Neurophysiol 103:155–163

Sabri A, Muske G, Zhang H, Pak E, Darrow A, Andrade-Gordon P, Steinberg SF (2000) Signaling properties and functions of two distinct cardiomyocyte protease-activated receptors. Circ Res 86:1054–1061

Saifeddine M, al-Ani B, Cheng CH, Wang L, Hollenberg MD (1996) Rat proteinase-activated receptor-2 (PAR-2): cDNA sequence and activity of receptor-derived peptides in gastric and vascular tissue. Br J Pharmacol 118:521–530

Seminario-Vidal L, Kreda S, Jones L, O'Neal W, Trejo J, Boucher RC, Lazarowski ER (2009) Thrombin promotes release of ATP from lung epithelial cells through coordinated activation of rho- and Ca^{2+}-dependent signaling pathways. J Biol Chem 284:20638–20648

Smith R, Ransjo M, Tatarczuch L, Song SJ, Pagel C, Morrison JR, Pike RN, Mackie EJ (2004) Activation of protease-activated receptor-2 leads to inhibition of osteoclast differentiation. J Bone Miner Res 19:507–516

Soh UJ, Dores MR, Chen B, Trejo J (2010) Signal transduction by protease-activated receptors. Br J Pharmacol 160:191–203

Song SJ, Pagel CN, Campbell TM, Pike RN, Mackie EJ (2005) The role of protease-activated receptor-1 in bone healing. Am J Pathol 166:857–868

Song XJ, Hu SJ, Greenquist KW, Zhang JM, LaMotte RH (1999) Mechanical and thermal hyperalgesia and ectopic neuronal discharge after chronic compression of dorsal root ganglia. J Neurophysiol 82:3347–3358

Steinberg SF (2005) The cardiovascular actions of protease-activated receptors. Mol Pharmacol 67:2–11

Steinhoff M, Vergnolle N, Young SH, Tognetto M, Amadesi S, Ennes HS, Trevisani M, Hollenberg MD, Wallace JL, Caughey GH, Mitchell SE, Williams LM, Geppetti P, Mayer EA, Bunnett NW (2000) Agonists of proteinase-activated receptor 2 induce inflammation by a neurogenic mechanism. Nat Med 6:151–158

Stoll G, Jander S, Schroeter M (1998) Inflammation and glial responses in ischemic brain lesions. Prog Neurobiol 56:149–171

Suckow SK, Anderson EM, Caudle RM (2012) Lesioning of TRPV1 expressing primary afferent neurons prevents PAR-2 induced motility, but not mechanical hypersensitivity in the rat colon. Neurogastroenterol Motil 24:e125–e135

Suen JY, Barry GD, Lohman RJ, Halili MA, Cotterell AJ, Le GT, Fairlie DP (2012) Modulating human proteinase activated receptor 2 with a novel antagonist (GB88) and agonist (GB110). Br J Pharmacol 165:1413–1423

Suidan HS, Niclou SP, Monard D (1996) The thrombin receptor in the nervous system. Semin Thromb Hemost 22:125–133

Takada M, Tanaka H, Yamada T, Ito O, Kogushi M, Yanagimachi M, Kawamura T, Musha T, Yoshida F, Ito M, Kobayashi H, Yoshitake S, Saito I (1998) Antibody to thrombin receptor inhibits neointimal smooth muscle cell accumulation without causing inhibition of platelet aggregation or altering hemostatic parameters after angioplasty in rat. Circ Res 82:980–987

Turk B (2006) Targeting proteases: successes, failures and future prospects. Nat Rev Drug Discov 5:785–799

Turnell AS, Brant DP, Brown GR, Finney M, Gallimore PH, Kirk CJ, Pagliuca TR, Campbell CJ, Michell RH, Grand RJ (1995) Regulation of neurite outgrowth from differentiated human neuroepithelial cells: a comparison of the activities of prothrombin and thrombin. Biochem J 308(Pt 3):965–973

Vecht CJ (1989) Nociceptive nerve pain and neuropathic pain. Pain 39(2):243–244
Vergnolle N (2000) Review article: proteinase-activated receptors—novel signals for gastrointestinal pathophysiology. Aliment Pharmacol Ther 14:257–266
Vergnolle N (2003) Proteinase-activated receptors and nociceptive pathways. Drug Dev Res 59:382–385
Vergnolle N, Bunnett NW, Sharkey KA, Brussee V, Compton SJ, Grady EF, Cirino G, Gerard N, Basbaum AI, Andrade-Gordon P, Hollenberg MD, Wallace JL (2001) Proteinase-activated receptor-2 and hyperalgesia: a novel pain pathway. Nat Med 7:821–826
Vu TK, Hung DT, Wheaton VI, Coughlin SR (1991) Molecular cloning of a functional thrombin receptor reveals a novel proteolytic mechanism of receptor activation. Cell 64(6):1057–1068
Wang H, Reiser G (2003) Thrombin signalling in the brain: the role of protease-activated receptors. Biol Chem 384:193–202
Wang S, Dai Y, Kobayashi K, Zhu W, Kogure Y, Yamanaka H, Wan Y, Zhang W, Noguchi K (2012) Potentiation of the P2X3 ATP receptor by PAR-2 in rat dorsal root ganglia neurons, through protein kinase-dependent mechanisms, contributes to inflammatory pain. Eur J Neurosci 36:2293–2301
Xiang Y, Masuko-Hongo K, Sekine T, Nakamura H, Yudoh K, Nishioka K, Kato T (2006) Expression of proteinase-activated receptors (PAR)-2 in articular chondrocytes is modulated by IL-1beta, TNF-alpha and TGF-beta. Osteoarthritis Cartilage 14:1163–1173
Zhu WJ, Yamanaka H, Obata K, Dai Y, Kobayashi K, Kozai T, Tokunaga A, Noguchi K (2005) Expression of mRNA for four subtypes of the proteinase-activated receptor in rat dorsal root ganglia. Brain Res 1041:205–211

Amygdala Pain Mechanisms

Volker Neugebauer

Contents

1 Pain-Related Amygdala Circuitry 262
2 Pain-Related Amygdala Plasticity 264
3 Pharmacology of Pain-Related Processing in the Amygdala 266
   3.1 Ionotropic Glutamate Receptors 266
   3.2 Metabotropic Glutamate Receptors 268
   3.3 GABA 272
   3.4 Neuropeptide CGRP 274
   3.5 Neuropeptide CRF 275
   3.6 Neuropeptide S 277
References 278

Abstract

A limbic brain area, the amygdala plays a key role in emotional responses and affective states and disorders such as learned fear, anxiety, and depression. The amygdala has also emerged as an important brain center for the emotional–affective dimension of pain and for pain modulation. Hyperactivity in the laterocapsular division of the central nucleus of the amygdala (CeLC, also termed the "nociceptive amygdala") accounts for pain-related emotional responses and anxiety-like behavior. Abnormally enhanced output from the CeLC is the consequence of an imbalance between excitatory and inhibitory mechanisms. Impaired inhibitory control mediated by a cluster of GABAergic interneurons in the intercalated cell masses (ITC) allows the development of glutamate- and neuropeptide-driven synaptic plasticity of excitatory inputs from

V. Neugebauer (✉)
Department of Pharmacology and Neuroscience, Center for Translational Neuroscience and Therapeutics, Texas Tech University Health Sciences Center, 3601 4th Street, Lubbock, TX 79430-6592, USA
e-mail: volker.neugebauer@ttuhsc.edu

© Springer-Verlag Berlin Heidelberg 2015
H.-G. Schaible (ed.), *Pain Control*, Handbook of Experimental Pharmacology 227,
DOI 10.1007/978-3-662-46450-2_13

the brainstem (parabrachial area) and from the lateral–basolateral amygdala network (LA-BLA, site of integration of polymodal sensory information). BLA hyperactivity also generates abnormally enhanced feedforward inhibition of principal cells in the medial prefrontal cortex (mPFC), a limbic cortical area that is strongly interconnected with the amygdala. Pain-related mPFC deactivation results in cognitive deficits and failure to engage cortically driven ITC-mediated inhibitory control of amygdala processing. Impaired cortical control allows the uncontrolled persistence of amygdala pain mechanisms.

Keywords
Amygdala • Pain • Plasticity • Neurotransmitter • mGluR • CGRP • CRF • NPS

Abbreviations

BLA	Basolateral amygdala
CB1	Cannabinoid receptor 1
CeA	Central nucleus of the amygdala
CeLC	Laterocapsular division of the central nucleus of the amygdala
ITC	Intercalated cell mass
LA	Lateral amygdala
mGluR	Metabotropic glutamate receptor
mPFC	Medial prefrontal cortex
NPS	Neuropeptide S
PB	Parabrachial area

Pain is a complex disorder with sensorimotor as well as emotional–affective and cognitive components (see Fig. 1). The amygdala, an almond-shaped brain area in the medial temporal lobe, plays an important role in the emotional–affective dimension of pain (Neugebauer et al. 2004, 2009) and, through interactions with cortical areas, also contributes to cognitive aspects such as pain-related decision-making deficits (Ji et al. 2010).

1 Pain-Related Amygdala Circuitry

The amygdala is closely interconnected with numerous cortical, subcortical, and brainstem areas. Figure 2 shows key amygdala nuclei and their connections relevant to sensory and pain-related processing. The lateral–basolateral nuclei (LA-BLA) form the input region for sensory, including nociceptive, information from thalamus (posterior areas) and cortical areas such as insula, anterior cingulate cortex, and

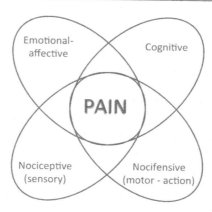

Fig. 1 Dimensions of pain

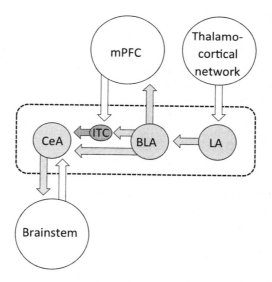

Fig. 2 Amygdala circuitry and interactions with cortical systems and brainstem. Input regions (lateral and basolateral amygdala, LA, BLA) process and transmit polymodal sensory and nociceptive information from thalamocortical systems to the amygdala output region (central nucleus, CeA) through direct excitatory projections or indirect feedforward inhibition involving interneurons in the intercalated cell mass (ITC). BLA forms close connections with mPFC that involve inhibitory interneurons in the cortex resulting in feedforward inhibition of mPFC principal cells. mPFC output neurons can engage ITC cells to control amygdala output. *Dashed line* indices area of the amygdala

other medial prefrontal cortical areas (Orsini and Maren 2012; Pape and Pare 2010; Marek et al. 2013; Price 2003). BLA projections of the medial prefrontal cortex (mPFC) provide emotion- and value-based information to guide executive functions such as decision-making and behavior control (McGaugh 2004; Holland

and Gallagher 2004; Laviolette and Grace 2006). The BLA contains neurons that respond preferentially to noxious stimuli (Ji et al. 2010).

Highly processed information generated in the LA-BLA network is transmitted to the central nucleus (CeA), which serves major amygdala output functions and projects to pain modulatory systems through forebrain and brainstem connections (Mason 2005; Neugebauer et al. 2004; Bourgeais et al. 2001; Price 2003) (Fig. 2). The laterocapsular division of the CeA (CeLC) receives nociceptive-specific information from the spinal cord and brainstem (external lateral parabrachial area) through the spino-parabrachio-amygdala pain pathway (Gauriau and Bernard 2002). The vast majority of CeLC neurons respond exclusively or predominantly to noxious stimuli and have large bilateral, mostly symmetrical receptive fields (Neugebauer et al. 2004, 2009). These CeLC neurons show non-accommodating spike firing properties characteristic of medium-size spine-laden peptidergic or GABAergic Type A projection neurons with targets in the brainstem, including PAG, and forebrain (Schiess et al. 1999; Jongen-Relo and Amaral 1998; Sun and Cassell 1993). Peptidergic (CRF or enkephalin containing) CeA projection neurons are innervated by calcitonin gene-related peptide (CGRP) containing terminals from the parabrachial area (Schwaber et al. 1988; Dobolyi et al. 2005; Harrigan et al. 1994), which is consistent with the peptidergic nature of the spino-parabrachio-amygdala pain pathway.

Interposed between LA-BLA and CeA is a cluster of inhibitory interneurons in the intercalated cell mass (ITC cells) which serve as a gate keeper to control amygdala output (Pape and Pare 2010; Likhtik et al. 2008; Marek et al. 2013; Jungling et al. 2008). ITC cells that inhibit CeA neurons are the target of excitatory projections from the infralimbic mPFC (McDonald 1998; Busti et al. 2011; Amir et al. 2011; Pinard et al. 2012) and are activated during behavioral extinction of negative emotional responses (Orsini and Maren 2012; Herry et al. 2010; Pape and Pare 2010; Sotres-Bayon and Quirk 2010). Pharmacological or mPFC-driven activation of ITC cells can also inhibit pain-related CeLC output and behaviors (Ren et al. 2013).

2 Pain-Related Amygdala Plasticity

Electrophysiological studies in anesthetized animals have consistently found increases in background and stimulus-evoked activity of individual CeLC neurons (Neugebauer and Li 2003; Li and Neugebauer 2004b, 2006; Ji and Neugebauer 2007, 2009; Ji et al. 2009) and BLA neurons (Ji et al. 2010) in a model of arthritic pain and in CeA neurons in a neuropathic pain model (Goncalves and Dickenson 2012) (Fig. 3). These "multireceptive" amygdala neurons are activated more strongly by noxious than innocuous stimuli and likely serve to integrate and evaluate sensory–affective information in the context of pain (Neugebauer et al. 2004). Biochemical and electrophysiological changes were observed only in the right amygdala in models of inflammatory pain (Ji and Neugebauer 2009; Carrasquillo and Gereau 2007, 2008). In neuropathic pain a transient activity

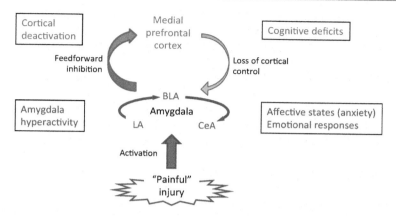

Fig. 3 Role of the amygdala in pain. Pain-producing events generate hyperactivity in the amygdala network of lateral, basolateral, and central nuclei (LA, BLA, CeA), which accounts for emotional–affective aspects of pain. Output from BLA deactivates medial prefrontal cortex through feedforward inhibition, resulting in cognitive deficits such as impaired decision-making. Decreased medial prefrontal cortical output to the amygdala allows the uncontrolled persistence of amygdala hyperactivity, hence persistence of pain

increase was observed in the left amygdala but a persistent change in the right amygdala (Goncalves and Dickenson 2012). Mechanisms of pain-related lateralization remain to be determined.

Neuronal activity changes in the amygdala are not simply a reflection of continued afferent input from spinal cord and other sources, but they result from an imbalance between excitatory and inhibitory synaptic mechanisms in the amygdala circuitry. Pain-related neuroplasticity in the amygdala has been established in electrophysiological and biochemical studies in brain slice preparations obtained from animals after the induction of different pain states, suggesting that brain changes persist at least in part independently of continued afferent input. Excitatory synaptic transmission to the CeLC is increased in acute models of arthritis (Neugebauer et al. 2003; Han et al. 2005; Bird et al. 2005; Fu and Neugebauer 2008; Fu et al. 2008), colitis (Han and Neugebauer 2004), and formalin-induced inflammation (Adedoyin et al. 2010) and in the spinal nerve ligation model of neuropathic pain (Ikeda et al. 2007; Nakao et al. 2012). Studies in the neuropathic pain model also reported right-hemispheric lateralization that depends on C-fiber-mediated inputs. Synaptic plasticity of parabrachial input to the CeLC is a consistent finding in various pain models, but there is also evidence for enhanced synaptic transmission at the LA-BLA (Ji et al. 2010) and BLA–CeLC synapses (Neugebauer et al. 2003; Fu and Neugebauer 2008; Ikeda et al. 2007; Ren and Neugebauer 2010; Ren et al. 2013).

Pain-related increased excitatory transmission and amygdala output can develop because inhibitory control mechanisms are impaired. Feedforward inhibition of CeLC neurons involves glutamatergic projections from BLA and mPFC to a cluster of GABAergic neurons in the intercalated cell masses (ITC cells). Decreased activation of this inhibitory gating mechanism in pain allows the development of glutamate- and neuropeptide-driven synaptic plasticity in the CeLC (Ren and Neugebauer 2010; Ren et al. 2013). A mechanism of impaired inhibition is the loss of cortical output (Ren et al. 2013) as the consequence of BLA hyperactivity that generates abnormally enhanced feedforward inhibition of principal cells in the mPFC (Ji et al. 2010; Ji and Neugebauer 2011). Failure to engage cortically driven ITC-mediated inhibitory control of amygdala processing may be a general mechanism of the abnormal persistence of emotional–affective states not just in pain (Apkarian et al. 2013; Ochsner and Gross 2005; Dalley et al. 2011).

Increased amygdala output as the result of neuroplasticity in the LA-BLA and CeLC has emerged as an important contributor to emotional–affective behaviors in animal pain models (Neugebauer et al. 2004, 2009). Decreasing amygdala activity with lesions or pharmacological interventions inhibits pain-related behaviors in different models (Han and Neugebauer 2005; Ren et al. 2013; Pedersen et al. 2007; Han et al. 2005; Manning 1998; Fu and Neugebauer 2008; Fu et al. 2008; Palazzo et al. 2008; Ji et al. 2010; Hebert et al. 1999). Importantly, increasing activity in the amygdala exogenously can exacerbate or generate pain responses under normal conditions in the absence of any tissue pathology (Han et al. 2010; Carrasquillo and Gereau 2007; Kolber et al. 2010; Qin et al. 2003; Myers et al. 2005; Myers and Greenwood-Van Meerveld 2010; Li et al. 2011; Ji et al. 2013). Increased amygdala activity is now also well documented in experimental and clinical pain conditions in humans (Liu et al. 2010; Baliki et al. 2008; Tillisch et al. 2010; Kulkarni et al. 2007; Simons et al. 2012).

3 Pharmacology of Pain-Related Processing in the Amygdala

The contributions of amino acid neurotransmitters (glutamate and GABA) and neuropeptides in the amygdala network relevant to pain-related processing are summarized in Fig. 4. A discussion of opioid function in the amygdala is beyond the scope of this article. There is strong evidence for a critical role of the amygdala, and the CeA in particular, in opioid-dependent pain modulation (Manning and Mayer 1995a, b; Zhang et al. 2013; Fields 2000; Manning 1998) and reward mechanisms (Bie et al. 2012; Cai et al. 2013), and μ-, κ-, and δ-opioid agonists can have direct effects on CeA amygdala neurons (Zhu and Pan 2004; Bie et al. 2009). However their role in pain-related amygdala processing and plasticity remains to be determined.

3.1 Ionotropic Glutamate Receptors

The CeLC receives glutamatergic inputs from the LA-BLA network and from brainstem areas (parabrachial input has been studied most extensively with regard to amygdala pain mechanisms). Activation of NMDA and non-NMDA receptors is

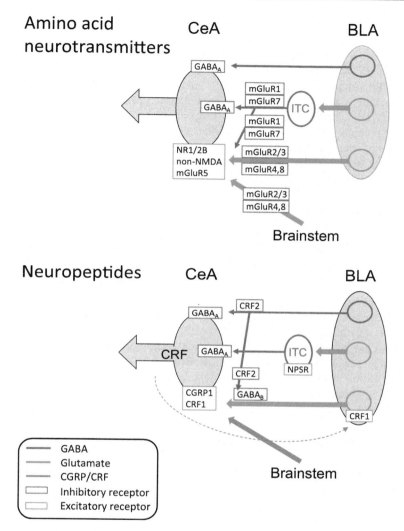

Fig. 4 Pharmacology of pain-related amygdala changes. Glutamatergic/GABAergic and peptidergic mechanisms are depicted in separate diagrams for the sake of clarity. *BLA* basolateral amygdala, *CeA* central nucleus of the amygdala, *ITC* intercalated cells, *mGluR* metabotropic glutamate receptor, *NR1/2B* NMDA receptor subunits NR1 and NR2B, *NPSR* neuropeptide S receptor. *Dashed blue line* indicates intra-amygdalar CRF release from CeA neurons

required for the generation of hyperactivity of CeLC neurons in the arthritis pain model, because administration of antagonists for NMDA (AP5) or non-NMDA (NBQX) receptors into the CeA 6 h postinduction inhibited the increased background activity and evoked responses of these neurons in anesthetized animals (Li and Neugebauer 2004a). In the same experimental in vivo preparation, an

NR2B-selective antagonist (Ro-256981) showed a different profile, inhibiting evoked but not background activity (Ji et al. 2009). In brain slices obtained from arthritic rats 6 h postinduction (acute stage), NMDA receptor-mediated transmission at the PB-CeLC synapse is increased through PKA-dependent phosphorylation of the NR1 subunit, and AP5 blocks the synaptic potentiation (Bird et al. 2005). In the subacute stage of the formalin pain model (24 h postinduction), NMDA receptors contribute to enhanced transmission at the PB-CeLC synapse (the NMDA receptor antagonist CPP inhibited transmission), but the NMDA/AMPA ratio is not increased compared to controls (Adedoyin et al. 2010). In contrast, synaptic plasticity of PB inputs to CeLC neurons in the spinal nerve ligation model of neuropathic pain (6–7 days postinduction) largely depends on non-NMDA but not NMDA receptors (Ikeda et al. 2007).

While the electrophysiological studies may suggest that NMDA receptors are more important for the induction rather than maintenance phase in models of inflammatory but not neuropathic pain, behavioral data argue against pain type- or stage-specific contributions. Bilateral intra-CeA injection of an NMDA receptor antagonist (MK-801) inhibited nocifensive (hind limb withdrawal reflex) and affective (place avoidance test) behaviors in the spared nerve injury model of neuropathic pain (Ansah et al. 2010). Even in normal animals, vocalization afterdischarges evoked by a noxious footshock (a measure of pain affect) were inhibited by antagonists for NMDA (AP5) or non-NMDA (CNQX) receptors administered into the CeA, whereas spinal reflexes (tail flick and hind limb movements) were unaffected (Spuz and Borszcz 2012). In this study bilateral administration of AP5 or CNQX was more effective than unilateral administration into either hemisphere.

3.2 Metabotropic Glutamate Receptors

The family of G-protein-coupled metabotropic glutamate receptors (mGluRs) comprises eight subtypes with splice variants, which can be classified into three groups (I–III) based on their sequence homology, signal transduction mechanism, and pharmacological profile.

Group I consists of mGluR1 and mGluR5 subtypes that couple to $G_{q/11}$ proteins to activate phospholipase C, resulting in the formation of IP3 to stimulate intracellular calcium release, generation of diacylglycerol (DAG) activating PKC, and activation of MAP kinases such as ERK (Nicoletti et al. 2011; Ferraguti et al. 2008; Niswender and Conn 2010). Electrophysiological, biochemical, and behavior data support an important role of mGluR1 and mGluR5 in pain-related amygdala neuroplasticity and amygdala-dependent behaviors. Antagonists for mGluR1 (CPCCOEt) and mGluR5 (MPEP) inhibited increased evoked responses of CeLC neurons recorded in anesthetized rats in a model of arthritis pain (Li and Neugebauer 2004b). MPEP, but not CPCCOEt, also inhibited enhanced background activity in the arthritis model and had inhibitory effects under normal conditions (Li and Neugebauer 2004b). Underlying mechanisms involve presynaptic mGluR1-mediated increase in excitatory and decrease in inhibitory transmission and postsynaptic mGluR5-mediated effects based on electrophysiological analyses in brain slices from arthritic rats (Ren and Neugebauer 2010; Neugebauer et al. 2003). In these studies selective mGluR1 antagonists (CPCCOEt and LY367385) inhibited excitatory but facilitated inhibitory transmission only in arthritis, whereas selective mGluR5 antagonists (MPEP and MTEP) inhibited synaptic transmission both under normal conditions and in arthritis.

Behavioral studies support an important role of amygdalar group I mGluRs in models of formalin-induced, arthritis, visceral, and neuropathic pain. Blockade of mGluR1 in the CeA with CPCCOEt inhibited stimulus-evoked audible and ultrasonic vocalizations and spinal reflexes that were increased in the kaolin/carrageenan arthritis pain model, whereas blockade of mGluR5 in the CeA with MPEP inhibited vocalizations but not spinal hind limb withdrawal reflexes. Vocalizations organized in limbic forebrain areas and particularly in the amygdala (measured as vocalization afterdischarges) were inhibited by both CPCCOEt and MPEP. Vocalizations during stimulation, which are organized at the brainstem level, were inhibited by CPCCOEt but not MPEP (Han and Neugebauer 2005). The results may suggest that mGluR1 and mGluR5 contribute to affective behaviors generated in the amygdala, whereas mGluR1, but not mGluR5, also regulate amygdala-dependent modulation of "sensory" nocifensive behaviors by descending brainstem systems. A similar pattern of differential contributions of mGluR1 and mGluR5 to amygdala functions was found in the spared nerve injury model of neuropathic pain. Application of antagonists for mGluR1 (CPCCOEt) or mGluR5 (MPEP) into the CeA inhibited aversive behaviors in the place avoidance test. CPCCOEt, but not MPEP, also inhibited hind limb withdrawal reflexes (Ansah et al. 2010). These studies were done in rats. Disruption of mGluR5 function in the CeLC of mice pharmacologically or with a conditional knockout approach inhibited paw withdrawal reflexes in the formalin pain model (Kolber et al. 2010) and visceromotor reflexes in the bladder distension pain model (Crock et al. 2012).

A possible conclusion emerging from these findings would be that mGluR1 contribute in a broader way to any type of pain behaviors generated or modulated by the amygdala, whereas mGluR5 function switches over time (in arthritis and neuropathic pain models) towards a preferential contribution to affective forebrain-mediated pain mechanisms. The electrophysiological data support this scenario because the presynaptic function of mGluR1 could affect the drive onto various amygdala output neurons, whereas the postsynaptic action of mGluR5 allows for the discrete activation of subsets of CeA neurons.

Under normal conditions, mGluR5 appears to mediate the facilitatory effects of group I mGluR activation in the CeA on vocalizations and spinal reflexes in the awake animal, on neuronal activity in the anesthetized preparation, and on synaptic transmission in amygdala brain slices (Neugebauer et al. 2003; Li et al. 2011; Ji and Neugebauer 2010; Li and Neugebauer 2004b). In these studies, the effects of an mGluR1/5 agonist (DHPG) were mimicked by a presumed mGluR5-selective agonist (CHPG) and/or blocked by a selective mGluR5 antagonist (MTEP). A selective mGluR1 antagonist (LY367385) inhibited the facilitatory effects of

DHPG only on neuronal responses to visceral stimuli (colorectal distention) (Ji and Neugebauer 2010). The facilitatory effects of mGluR5 involved IP3-mediated calcium release to increase mitochondrial ROS production resulting in the activation of ERK1/2 and PKA, but not PKC. Inhibition of ERK (U0126) and PKA (KT5720) was necessary to block completely the excitatory effects of a ROS donor (tBOOH); a PKC inhibitor (GF109203X) had no effect (Li et al. 2011). ERK activation downstream of mGluR5 is also supported by studies in mice. DHPG administered into the CeA induced mechanical (decreased paw withdrawal thresholds) and visceral (response to bladder distension) hypersensitivity under normal conditions. The facilitatory effect was reduced by an mGluR5 antagonist (MPEP) or mGluR5 knockdown and was accompanied by ERK1/2 activation in the CeA (Kolber et al. 2010) and in the spinal cord (Crock et al. 2012).

Group II comprises mGluR2 and mGluR3 subtypes that couple negatively to adenylyl cyclase, cAMP, and PKA activation via G_i proteins (Nicoletti et al. 2011; Niswender and Conn 2010). Typically localized extrasynaptically they serve as presynaptic auto- or hetero-receptors (Niswender and Conn 2010). Pharmacological activation of group II mGluRs in the CeA with mGluR2/3 agonists (LCCG1 or LY354740) decreased the responses of CeLC neurons under normal conditions and in the kaolin-/carrageenan-induced knee joint arthritis pain model (Li and Neugebauer 2006). There was an increase in potency for inhibition of responses to noxious stimulation of the arthritic knee in the pain model but not for inhibition of responses to innocuous stimulation of the knee or of the intact ankle or background activity, suggesting perhaps an input- and activity-dependent effect, because noxious stimulation of the arthritic knee would likely generate the strongest input to the amygdala. A group II mGluR antagonist (EGLU) had no effect under normal conditions but increased the responses to stimulation of the knee in the arthritis pain state. This would be consistent with activity-dependent endogenous activation and gain of function of group II mGluRs in pain.

Electrophysiological analysis in brain slices showed that group II mGluRs act presynaptically to modulate synaptic plasticity in the amygdala in a model of arthritic pain (Han et al. 2006). A selective group II mGluR agonist (LY354740) inhibited excitatory transmission at the PB-CeLC synapse in slices from arthritic rats more potently than in controls without affecting neuronal excitability. A group II mGluR antagonist (EGLU) but not a $GABA_A$ receptor antagonist (bicuculline) reversed the inhibitory effect of LY354740. EGLU had no effect on its own, suggesting that the endogenous activation of group II mGluRs in the in vivo condition (see previous paragraph) may be due to inputs not preserved or active in the brain slice preparation. The pain-related function of group II mGluRs at the BLA–CeLC synapse remains to be determined, but group II mGluR agonists (LCCG1 and LY354740) inhibited excitatory transmission from BLA to CeLC under normal conditions, and the inhibitory effect persisted in the presence of antagonists for mGluR1 (CPCCOEt), mGluR5 (SIB-1893), $GABA_A$ (picrotoxin), and $GABA_B$ (CGP 55845) (Neugebauer et al. 2000).

The relative contribution of mGluR2 and 3 was examined in amygdala brain slices using an inhibitor (ZJ43) of the peptide neurotransmitter

N-acetylaspartylglutamate (NAAG) that activates preferentially mGluR3 (Adedoyin et al. 2010). ZJ43 inhibited excitatory transmission at the PB-CeLC synapse more strongly than a group II mGluR agonist that does not distinguish mGluR2 and 3 (SLx-3095-1), suggesting a role for endogenous NAAG and perhaps a predominant contribution of mGluR3 under normal conditions. Both effects were blocked by a group II mGluR antagonist (LY341495). In the formalin pain model (24 h postinduction), ZJ43 was much less effective than SLx-3095-1 in reducing excitatory synaptic transmission, which would be consistent with a loss of function of NAAG or an increase in the contribution of mGluR2 relative to mGluR3. Electrophysiological effects correlated with behavior because systemic injection of ZJ43 blocked formalin pain-related synaptic plasticity in the CeLC as well as mechanical allodynia (Adedoyin et al. 2010).

Group III consists of mGluR4, 6, 7, and 8 subtypes. With the exception of mGluR6, which is found only in retinal ON-bipolar cells, they couple negatively to adenylyl cyclase, cAMP and PKA activation via G_i proteins like group II mGluRs (Nicoletti et al. 2011; Niswender and Conn 2010). Predominantly presynaptic receptors, mGluR4 and mGluR8 have high affinity for glutamate and are localized extrasynaptically on glutamatergic terminals, whereas low affinity mGluR7 is found in or near the active zone of the synapse on GABAergic terminals (Niswender and Conn 2010).

Evidence from electrophysiological studies using the prototypical broad spectrum agonist (LAP4) suggests that group III mGluRs can inhibit pain-related amygdala neuroplasticity. LAP4 inhibited evoked responses of CeLC neurons recorded in anesthetized rats more potently in the arthritis pain state than under normal conditions, whereas the inhibitory effect on background activity did not change (Li and Neugebauer 2006). A group III mGluR antagonist (UBP1112) had no effect under normal conditions but facilitated the evoked responses in arthritic rats. Electrophysiological analysis in brain slices showed that LAP4 decreased excitatory transmission at the PB-CeLC synapse more potently in the arthritis pain model than in control slices, and the inhibitory effects involved a presynaptic site of action (Han et al. 2004). UBP1112 reversed the inhibitory effect of LAP4 but had no effect on its own, which is different from the in vivo situation where the facilitatory effects of the antagonist suggest the possibility of endogenous receptor activation in the pain model. A likely explanation would be the requirement of amygdala inputs not present or active in the slice preparation. A group III agonist (LAP4) also inhibited transmission at the BLA–CeLC synapse under normal conditions, but pain-related function of group III mGluRs at this synapse remains to be determined.

The recent availability of subtype selective agonists allowed a more detailed analysis that showed opposing functions of mGluR7 and mGluR8 in the amygdala related to pain processing and modulation (Palazzo et al. 2008; Ren et al. 2011). Activation of mGluR7 with AMN082 increased CeLC output (synaptically evoked spiking) in brain slices from control rats but not from arthritic rats (Ren et al. 2011). AMN082 acted presynaptically to inhibit glutamate-driven synaptic inhibition of CeLC neurons (feedforward inhibition from BLA), but not monosynaptic inhibitory transmission, implicating BLA-activated GABAergic interneurons such as intercalated cells (ITC; see Sect. 4.2). In support of this interpretation, the effect of AMN082 was occluded in the presence of bicuculline to block $GABA_A$ receptors. The mGluR7-mediated disinhibition facilitated the excitatory drive of CeLC neurons. Thus, mGluR7 would act as a gatekeeper to regulate information flow to the CeLC by releasing GABAergic control, hence permitting excitatory inputs from the LA-BLA network to reach the CeLC under normal conditions. In contrast, a selective mGluR8 agonist (DCPG) inhibited excitatory transmission at the BLA–CeLC synapse and output of CeLC neurons more potently in brain slices from arthritic rats than under normal conditions. The mechanism was presynaptic on glutamatergic terminals and did not involve GABAergic modulation; DCPG had no effect on inhibitory transmission (Ren et al. 2011). The direct inhibitory effect of mGluR8 and the disinhibition by mGluR7 could be blocked with a group III antagonist (MAP4).

Behavioral data are consistent with these electrophysiological findings (Palazzo et al. 2008). Activation of mGluR7 in the CeA of normal rats with AMN082 facilitated spinal reflexes (withdrawal thresholds) and supraspinally organized affective responses (vocalizations); AMN082 also increased anxiety-like behavior (decreased open-arm preference in the elevated plus maze). The facilitatory effects of AMN082 were not detected in arthritic rats showing increased pain behaviors. Intra-CeA administration of an mGluR8 agonist (DCPG) had no effect in normal animals but inhibited the increased spinal reflexes and vocalizations of arthritic rats and had anxiolytic-like effects in these animals tested in the elevated plus maze (Palazzo et al. 2008). DCPG also decreased thermal hypersensitivity in a carrageenan-induced hindpaw inflammation model and the antinociceptive effect was blocked with a group III mGluR antagonist (MSOP) (Palazzo et al. 2011). Furthermore, an increase in mGluR8 gene, protein, and staining, the latter being associated with vesicular GABA transporter-positive profiles, has been found in the CeA after carrageenan-induced inflammatory pain. These results show that stimulation of mGluR8, which was overexpressed within the CeA in inflammatory pain conditions, inhibits nociceptive behavior. Such an effect is associated with an increase in 5-HT and Glu release, a decrease in GABA, and the inhibition of ON- and the stimulation of OFF-cell activities within RVM.

The results suggest that under normal conditions mGluR7, but not mGluR8, facilitates pain responses and has anxiogenic-like properties. In pain models, however, mGluR8, but not mGluR7, has inhibitory behavioral effects (Palazzo et al. 2008, 2011), and this functional change may involve upregulation of mGluR8 gene and protein expression (Palazzo et al. 2011).

3.3 GABA

The amygdala is rich in GABAergic neurons and GABA receptors, allowing the control of amygdala output through direct inhibition, feedforward inhibition, and disinhibition. Recent work has focused on a cluster of inhibitory interneurons in the intercalated cell mass (ITC cells) positioned between BLA and CeLC (Ren et al. 2013).

Evidence from electrophysiological studies in brain slices suggests that GABAergic transmission can control excitatory inputs and synaptically evoked outputs of CeLC neurons (Fu and Neugebauer 2008; Ren et al. 2011, 2013; Ren and Neugebauer 2010). However, $GABA_A$ receptor-mediated synaptic inhibition is lost or impaired in a model of arthritic pain. Decreased monosynaptic (Ren et al. 2011) and glutamate-driven disynaptic inhibition (feedforward inhibition sensitive to NBQX) (Ren and Neugebauer 2010; Ren et al. 2013) of CeLC neurons was found in brain slices from arthritis rats compared to controls. Feedforward inhibition can be evoked by stimulation of BLA output (Ren and Neugebauer 2010) or of medial prefrontal cortical fibers in the external capsule (Ren et al. 2013). ITC cells have been implicated in feedforward inhibition of CeLC neurons because pharmacological activation of ITC cells with neuropeptide S (NPS) (see Sect. 5) or high-frequency stimulation of excitatory external capsule input to ITC cells inhibited synaptic activation of CeLC neurons (Ren et al. 2013). These inhibitory inputs exert a tonic GABAergic tone on CeLC neurons under normal conditions but not in the arthritis pain model. Blockade of $GABA_A$ receptors with bicuculline in the CeLC under normal conditions facilitated excitatory transmission at the BLA–CeLC synapse through an indirect action in the network (Ren and Neugebauer 2010) and also increased synaptically evoked action potential firing, a measure of neuronal output (Ren et al. 2011). In brain slices from arthritic rats, however, bicuculline had no significant effect on excitatory synaptic responses of CeLC neurons, suggesting a loss of GABAergic inhibitory control that may contribute to the pain-related increase of excitatory transmission (Ren and Neugebauer 2010).

Behavioral data somewhat agree with the electrophysiological findings. In a model of neuropathic pain chronic constriction injury, bicuculline administered into the CeA had no effect on mechanical allodynia and hyperalgesia, but attenuated, rather than facilitated, affective pain behaviors (escape/avoidance test). The data would be consistent with a loss, and possibly even reversal, of tonic GABAergic inhibitory control (Pedersen et al. 2007). Exogenous activation of $GABA_A$ receptors in the CeA with muscimol, however, had antinociceptive effects and attenuated escape/avoidance behaviors, which would indicate the presence of functional $GABA_A$ receptors. Neurochemical studies found no evidence for significant changes of extracellular GABA levels in the BLA in the formalin pain model (Rea et al. 2009) and in the CeA in a model of carrageenan-induced hindpaw inflammation (Palazzo et al. 2011). In these studies, baseline rather than evoked release was studied. Electrophysiological data implicated mGluR1 in the reduced or lost GABAergic inhibition (see Sect. 4.1 Group I). The mechanisms of impaired GABAergic control of amygdala function remain to be determined.

3.4 Neuropeptide CGRP

CGRP is a 37-amino-acid peptide that binds to G-protein-coupled receptors, including CGRP1, which couple positively to adenylyl cyclase, cyclic AMP formation, and protein kinase A (PKA) activation (Wimalawansa 1996; Poyner et al. 2002; Van Rossum et al. 1997). Functional CGRP1 receptors are formed by a heterodimeric complex of the calcitonin receptor-like receptor (CRLR) and receptor activity-modifying protein 1 (RAMP1) (Robinson et al. 2009; McLatchie et al. 1998). Particularly high levels of CGRP (de Lacalle and Saper 2000; Schwaber et al. 1988; Kruger et al. 1988; Dobolyi et al. 2005), CGRP binding sites (Wimalawansa 1996; Van Rossum et al. 1997), and proteins (CRLR and RAMP1) required for functional CGRP1 receptors (Ma et al. 2003; Oliver et al. 2001) have been described in the amygdala (CeA). CGRP can interact with other receptors (Robinson et al. 2009; Hay 2007; Poyner et al. 2002) such as a putative CGRP2 receptor that may include RAMP2 or RAMP3 rather than RAMP1 (Hay 2007). Whereas RAMP1 dominates in the CeA, the BLA contains relatively more RAMP2 than RAMP1 (Oliver et al. 2001). The exclusive source of CGRP in the amygdala is the lateral parabrachial area, and CGRP-immunoreactive parabrachial fibers essentially delineate the CeLC (Schwaber et al. 1988; de Lacalle and Saper 2000; Kruger et al. 1988; Dobolyi et al. 2005), making CGRP a marker of parabrachial inputs to the "nociceptive amygdala."

Electrophysiological and behavioral data show an important role of CGRP and CGRP1 receptors in the CeA in pain-related neuroplasticity and behaviors. Administration of selective CGRP1 receptor antagonists (CGRP$_{8-37}$ and BIBN4096BS) into the CeA inhibited increased responses of CeLC neurons to mechanical stimulation of the arthritic knee and non-injured ankle in anesthetized rats with a kaolin-/carrageenan-induced knee joint arthritis (Han et al. 2005). The antagonists were more efficacious in the pain model than under normal conditions. In amygdala brain slices from arthritic rats, CGRP1 receptor antagonists inhibited synaptic plasticity of parabrachial inputs to the CeLC but had no significant effect under normal conditions (Han et al. 2005). Detailed electrophysiological analyses showed that CGRP1 receptors contribute to pain-related plasticity through a protein kinase A (PKA)-dependent postsynaptic mechanism that involves NMDA, but not AMPA, receptors. Pharmacological blockade of CGRP1 receptors in the CeA with CGRP$_{8-37}$ inhibited spinal hind limb withdrawal reflexes and supraspinally organized pain behaviors (audible and ultrasonic vocalizations) of awake arthritic rats (Han et al. 2005). CGRP$_{8-37}$ had no effect on these behaviors in normal animals without arthritis.

Under normal conditions CGRP in the CeA can facilitate synaptic transmission and generate nocifensive and affective pain behaviors (Han et al. 2010). In brain slices from normal rats, CGRP increased excitatory transmission of parabrachial inputs to CeLC neurons through a postsynaptic mechanism and also increased neuronal excitability. CGRP-induced synaptic facilitation was reversed by an NMDA receptor antagonist (AP5) or a PKA inhibitor (KT5720), but not by a PKC inhibitor (GF109203X). Stereotaxic administration of CGRP into the CeLC of awake rats increased audible and ultrasonic vocalizations and decreased hind limb withdrawal thresholds (Han et al. 2010). Behavioral effects of CGRP were largely blocked by KT5720 but not GF109203X. Electrophysiological and behavioral effects of CGRP were blocked by a CGRP1 receptor antagonist (CGRP$_{8-37}$). The data suggest that CGRP in the amygdala exacerbates pain behaviors under normal conditions through a PKA and NMDA receptor-dependent direct action on CeLC neurons.

In contrast, antinociceptive effects of CGRP have been reported in the BLA of normal animals (Li et al. 2008). Hindpaw withdrawal latencies to noxious thermal and mechanical stimulations increased significantly after intra-BLA administration of CGRP, and the antinociceptive effect was blocked by CGRP$_{8-37}$. CGRP receptor composition in different nuclei of the amygdala may explain the differential effects of CGRP. CGRP could also activate inhibitory projections from the BLA to the CeLC (see Sect. 4.2). Neuronal effects of CGRP in the BLA remain to be determined.

3.5 Neuropeptide CRF

Corticotropin-releasing factor (CRF) is not only a "stress hormone" but also a neuromodulator outside the hypothalamic-pituitary-adrenocortical (HPA) axis, and the amygdala is a major site of extrahypothalamic expression of CRF and its receptors (Charney 2003; Gray 1993; Koob 2010; Asan et al. 2005; Sanchez et al. 1999; Takahashi 2001; Tache and Bonaz 2007; Bale and Vale 2004; Reul and Holsboer 2002; Hauger et al. 2009). CRF receptors can couple to a number of G-proteins to activate a variety of intracellular signaling pathways, and PKA and PKC appear to play particular important roles (Blank et al. 2003). Sources of CRF in the amygdala are CeA neurons as well as afferents from the lateral hypothalamic area and dorsal raphe nucleus (Uryu et al. 1992; Commons et al. 2003). CRF containing neurons in the CeA are innervated by CGRP containing terminals from the parabrachial area (Schwaber et al. 1988; Dobolyi et al. 2005; Harrigan et al. 1994) and project to widespread regions of the basal forebrain and brain stem (Gray 1993).

Electrophysiological studies showed an important contribution of endogenously activated CRF1 receptors to neuroplasticity in the CeA and BLA in a model of arthritic pain (Ji and Neugebauer 2007, 2008; Fu and Neugebauer 2008). In anesthetized rats, a selective CRF1 receptor antagonist (NBI27914) administered into the CeA inhibited the evoked responses and background activity of CeLC neurons in an arthritis pain model (Ji and Neugebauer 2007). Administration of a NBI27914 into the BLA, but not CeA, also inhibited the increased background and evoked activity of BLA neurons in the arthritis model (Ji et al. 2010). In contrast, a selective CRF2 receptor antagonist (astressin-2B) had facilitatory effects on CeLC neurons under normal conditions but not in the pain model (Ji and Neugebauer 2007).

Patch-clamp analysis in brain slices found differential pre- and postsynaptic mechanisms of CRF1 and CRF2 receptor-mediated effects in the amygdala. In brain slices from arthritic rats, NBI27914 inhibited enhanced synaptic inputs to CeLC neurons from the parabrachial area and the BLA (Fu and Neugebauer 2008) as well as synaptic plasticity in BLA neurons (Ji et al. 2010). NBI27914 had no significant effect on CeLC or BLA neurons under normal conditions. The synaptic effects involved a postsynaptic mechanism and PKA-dependent inhibition of an NMDA receptor-mediated synaptic component. NBI27914 also decreased neuronal excitability by inhibiting ion channels important for action potential repolarization such as Kv3-type potassium channels. In contrast, a CRF2 receptor antagonist (astressin-2B) had no effect on neuronal excitability but facilitated excitatory transmission at the parabrachial–CeLC and BLA–CeLC synapses through presynaptic inhibition of GABAergic transmission (disinhibition) (Fu and Neugebauer 2008). Astressin-2B inhibited $GABA_A$ receptor-mediated inhibitory monosynaptic transmission from BLA to CeLC, whereas NBI27914 had no effect on inhibitory transmission. In brain slices from normal rats neither antagonist had no effects on basal synaptic transmission (Fu and Neugebauer 2008).

In conclusion, endogenous CRF1 receptor activation in the amygdala contributes to pain-related synaptic plasticity and hyperactivity through a postsynaptic mechanism, whereas presynaptic CRF2 receptor-mediated inhibitory function is lost in the arthritis pain model. Studies in whole animals (Ji and Neugebauer 2007) and in brain slice preparations (Fu and Neugebauer 2008) are largely in agreement. Discrepancies regarding CRF2 receptor function under normal conditions may be due to differences in the availability of extra-amygdaloid inputs providing endogenous ligands and tonic inhibition required for CRF2 receptor blockade to have a detectable effect.

Behavioral evidence suggests that pain-related endogenous activation of CRF1 receptors in the amygdala contributes to pain modulation and pain affect. Blockade of CRF1 receptors in the CeA with NBI27914 inhibited pain-related behaviors (audible and ultrasonic vocalizations and hind limb withdrawal reflexes) and anxiety-like behaviors in a model of arthritic pain but had no effect in normal animals (Ji et al. 2007; Fu and Neugebauer 2008). CRF2 receptor-mediated inhibition does not reach behavioral significance since astressin-2B had no significant effect on pain behaviors (Fu and Neugebauer 2008). In a neuropathic pain model (spared nerve injury) a nonselective CRF receptor antagonist (CRF_{9-41}) had no effect on emotional–affective (aversive place-conditioning test) and nocifensive (hind limb withdrawal thresholds) pain behaviors (Bourbia et al. 2010). Increasing endogenous CRF in the CeA with a CRF-binding protein inhibitor (CRF_{6-33}) had mixed effects, facilitating nocifensive responses while attenuating emotional–affective behaviors (Bourbia et al. 2010). Microinjections of a CRF receptor antagonist (CRF_{9-41}) into the CeA reduced hyperalgesia (tail flick test) associated with morphine withdrawal but had no effect in normal controls (McNally and Akil 2002).

Pain-related changes of CRF function in the amygdala are supported by biochemical data. CRF and CRF mRNA are increased in CeA neurons in the chronic constriction injury model of neuropathic pain independently of HPA axis activation (Rouwette et al. 2011; Ulrich-Lai et al. 2006). Increased CRF mRNA expression in the CeA was also found in a colitis model of visceral pain (Greenwood-Van Meerveld et al. 2006).

While there is little evidence for endogenous activation of the CRF system in the amygdala under normal conditions (Ji and Neugebauer 2007; Fu and Neugebauer 2008), exogenously administered CRF can have CRF1 receptor-mediated facilitatory and CRF2 receptor-mediated inhibitory effects in the CeA (Ji and Neugebauer 2008; Ji et al. 2013). In anesthetized normal rats CRF increased background and evoked activity of CeLC neurons but decreased neuronal activity at higher concentrations (Ji and Neugebauer 2008). Facilitatory effects of CRF were blocked by a selective CRF1 receptor antagonist (NBI27914) but not a CRF2 receptor antagonist (astressin-2B) and by a PKA, but not PKC, inhibitor. Inhibitory effects of CRF were reversed by astressin-2B. In brain slices from normal rats without injury, CRF increased excitatory transmission at the parabrachial–CeLC synapse and also neuronal output (synaptically evoked spiking) through a postsynaptic PKA-dependent action (Ji et al. 2013). The CRF effects were blocked by NBI27914, but not astressin-2B, and by an inhibitor of PKA (KT5720), but not PKC (GF109203x). CRF increased a latent NMDA receptor-mediated synaptic component through CRF1 receptor-mediated PKA activation. Thus CRF can induce changes resembling pain-related plasticity of CeLC neurons that involves PKA-dependent NR1 subunit phosphorylation and increased NMDA receptor-mediated synaptic transmission (Bird et al. 2005; Han et al. 2005).

Behavioral consequences of non-pain-related activation of CRF1 receptors in the amygdala (CeA) are increased audible and ultrasonic vocalizations and decreased hind limb withdrawal thresholds (Ji et al. 2013). In agreement with the electrophysiological findings, behavioral effects of CRF were blocked by NBI27914 and KT5720 but not GF109203x. Importantly, CRF effects persisted when HPA axis function was suppressed by pretreatment with dexamethasone (subcutaneously). It is conceivable that conditions of increased amygdala CRF levels can contribute to pain in the absence of tissue pathology or disease state.

3.6 Neuropeptide S

Recently discovered NPS selectively enhances ITC-dependent feedforward inhibition of CeA neurons to produce powerful anxiolytic effects (Jungling et al. 2008). NPS acts on a Gq-/Gs-coupled receptor (NPSR) to increase intracellular calcium and cAMP-PKA signaling (Reinscheid 2008; Guerrini et al. 2010). The amygdala is one of the brain areas with the strongest expression of NPSR (Leonard and Ring 2011). In the rat amygdala, the highest level of NPSR mRNA is found in and around ITC cells but not in other elements of the pain-related amygdala circuitry (Xu et al. 2007).

Electrophysiological and behavioral data suggest an important role for NPS in the control of pain-related amygdala output and affective behaviors through a direct action on inhibitory ITC cells (Ren et al. 2013). In brain slices from arthritic rats, NPS inhibited the enhanced excitatory drive of CeLC neurons from BLA. The inhibitory effect of NPS was not due to a direct postsynaptic action on CeLC neurons but involved a presynaptic, action potential-dependent network mechanism. In fact, NPS increased feedforward inhibition of excitatory drive and output of CeLC neurons by activating GABAergic ITC neurons. The cellular mechanisms by which feedforward inhibition controls CeLC output remain to be determined. Feedforward inhibition was generated by cortical (mPFC) fiber activation in the external capsule. NPS increased excitatory drive and synaptically evoked output of ITC cells through a PKA-dependent facilitatory postsynaptic NPSR-mediated action. A selective NPSR antagonist ([D-Cys(tBu)^5]NPS) blocked the electrophysiological effects of NPS but had no effect on its own (Ren et al. 2013).

As a consequence of synaptic inhibition of amygdala output NPS inhibited pain behaviors in an arthritis pain model (Ren et al. 2013). Administration of NPS into the ITC, but not CeLC, inhibited vocalizations and anxiety-like behavior in arthritic rats. A selective NPS receptor antagonist ([D-Cys(tBu)^5]NPS) blocked the behavioral effects of NPS but had no effect on its own. These findings are in line with previous reports that intracerebroventricular administration of NPS had anxiolytic (Ruzza et al. 2012; Xu et al. 2004; Jungling et al. 2008) and antinociceptive (Li et al. 2009; Peng et al. 2010) effects. Intracerebroventricular NPS attenuated nociceptive behaviors (paw licking) in the mouse formalin pain model (Peng et al. 2010) but also had antinociceptive effects on the tail withdrawal and hot-plate tests in normal mice (Li et al. 2009). NPSR antagonists ([D-Val^5]NPS and [D-Cys(tBu)^5]NPS) blocked the effects of NPS but had no effect on their own. In conclusion, the amygdala is an important site for pain-inhibiting effects of NPS through a mechanism that involves activation of feedforward inhibition of CeA neurons through a PKA-dependent postsynaptic action on inhibitory ITC cells.

Acknowledgements Work in the author's lab is supported by National Institute of Neurological Disorders and Stroke Grants NS-081121, NS-38261, and NS-11255.

References

Adedoyin MO, Vicini S, Neale JH (2010) Endogenous N-acetylaspartylglutamate (NAAG) inhibits synaptic plasticity/transmission in the amygdala in a mouse inflammatory pain model. Mol Pain 6:60–77

Amir A, Amano T, Pare D (2011) Physiological identification and infralimbic responsiveness of rat intercalated amygdala neurons. J Neurophysiol 105:3054–3066

Ansah OB, Bourbia N, Goncalves L, Almeida A, Pertovaara A (2010) Influence of amygdaloid glutamatergic receptors on sensory and emotional pain-related behavior in the neuropathic rat. Behav Brain Res 209:174–178

Apkarian AV, Neugebauer V, Koob G, Edwards S, Levine JD, Ferrari L, Egli M, Regunathan S (2013) Neural mechanisms of pain and alcohol dependence. Pharmacol Biochem Behav 112C:34–41

Asan E, Yilmazer-Hanke DM, Eliava M, Hantsch M, Lesch KP, Schmitt A (2005) The corticotropin-releasing factor (CRF)-system and monoaminergic afferents in the central amygdala: investigations in different mouse strains and comparison with the rat. Neuroscience 131:953–967

Bale TL, Vale WW (2004) CRF and CRF receptors: role in stress responsivity and other behaviors. Annu Rev Pharmacol Toxicol 44:525–557

Baliki MN, Geha PY, Jabakhanji R, Harden N, Schnitzer TJ, Apkarian AV (2008) A preliminary fMRI study of analgesic treatment in chronic back pain and knee osteoarthritis. Mol Pain 4:47

Bie B, Zhu W, Pan ZZ (2009) Rewarding morphine-induced synaptic function of delta-opioid receptors on central glutamate synapses. J Pharmacol Exp Ther 329:290–296

Bie B, Wang Y, Cai YQ, Zhang Z, Hou YY, Pan ZZ (2012) Upregulation of nerve growth factor in central amygdala increases sensitivity to opioid reward. Neuropsychopharmacology 37:2780–2788

Bird GC, Lash LL, Han JS, Zou X, Willis WD, Neugebauer V (2005) Protein kinase A-dependent enhanced NMDA receptor function in pain-related synaptic plasticity in rat amygdala neurones. J Physiol 564:907–921

Blank T, Nijholt I, Grammatopoulos DK, Randeva HS, Hillhouse EW, Spiess J (2003) Corticotropin-releasing factor receptors couple to multiple G-proteins to activate diverse intracellular signaling pathways in mouse hippocampus: role in neuronal excitability and associative learning. J Neurosci 23:700–707

Bourbia N, Ansah OB, Pertovaara A (2010) Corticotropin-releasing factor in the rat amygdala differentially influences sensory-discriminative and emotional-like pain response in peripheral neuropathy. J Pain 11:1461–1471

Bourgeais L, Gauriau C, Bernard J-F (2001) Projections from the nociceptive area of the central nucleus of the amygdala to the forebrain: a PHA-L study in the rat. Eur J Neurosci 14:229–255

Busti D, Geracitano R, Whittle N, Dalezios Y, Manko M, Kaufmann W, Satzler K, Singewald N, Capogna M, Ferraguti F (2011) Different fear states engage distinct networks within the intercalated cell clusters of the amygdala. J Neurosci 31:5131–5144

Cai YQ, Wang W, Hou YY, Zhang Z, Xie J, Pan ZZ (2013) Central amygdala GluA1 facilitates associative learning of opioid reward. J Neurosci 33:1577–1588

Carrasquillo Y, Gereau RW (2007) Activation of the extracellular signal-regulated kinase in the amygdala modulates pain perception. J Neurosci 27:1543–1551

Carrasquillo Y, Gereau RW (2008) Hemispheric lateralization of a molecular signal for pain modulation in the amygdala. Mol Pain 4:24

Charney DS (2003) Neuroanatomical circuits modulating fear and anxiety behaviors. Acta Psychiatr Scand Suppl (417):38–50

Commons KG, Connolley KR, Valentino RJ (2003) A neurochemically distinct dorsal raphe-limbic circuit with a potential role in affective disorders. Neuropsychopharmacology 28:206–215

Crock LW, Kolber BJ, Morgan CD, Sadler KE, Vogt SK, Bruchas MR, Gereau RW (2012) Central amygdala metabotropic glutamate receptor 5 in the modulation of visceral pain. J Neurosci 32:14217–14226

Dalley JW, Everitt BJ, Robbins TW (2011) Impulsivity, compulsivity, and top-down cognitive control. Neuron 69:680–694

de Lacalle S, Saper CB (2000) Calcitonin gene-related peptide-like immunoreactivity marks putative visceral sensory pathways in human brain. Neuroscience 100:115–130

Dobolyi A, Irwin S, Makara G, Usdin TB, Palkovits M (2005) Calcitonin gene-related peptide-containing pathways in the rat forebrain. J Comp Neurol 489:92–119

Ferraguti F, Crepaldi L, Nicoletti F (2008) Metabotropic glutamate 1 receptor: current concepts and perspectives. Pharmacol Rev 60:536–581

Fields HL (2000) Pain modulation: expectation, opioid analgesia and virtual pain. Prog Brain Res 122:245–253

Fu Y, Neugebauer V (2008) Differential mechanisms of CRF1 and CRF2 receptor functions in the amygdala in pain-related synaptic facilitation and behavior. J Neurosci 28:3861–3876

Fu Y, Han J, Ishola T, Scerbo M, Adwanikar H, Ramsey C, Neugebauer V (2008) PKA and ERK, but not PKC, in the amygdala contribute to pain-related synaptic plasticity and behavior. Mol Pain 4:26–46

Gauriau C, Bernard J-F (2002) Pain pathways and parabrachial circuits in the rat. Exp Physiol 87 (2):251–258

Goncalves L, Dickenson AH (2012) Asymmetric time-dependent activation of right central amygdala neurones in rats with peripheral neuropathy and pregabalin modulation. Eur J Neurosci 36:3204–3213

Gray TS (1993) Amygdaloid CRF pathways. Role in autonomic, neuroendocrine, and behavioral responses to stress. Ann N Y Acad Sci 697:53–60

Greenwood-Van Meerveld B, Johnson AC, Schulkin J, Myers DA (2006) Long-term expression of corticotropin-releasing factor (CRF) in the paraventricular nucleus of the hypothalamus in response to an acute colonic inflammation. Brain Res 1071:91–96

Guerrini R, Salvadori S, Rizzi A, Regoli D, Calo' G (2010) Neurobiology, pharmacology, and medicinal chemistry of neuropeptide S and its receptor. Med Res Rev 30:751–777

Han JS, Neugebauer V (2004) Synaptic plasticity in the amygdala in a visceral pain model in rats. Neurosci Lett 361:254–257

Han JS, Neugebauer V (2005) mGluR1 and mGluR5 antagonists in the amygdala inhibit different components of audible and ultrasonic vocalizations in a model of arthritic pain. Pain 113:211–222

Han JS, Bird GC, Neugebauer V (2004) Enhanced group III mGluR-mediated inhibition of pain-related synaptic plasticity in the amygdala. Neuropharmacology 46:918–926

Han JS, Li W, Neugebauer V (2005) Critical role of calcitonin gene-related peptide 1 receptors in the amygdala in synaptic plasticity and pain behavior. J Neurosci 25:10717–10728

Han JS, Fu Y, Bird GC, Neugebauer V (2006) Enhanced group II mGluR-mediated inhibition of pain-related synaptic plasticity in the amygdala. Mol Pain 2:18–29

Han JS, Adwanikar H, Li Z, Ji G, Neugebauer V (2010) Facilitation of synaptic transmission and pain responses by CGRP in the amygdala of normal rats. Mol Pain 6:10–23

Harrigan EA, Magnuson DJ, Thunstedt GM, Gray TS (1994) Corticotropin releasing factor neurons are innervated by calcitonin gene-related peptide terminals in the rat central amygdaloid nucleus. Brain Res Bull 33:529–534

Hauger RL, Risbrough V, Oakley RH, Olivares-Reyes JA, Dautzenberg FM (2009) Role of CRF receptor signaling in stress vulnerability, anxiety, and depression. Ann N Y Acad Sci 1179:120–143

Hay DL (2007) What makes a CGRP2 receptor? Clin Exp Pharmacol Physiol 34:963–971

Hebert MA, Ardid D, Henrie JA, Tamashiro K, Blanchard DC, Blanchard RJ (1999) Amygdala lesions produce analgesia in a novel, ethologically relevant acute pain test. Physiol Behav 67 (1):99–105

Herry C, Ferraguti F, Singewald N, Letzkus JJ, Ehrlich I, Luthi A (2010) Neuronal circuits of fear extinction. Eur J Neurosci 31:599–612

Holland PC, Gallagher M (2004) Amygdala-frontal interactions and reward expectancy. Curr Opin Neurobiol 14:148–155

Ikeda R, Takahashi Y, Inoue K, Kato F (2007) NMDA receptor-independent synaptic plasticity in the central amygdala in the rat model of neuropathic pain. Pain 127:161–172

Ji G, Neugebauer V (2007) Differential effects of CRF1 and CRF2 receptor antagonists on pain-related sensitization of neurons in the central nucleus of the amygdala. J Neurophysiol 97:3893–3904

Ji G, Neugebauer V (2008) Pro- and anti-nociceptive effects of corticotropin-releasing factor (CRF) in central amygdala neurons are mediated through different receptors. J Neurophysiol 99:1201–1212

Ji G, Neugebauer V (2009) Hemispheric lateralization of pain processing by amygdala neurons. J Neurophysiol 1102:2253–2264

Ji G, Neugebauer V (2010) Reactive oxygen species are involved in group I mGluR-mediated facilitation of nociceptive processing in amygdala neurons. J Neurophysiol 104:218–229

Ji G, Neugebauer V (2011) Pain-related deactivation of medial prefrontal cortical neurons involves mGluR1 and GABAA receptors. J Neurophysiol 106:2642–2652

Ji G, Fu Y, Ruppert KA, Neugebauer V (2007) Pain-related anxiety-like behavior requires CRF1 receptors in the amygdala. Mol Pain 3:13–17

Ji G, Horvath C, Neugebauer V (2009) NR2B receptor blockade inhibits pain-related sensitization of amygdala neurons. Mol Pain 5:21–26

Ji G, Sun H, Fu Y, Li Z, Pais-Vieira M, Galhardo V, Neugebauer V (2010) Cognitive impairment in pain through amygdala-driven prefrontal cortical deactivation. J Neurosci 30:5451–5464

Ji G, Fu Y, Adwanikar H, Neugebauer V (2013) Non-pain-related CRF1 activation in the amygdala facilitates synaptic transmission and pain responses. Mol Pain 9:2

Jongen-Relo AL, Amaral DG (1998) Evidence for a GABAergic projection from the central nucleus of the amygdala to the brainstem of the macaque monkey: a combined retrograde tracing and in situ hybridization study. Eur J Neurosci 10:2924–2933

Jungling K, Seidenbecher T, Sosulina L, Lesting J, Sangha S, Clark SD, Okamura N, Duangdao DM, Xu YL, Reinscheid RK, Pape HC (2008) Neuropeptide S-mediated control of fear expression and extinction: role of intercalated GABAergic neurons in the amygdala. Neuron 59:298–310

Kolber BJ, Montana MC, Carrasquillo Y, Xu J, Heinemann SF, Muglia LJ, Gereau RW (2010) Activation of metabotropic glutamate receptor 5 in the amygdala modulates pain-like behavior. J Neurosci 30:8203–8213

Koob GF (2010) The role of CRF and CRF-related peptides in the dark side of addiction. Brain Res 1314:3–14

Kruger L, Sternini C, Brecha NC, Mantyh PW (1988) Distribution of calcitonin gene-related peptide immunoreactivity in relation to the rat central somatosensory projection. J Comp Neurol 273:149–162

Kulkarni B, Bentley DE, Elliott R, Julyan PJ, Boger E, Watson A, Boyle Y, El-Deredy W, Jones AK (2007) Arthritic pain is processed in brain areas concerned with emotions and fear. Arthritis Rheum 56:1345–1354

Laviolette SR, Grace AA (2006) Cannabinoids potentiate emotional learning plasticity in neurons of the medial prefrontal cortex through basolateral amygdala inputs. J Neurosci 26:6458–6468

Leonard SK, Ring RH (2011) Immunohistochemical localization of the neuropeptide S receptor in the rat central nervous system. Neuroscience 172:153–163

Li W, Neugebauer V (2004a) Block of NMDA and non-NMDA receptor activation results in reduced background and evoked activity of central amygdala neurons in a model of arthritic pain. Pain 110:112–122

Li W, Neugebauer V (2004b) Differential roles of mGluR1 and mGluR5 in brief and prolonged nociceptive processing in central amygdala neurons. J Neurophysiol 91:13–24

Li W, Neugebauer V (2006) Differential changes of group II and group III mGluR function in central amygdala neurons in a model of arthritic pain. J Neurophysiol 96:1803–1815

Li N, Liang J, Fang CY, Han HR, Ma MS, Zhang GX (2008) Involvement of CGRP and CGRPl receptor in nociception in the basolateral nucleus of amygdala of rats. Neurosci Lett 443:184–187

Li W, Chang M, Peng YL, Gao YH, Zhang JN, Han RW, Wang R (2009) Neuropeptide S produces antinociceptive effects at the supraspinal level in mice. Regul Pept 156:90–95

Li Z, Ji G, Neugebauer V (2011) Mitochondrial reactive oxygen species are activated by mGluR5 through IP3 and activate ERK and PKA to increase excitability of amygdala neurons and pain behavior. J Neurosci 31:1114–1127

Likhtik E, Popa D, Apergis-Schoute J, Fidacaro GA, Pare D (2008) Amygdala intercalated neurons are required for expression of fear extinction. Nature 454:642–645

Liu CC, Ohara S, Franaszczuk P, Zagzoog N, Gallagher M, Lenz FA (2010) Painful stimuli evoke potentials recorded from the medial temporal lobe in humans. Neuroscience 165:1402–1411

Ma W, Chabot J-G, Powell KJ, Jhamandas K, Dickerson IM, Quirion R (2003) Localization and modulation of calcitonin gene-related peptide-receptor component protein-immunoreactive cells in the rat central and peripheral nervous systems. Neuroscience 120:677–694

Manning BH (1998) A lateralized deficit in morphine antinociception after unilateral inactivation of the central amygdala. J Neurosci 18:9453–9470

Manning BH, Mayer DJ (1995a) The central nucleus of the amygdala contributes to the production of morphine antinociception in the formalin test. Pain 63:141–152

Manning BH, Mayer DJ (1995b) The central nucleus of the amygdala contributes to the production of morphine antinociception in the rat tail-flick test. J Neurosci 15(12):8199–8213

Marek R, Strobel C, Bredy TW, Sah P (2013) The amygdala and medial prefrontal cortex: partners in the fear circuit. J Physiol 591:2381–2391

Mason P (2005) Deconstructing endogenous pain modulations. J Neurophysiol 94:1659–1663

McDonald AJ (1998) Cortical pathways to the mammalian amygdala. Prog Neurobiol 55:257–332

McGaugh JL (2004) The amygdala modulates the consolidation of memories of emotionally arousing experiences. Annu Rev Neurosci 27:1–28

McLatchie LM, Fraser NJ, Main MJ, Wise A, Brown J, Thompson N, Solari R, Lee MG, Foord SM (1998) RAMPs regulate the transport and ligand specificity of the calcitonin-receptor-like receptor. Nature 393:333–339

McNally GP, Akil H (2002) Role of corticotropin-releasing hormone in the amygdala and bed nucleus of the stria terminalis in the behavioral, pain modulatory, and endocrine consequences of opiate withdrawal. Neuroscience 112:605–617

Myers B, Greenwood-Van Meerveld B (2010) Divergent effects of amygdala glucocorticoid and mineralocorticoid receptors in the regulation of visceral and somatic pain. Am J Physiol Gastrointest Liver Physiol 298:G295–G303

Myers DA, Gibson M, Schulkin J, Greenwood Van-Meerveld B (2005) Corticosterone implants to the amygdala and type 1 CRH receptor regulation: effects on behavior and colonic sensitivity. Behav Brain Res 161:39–44

Nakao A, Takahashi Y, Nagase M, Ikeda R, Kato F (2012) Role of capsaicin-sensitive C-fiber afferents in neuropathic pain-induced synaptic potentiation in the nociceptive amygdala. Mol Pain 8:51

Neugebauer V, Li W (2003) Differential sensitization of amygdala neurons to afferent inputs in a model of arthritic pain. J Neurophysiol 89:716–727

Neugebauer V, Zinebi F, Russell R, Gallagher JP, Shinnick-Gallagher P (2000) Cocaine and kindling alter the sensitivity of group II and III metabotropic glutamate receptors in the central amygdala. J Neurophysiol 84:759–770

Neugebauer V, Li W, Bird GC, Bhave G, Gereau RW (2003) Synaptic plasticity in the amygdala in a model of arthritic pain: differential roles of metabotropic glutamate receptors 1 and 5. J Neurosci 23:52–63

Neugebauer V, Li W, Bird GC, Han JS (2004) The amygdala and persistent pain. Neuroscientist 10:221–234

Neugebauer V, Galhardo V, Maione S, Mackey SC (2009) Forebrain pain mechanisms. Brain Res Rev 60:226–242

Nicoletti F, Bockaert J, Collingridge GL, Conn PJ, Ferraguti F, Schoepp DD, Wroblewski JT, Pin JP (2011) Metabotropic glutamate receptors: from the workbench to the bedside. Neuropharmacology 60:1017–1041

Niswender CM, Conn PJ (2010) Metabotropic glutamate receptors: physiology, pharmacology, and disease. Annu Rev Pharmacol Toxicol 50:295–322

Ochsner KN, Gross JJ (2005) The cognitive control of emotion. Trends Cogn Sci 9:242–249

Oliver KR, Kane SA, Salvatore CA, Mallee JJ, Kinsey AM, Koblan KS, Keyvan-Fouladi N, Heavens RP, Wainwright A, Jacobson M, Dickerson IM, Hill RG (2001) Cloning, characterization and central nervous system distribution of receptor activity modifying proteins in the rat. Eur J Neurosci 14:618–628

Orsini CA, Maren S (2012) Neural and cellular mechanisms of fear and extinction memory formation. Neurosci Biobehav Rev 36:1773–1802

Palazzo E, Fu Y, Ji G, Maione S, Neugebauer V (2008) Group III mGluR7 and mGluR8 in the amygdala differentially modulate nocifensive and affective pain behaviors. Neuropharmacology 55:537–545

Palazzo E, Marabese I, Soukupova M, Luongo L, Boccella S, Giordano C, de Novellis V, Rossi F, Maione S (2011) Metabotropic glutamate receptor subtype 8 in the amygdala modulates thermal threshold, neurotransmitter release, and rostral ventromedial medulla cell activity in inflammatory pain. J Neurosci 31:4687–4697

Pape HC, Pare D (2010) Plastic synaptic networks of the amygdala for the acquisition, expression, and extinction of conditioned fear. Physiol Rev 90:419–463

Pedersen LH, Scheel-Kruger J, Blackburn-Munro G (2007) Amygdala GABA-A receptor involvement in mediating sensory-discriminative and affective-motivational pain responses in a rat model of peripheral nerve injury. Pain 127:17–26

Peng YL, Zhang JN, Chang M, Li W, Han RW, Wang R (2010) Effects of central neuropeptide S in the mouse formalin test. Peptides 31:1878–1883

Pinard CR, Mascagni F, McDonald AJ (2012) Medial prefrontal cortical innervation of the intercalated nuclear region of the amygdala. Neuroscience 205:112–124

Poyner DR, Sexton PM, Marshall I, Smith DM, Quirion R, Born W, Muff R, Fischer JA, Foord SM (2002) International Union of Pharmacology. XXXII. The mammalian calcitonin gene-related peptides, adrenomedullin, amylin, and calcitonin receptors. Pharmacol Rev 54:233–246

Price JL (2003) Comparative aspects of amygdala connectivity. In: Shinnick-Gallagher P, Pitkanen A, Shekhar A, Cahill L (eds) The amygdala in brain function. Basic and clinical approaches, vol 985. The New York Academy of Sciences, New York, pp 50–58

Qin C, Greenwood-Van Meerveld B, Foreman RD (2003) Visceromotor and spinal neuronal responses to colorectal distension in rats with aldosterone onto the amygdala. J Neurophysiol 90:2–11

Rea K, Lang Y, Finn DP (2009) Alterations in extracellular levels of gamma-aminobutyric acid in the rat basolateral amygdala and periaqueductal gray during conditioned fear, persistent pain and fear-conditioned analgesia. J Pain 10:1088–1098

Reinscheid RK (2008) Neuropeptide S: anatomy, pharmacology, genetics and physiological functions. Results Probl Cell Differ 46:145–158

Ren W, Neugebauer V (2010) Pain-related increase of excitatory transmission and decrease of inhibitory transmission in the central nucleus of the amygdala are mediated by mGluR1. Mol Pain 6:93–106

Ren W, Palazzo E, Maione S, Neugebauer V (2011) Differential effects of mGluR7 and mGluR8 activation on pain-related synaptic activity in the amygdala. Neuropharmacology 61:1334–1344

Ren W, Kiritoshi T, Gregoire S, Ji G, Guerrini R, Calo G, Neugebauer V (2013) Neuropeptide S: a novel regulator of pain-related amygdala plasticity and behaviors. J Neurophysiol 110:1765–1781

Reul JM, Holsboer F (2002) Corticotropin-releasing factor receptors 1 and 2 in anxiety and depression. Curr Opin Pharmacol 2:23–33

Robinson SD, Aitken JF, Bailey RJ, Poyner DR, Hay DL (2009) Novel peptide antagonists of adrenomedullin and calcitonin gene-related peptide receptors: identification, pharmacological characterization, and interactions with position 74 in receptor activity-modifying protein 1/3. J Pharmacol Exp Ther 331:513–521

Rouwette T, Vanelderen P, Reus MD, Loohuis NO, Giele J, van Egmond J, Scheenen W, Scheffer GJ, Roubos E, Vissers K, Kozicz T (2011) Experimental neuropathy increases limbic forebrain CRF. Eur J Pain 16(1):61–71

Ruzza C, Pulga A, Rizzi A, Marzola G, Guerrini R, Calo' G (2012) Behavioural phenotypic characterization of CD-1 mice lacking the neuropeptide S receptor. Neuropharmacology 62:1999–2009

Sanchez MM, Young LJ, Plotsky PM, Insel TR (1999) Autoradiographic and in situ hybridization localization of corticotropin-releasing factor 1 and 2 receptors in nonhuman primate brain. J Comp Neurol 408(3):365–377

Schiess MC, Callahan PM, Zheng H (1999) Characterization of the electrophysiological and morphological properties of rat central amygdala neurons in vitro. J Neurosci Res 58:663–673

Schwaber JS, Sternini C, Brecha NC, Rogers WT, Card JP (1988) Neurons containing calcitonin gene-related peptide in the parabrachial nucleus project to the central nucleus of the amygdala. J Comp Neurol 270:416–426

Simons LE, Moulton EA, Linnman C, Carpino E, Becerra L, Borsook D (2012) The human amygdala and pain: evidence from neuroimaging. Hum Brain Mapp 35(2):527–538. doi:10.1002/hbm.22199

Sotres-Bayon F, Quirk GJ (2010) Prefrontal control of fear: more than just extinction. Curr Opin Neurobiol 20:231–235

Spuz CA, Borszcz GS (2012) NMDA or non-NMDA receptor antagonism within the amygdaloid central nucleus suppresses the affective dimension of pain in rats: evidence for hemispheric synergy. J Pain 13:328–337

Sun N, Cassell MD (1993) Intrinsic GABAergic neurons in the rat central extended amygdala. J Comp Neurol 330:381–404

Tache Y, Bonaz B (2007) Corticotropin-releasing factor receptors and stress-related alterations of gut motor function. J Clin Invest 117:33–40

Takahashi LK (2001) Role of CRF(1) and CRF(2) receptors in fear and anxiety. Neurosci Biobehav Rev 25:627–636

Tillisch K, Mayer EA, Labus JS (2010) Quantitative meta-analysis identifies brain regions activated during rectal distension in irritable bowel syndrome. Gastroenterology 140:91–100

Ulrich-Lai YM, Xie W, Meij JTA, Dolgas CM, Yu L, Herman JP (2006) Limbic and HPA axis function in an animal model of chronic neuropathic pain. Physiol Behav 88:67–76

Uryu K, Okumura T, Shibasaki T, Sakanaka M (1992) Fine structure and possible origins of nerve fibers with corticotropin-releasing factor-like immunoreactivity in the rat central amygdaloid nucleus. Brain Res 577:175–179

Van Rossum D, Hanish U-K, Quirion R (1997) Neuroanatomical localization, pharmacological characterization and functions of CGRP, related peptides and their receptors. Neurosci Biobehav Rev 21:649–678

Wimalawansa SJ (1996) Calcitonin gene-related peptide and its receptors: molecular genetics, physiology, pathophysiology, and therapeutic potentials. Endocr Rev 17:533–585

Xu YL, Reinscheid RK, Huitron-Resendiz S, Clark SD, Wang Z, Lin SH, Brucher FA, Zeng J, Ly NK, Henriksen SJ, De LL, Civelli O (2004) Neuropeptide S: a neuropeptide promoting arousal and anxiolytic-like effects. Neuron 43:487–497

Xu YL, Gall CM, Jackson VR, Civelli O, Reinscheid RK (2007) Distribution of neuropeptide S receptor mRNA and neurochemical characteristics of neuropeptide S-expressing neurons in the rat brain. J Comp Neurol 500:84–102

Zhang RX, Zhang M, Li A, Pan L, Berman BM, Ren K, Lao L (2013) DAMGO in the central amygdala alleviates the affective dimension of pain in a rat model of inflammatory hyperalgesia. Neuroscience 252:359–366

Zhu W, Pan ZZ (2004) Synaptic properties and postsynaptic opioid effects in rat central amygdala neurons. Neuroscience 127:871–879

Itch and Pain Differences and Commonalities

Martin Schmelz

Contents

1 Differentiation Between Pain and Itch 286
   1.1 Specificity for Itch 286
   1.2 Intensity and Pattern Theory of Itch 290
2 Central and Peripheral Sensitization in Itch and Pain 293
   2.1 Central Sensitization 293
   2.2 Peripheral Sensitization 295
3 Perspectives: Mechanisms for Itch or Pain in Neuropathy and Chronic Inflammation .. 295
References 296

Abstract

Pain and itch are generally regarded antagonistic as painful stimuli such as scratching suppresses itch. Moreover, inhibition of pain processing by opioids generates itch further supporting their opposing role. Separate specific pathways for itch and pain processing have been uncovered, and several molecular markers have been established in mice that identify neurons involved in the processing of histaminergic and non-histaminergic itch on primary afferent and spinal level. These results are in agreement with the specificity theory for itch and might suggest that pain and itch should be investigated separately on the level of neurons, mediators, and mechanisms. However, in addition to broadly overlapping mediators of itch and pain, there is also evidence for overlapping functions in primary afferents: nociceptive primary afferents can provoke itch when activated very locally in the epidermis, and sensitization of both nociceptors and pruriceptors has been found following local nerve growth factor application in volunteers. Thus, also mechanisms that underlie the development

M. Schmelz (✉)
Department of Anesthesiology and Intensive Care Medicine, Faculty of Medicine Mannheim, University of Heidelberg, Theodor-Kutzer Ufer 1-3, Mannheim 68167, Germany
e-mail: martin.schmelz@medma.uni-heidelberg.de

of chronic itch and pain including spontaneous activity and sensitization of primary afferents as well as spinal cord sensitization may well overlap to a great extent. Rather than separating itch and pain, research concepts should therefore address the common mechanisms. Such an approach appears most appropriate for clinical conditions of neuropathic itch and pain and also chronic inflammatory conditions. While itch researchers can benefit from the large body of information of the pain field, pain researchers will find behavioral readouts of spontaneous itch much simpler than those for spontaneous pain in animals and the skin as source of the pruritic activity much more accessible even in patients.

Keywords
Specificity • Intensity theory • Pattern theory • Sensitization

1 Differentiation Between Pain and Itch

Itch and pain can be clearly separated by their distinct sensations and their characteristic reflex patterns. Acute pain evokes withdrawal of the stimulated limb which enables escape from a potentially damaging external stimulus that threatens the organism. In contrast, the scratch reflex directs attention to the stimulated site, and scratching provides the means to remove a potentially damaging stimulus that already has invaded the skin and now poses a threat from inside the body. While virtually all organs of the human body except the brain itself are innervated by nociceptors, itch can be induced only from skin and adjoining mucosae. Actually, only in these locations scratching appears a reasonable approach to remove superficially localized agents. In the respiratory tract, coughing has a very similar protective role and instructively has been termed "airway itch" (Gibson 2004). The clear functional separation between itch and pain could be explained most easily by two specific sensory pathways.

1.1 Specificity for Itch

Specific sets of primary afferent dedicated to pain ("nociceptors") and to itch ("pruriceptors") have been hypothesized in the late nineteenth and early twentieth century (Handwerker 2014). Unmyelinated primary afferents that responded to histamine iontophoresis in parallel to the itch ratings of subjects were finally discovered among the group of mechano-insensitive C-fibers (Schmelz et al. 1997). In contrast, the most common type of C-fibers, mechano-heat nociceptors ("polymodal nociceptors"), is either insensitive to histamine or only weakly activated by this stimulus (Schmelz et al. 2003b). Hence, this fiber type cannot account for the prolonged itch induced by the iontophoretic application of histamine. Yet, when histamine is injected intracutaneously, also polymodal nociceptors are activated for several minutes (Johanek et al. 2008). Thus, a contribution of this fiber class to histamine-induced itch cannot be entirely ruled out.

The histamine-sensitive pruriceptors among the mechano-insensitive C-nociceptors are characterized by a particular low conduction velocity, large innervation territories, mechanical unresponsiveness, and high transcutaneous electrical thresholds (Schmelz et al. 1997, 2003b; Schmidt et al. 2002). Interestingly, only these histamine-sensitive afferents were activated by prostaglandin E2 injected intracutaneously in their innervation territory (Schmelz et al. 2003b). In line with the large innervation territories of these fibers, two-point discrimination for histamine-induced itch is poor (15 cm in the upper arm) (Wahlgren and Ekblom 1996). The excellent locognosia for histamine-induced itch in the hand (Koltzenburg et al. 1993) might therefore be based on central processing compensating for low spatial resolution in the periphery. Among the mechano-insensitive afferent C-fibers, only a subset of units shows a strong and sustained response to histamine. They comprise about 20 % of the mechano-insensitive class of C-fibers, i.e., about 5 % of all C-fibers in the superficial peroneal nerve.

In accordance with the existence of dedicated histamine-sensitive primary afferents, cat spinal cord recordings provided evidence for a specific class of dorsal horn neurons projecting to the thalamus which respond strongly to histamine administered to the skin by iontophoresis (Andrew and Craig 2001). These neurons were also unresponsive to mechanical stimulation, and their axons had a lower conduction velocity and anatomically distinct projections to the thalamus. The itch-selective units in lamina I of the spinal cord form a distinct pathway projecting to the posterior part of the ventromedial thalamic nucleus which projects to the dorsal insular cortex (Craig 2002), a region which has been shown to be involved in a variety of interoceptive modalities like thermoception, visceral sensations, thirst, and hunger.

Thus, the combination of dedicated peripheral and central neurons with a unique response pattern to pruritogenic mediators and anatomically distinct projections to the thalamus provides the basis for a specific neuronal pathway for itch.

1.1.1 Molecular Markers for Itch-Processing Neurons

Functional classes of primary afferent neurons are defined primarily on the basis of their response characteristics. However, functional markers are required to identify the neuronal classes also in vitro. For the separation of functional classes among primary afferents, marker proteins have been established that are involved in sensory transduction such as vanilloid receptors (TRPV1, TRPA1) and purinergic receptors (P2X3). Moreover, neuropeptides such as substance P and calcitonin gene-related peptide, receptors for growth factors, and also receptors of yet unknown function such as the family of Mas-related G protein-coupled receptors (Mrgpr) are used. Markers that have been used to characterize neurons involved in itch processing (Akiyama and Carstens 2013) include histamine H1 receptors, the neuropeptides gastrin-releasing peptide and B-type natriuretic peptide, and the several members of the Mrgpr family (A3, D, C11) (LaMotte et al. 2014; Bautista et al. 2014; Braz et al. 2014b). Unfortunately, there are only few examples for a convincing link between the rodent marker and functional neuronal class in primates. For a very special subtype of afferent C-fiber, the very low-threshold so-called C-touch fibers (CT afferents) (Ackerley et al. 2014), links to the expression of MrgprB4 (Vrontou et al. 2013) and to the expression of the glutamate transporter VGLUT3 (Seal et al. 2009) have been described.

In the realm of itch processing, however, we do not have such convincing ties between molecular markers used in rodents and fiber classes in the primate. There is evidence that cowhage induces itch via activation of proteinase-activated receptors (Reddy et al. 2008). Thus, the activation of QC-type mechano-sensitive nociceptors by cowhage (Johanek et al. 2008) might be a possible link to MrgprC11 (Akiyama and Carstens 2013). Beta-alanine, the activator of MrgprD, does provoke itch in humans (Qu et al. 2014; Han et al. 2012; Liu et al. 2012) and activates primarily QC-type mechano-sensitive nociceptors in the monkey (Wooten et al. 2014), but the corresponding fiber type in human is yet unclear. This is similarly true for BAM8-22, activator of MrgprC11, that also provokes histamine-independent itch in humans (Sikand et al. 2011) probably via activating MrgprX1, the human homologue of rodent MrgprC11. Activation of polymodal nociceptors by agonists of receptors thought to be itch specific poses some problems to the concept of specificity of itch and pain (see pattern theory below).

1.1.2 Antagonistic Interaction Between Itch and Pain

Our common experience tells us that pain inhibits itch. Also experimentally, the inhibition of itch by painful stimuli has been demonstrated by the use of various painful thermal, mechanical, and chemical stimuli. Electrical stimulation via an array of pointed electrodes ("cutaneous field stimulation") has also been successfully used to inhibit histamine-induced itch for several hours in an area around a stimulated site of 20 cm in diameter. The large area of inhibition suggests a central mode of action (Nilsson et al. 1997). Consistent with these results, itch is suppressed inside the secondary zone of capsaicin-induced mechanical hyperalgesia (Brull et al. 1999). This central effect of nociceptor excitation by capsaicin should be clearly distinguished from the neurotoxic effect of higher concentrations of capsaicin which destroy most C-fiber terminals, including fibers that mediate itch (Simone et al. 1998). The latter mechanism, therefore, also abolishes pruritus locally, until the nerve terminals are regenerated.

Not only is itch inhibited by enhanced input of pain stimuli, but vice versa, inhibition of pain processing may reduce its inhibitory effect and thus enhance itch (Atanassoff et al. 1999). This phenomenon is particularly relevant to spinally administered μ-opioid receptor agonists which induce segmental analgesia often combined with segmental pruritus (Andrew et al. 2003), but has also been confirmed in animal experiments (Nojima et al. 2004). Conversely, kappa-opioid antagonists have been found to enhance itch (Kamei and Nagase 2001). In line with these results, the κ-opioid agonist nalbuphine has been shown to reduce μ-opioid-induced pruritus in a meta-analysis (Kjellberg and Tramer 2001). This therapeutic concept has already been tested successfully in chronic itch patients using a newly developed κ-opioid agonist (Kumagai et al. 2010). Recently,

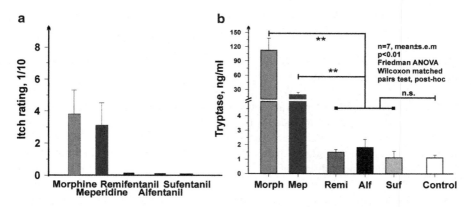

Fig. 1 Opioids were applied in the volar forearm of volunteers by dermal microdialysis. Intensity of opioid-induced maximum itch is shown in (**a**) (visual analog scale 0–10, mean ± SEM). (**b**) Peak mast cell tryptase release during stimulation with the opioids is shown (mean ± SEM). Only low affinity opioids meperidine (40.4 mM) and morphine (3.11 mM) caused tryptase release from mast cells. The potent opioids alfentanil (1.2 mM), sufentanil (0.12 mM), and remifentanil (2.65 mM) provoked neither itch nor tryptase release (modified from Blunk et al. 2004)

GABAergic interneurons harboring the transcription factor Bhlhb5 in the dorsal horn have been found to be crucial to inhibit itch (Ross et al. 2010; Braz et al. 2014a). Most interestingly, these neurons appear to mediate spinal suppression of itch by releasing the κ-opioid agonist dynorphin (Kardon et al. 2014).

Opioid-induced itch has often been linked to peripheral release of histamine from mast cells as intradermally injected opioids can activate mast cells by a non-receptor-mediated mechanism (Ferry et al. 2002). Accordingly, weak opioids, such as codeine, have been used as a positive control in skin prick tests. The consecutive release of histamine and mast cell tryptase can be specifically monitored by measuring tryptase concentration with dermal microdialysis following intraprobe delivery (Blunk et al. 2003). In contrast to morphine, the highly potent μ-opioid agonist, fentanyl, does not provoke any mast cell degranulation, even if applied at concentrations having μ-agonistic effects exceeding those of morphine (Fig. 1). Thus, high local concentrations of opioids are required to degranulate mast cells, and therefore, itch induced by systemic administration of potent μ-opioid agonists in therapeutic doses is based on central mechanisms.

Central inhibition of itch can also be achieved by cold stimulation (Bromm et al. 1995). In addition, cooling has a peripheral inhibitory effect: histamine-induced activation of nociceptors can be reduced by cooling (Mizumura and Koda 1999). Also in humans, cooling of a histamine-treated skin site reduced the activity of the primary afferents and decreased the area of "itchy skin" or "hyperknesis" around the application site (Heyer et al. 1995). Unexpectedly, there is an initial increase of itch intensity upon cooling the histamine application site (Pfab et al. 2006) that can be used as experimental model for central imaging (Napadow et al. 2014). Conversely, tonic warming the skin would lead to an exacerbation of itch. However, as soon as the heating becomes painful, central inhibition of pruritus will counteract this effect (Schmelz 2002).

Recent work on antipruritic effects of subpopulations of primary nociceptive afferents indicates that especially the input from the VGLUT2-positive subpopulation is crucial to explain inhibition of itch behavior by painful stimuli (Lagerstrom et al. 2010; Liu et al. 2010). When VGLUT2 and, thereby, glutamate release in NaV1.8-positive nociceptors was deficient, inflammatory and neuropathic pain responses were grossly abolished, but spontaneous scratching behavior and increased experimental itch were massively enhanced (Liu et al. 2010). Most interestingly, capsaicin-induced pain behavior was changed into scratching behavior in these mice suggesting that the lack of noxious input via VGLUT2-positive nociceptors disinhibited itch (Liu et al. 2010). The exact nature of the crucial nociceptor class is still unclear, as another group did not find increased scratching when VGLUT2 was knocked out in NaV1.8-positive but in TRPV1-positive primary afferent neurons (Lagerstrom et al. 2010). It will be of major interest to further characterize the nociceptor class having a crucial itch-inhibiting spinal effect.

It is important to note that we have so far followed the ideas of two separate populations for itch and pain. This segregation is actually required for the genetic approaches described above. Unfortunately, several aspects of itch generation such as pruritus induced by activation of polymodal nociceptors or spatial stimulus characteristics switching pain to itch cannot be sufficiently explained by this approach.

1.2 Intensity and Pattern Theory of Itch

Numerous neuronal markers that were linked to itch, but not pain processing, support the specificity theory of itch. However, as indicated above, there is evidence that itch can also be induced by activation of nociceptors. Nociceptors could provoke itch either by an intensity coding ("intensity theory") or by a particular population encoding ("pattern theory") (Namer and Reeh 2013; Handwerker 2014). Albeit this general question might appear purely academic, it is crucial to provide the most promising research approach to identify pharmacological targets in chronic itch and pain.

1.2.1 Non-histaminergic Itch

Histamine application causes a local wheal response surrounded by an area of vasodilation ("axon reflex erythema") (Lewis et al. 1927). This vasodilation is induced by neuropeptide release from mechano-insensitive C-fibers (Schmelz et al. 2000; Geppetti and Holzer 1996). The absence of an axon reflex flare therefore suggests that the itch is independent of histamine-sensitive C-fibers. Indeed, itch was induced by papain in an early study in the absence of a flare response indicating a histamine-independent action (Hägermark 1973). Itch without axon reflex flare can also be elicited by weak electrical stimulation (Shelley and Arthur 1957; Ikoma et al. 2005), providing further evidence that the sensation of itch can be dissociated from cutaneous vasodilation.

Cowhage spicules inserted into human skin produce itch in an intensity which is comparable to that following histamine application (LaMotte et al. 2009; Sikand et al. 2009), but is not accompanied by an axon reflex erythema and unresponsive to histamine (H1) blocker (Johanek et al. 2007). The active compound, the cysteine protease muconain, has been identified lately and has shown to activate proteinase-activated receptor 2 (PAR 2) and even more potently PAR 4 (Reddy et al. 2008). Interestingly, mechano-responsive "polymodal" C-fiber afferents, the most common type of afferent C-nociceptors in human skin (Schmidt et al. 1995), can be activated by cowhage in the cat (Tuckett and Wei 1987), in nonhuman primates (Johanek et al. 2007, 2008), and in human volunteers (Namer et al. 2008) (Fig. 2).

Given that cowhage spicules can activate a large proportion of polymodal nociceptors, we face a major problem to explain why activation of these fibers by heat or by scratching actually inhibits itch, whereas activation by cowhage produces it. On the other hand, data from monkey suggest that mechano-heat-sensitive C-nociceptors with a fast response to heating ("QC") might play a more important role in mediating cowhage-induced itch (Johanek et al. 2008). One might therefore still hypothesize that there is a certain selectivity among mechano-sensitive C-nociceptors for cowhage that would allow the central nervous system to separate nociceptive from pruriceptive stimuli (LaMotte et al. 2014). Along the same lines, in particular QC-nociceptors, but not mechano-insensitive nociceptors, were activated by beta-alanine (Wooten et al. 2014), the activator of MrgprD that provokes itch in humans.

1.2.2 Encoding Itch by Patterns of Activated Nociceptors

Considering nociceptors being involved in generating itch, a population code has been postulated ("pattern theory") (Handwerker 2014; Akiyama and Carstens 2013; McMahon and Koltzenburg 1992) in which only a subpopulation of nociceptors can also be activated by pruritic stimuli, whereas pure nociceptors are only responsive to algogens. Accordingly, itch will be felt when only the first subpopulation is responding, but pain when both populations are active.

The encoding of itch by nociceptors has also been proposed to be based on a spatial code (Schmelz and Handwerker 2013) based on the itch induced by capsaicin being applied very localized on a cowhage spicule into the epidermis (Sikand et al. 2009). The highly localized stimulation in the epidermis strongly activates some of the local nociceptors, while their immediate neighbors remain silent resulting in a mismatch signal of activation and absence of activation from this site. It has thus been hypothesized that this mismatch might be perceived by the central nervous system as itch (Schmelz and Handwerker 2013; Namer et al. 2008). Interestingly, this result also speaks against capsaicin's pain specificity (Ross 2011). Teleologically, it is obvious that scratching behavior in the case of a highly localized superficial noxious focus is an adequate response as it can eliminate the presumed cause. Moreover, scratching activates all the mechano-sensitive nociceptors in the stimulated area, and thus, the mismatch signal of activated and

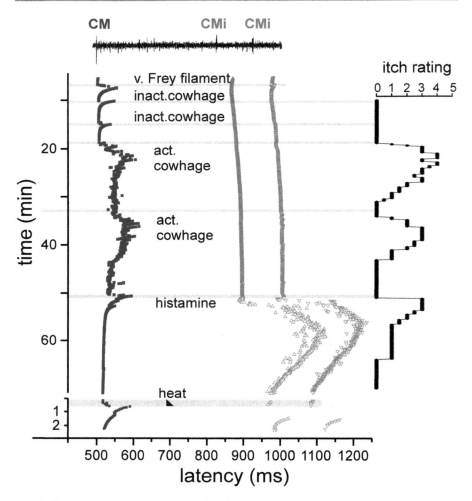

Fig. 2 Specimen of a multifiber recording from a mechano-responsive (CM, *blue*) and two mechano-insensitive nociceptors (CMi, *red*) in human (raw signal with marked action potential on top). Conduction latencies of these three marked fibers (*filled square, open triangles*) in response to successive electrical stimulation at the receptive field are plotted from top to bottom. When activated by mechanical (v. Frey filament, inactivated cowhage spicules), chemical (active cowhage, histamine), or heat test stimuli (*black triangle*), C-fibers exhibit activity-dependent increase of response latency followed by a gradual normalization ("marking"). The mechano-responsive fiber is activated during mechanical stimulation with the v. Frey filament and during application of inactive cowhage, but lasting activation is only seen after application of active cowhage. In contrast, the mechano-insensitive fibers do not respond to cowhage stimulation, but are activated following histamine iontophoresis. At the right side of the panel, the itch ratings of the subject are depicted which were assessed during this experiment. Ratings are given on a numerical rating scale from 0 (0 = no itch) to 10 (10 = maximal imaginable itch). Inactive cowhage does not evoke any itch, whereas active cowhage and histamine evoke itch similarly in time course and maximum mirroring nicely the activation pattern of the fibers. Modified from Namer et al. (2008)

nonactivated nociceptors at this site is terminated. Therefore, it needs to be pointed out that pruritus cannot only be explained by itch-specific or itch-selective neurons (LaMotte et al. 2014) along the specificity theory. In addition, the pure spatial pattern of activated nociceptors might similarly underlie the itch sensation without any requirement of itch-specific primary afferent neurons.

2 Central and Peripheral Sensitization in Itch and Pain

Spontaneous itch and pain are of paramount clinical relevance as they correlate to the main complaint of chronic itch and pain patients. It is highly interesting that the patterns of peripheral and central sensitization linked to chronic pain and itch are remarkably similar.

2.1 Central Sensitization

Activity in chemo-nociceptors leads not only to acute pain but, in addition, can sensitize second-order neurons in the dorsal horn, thereby leading to increased sensitivity to pain (hyperalgesia). Two types of mechanical hyperalgesia can be differentiated. Normally painless touch sensations in the uninjured surroundings of a trauma are felt as painful "touch- or brush-evoked hyperalgesia" or allodynia. Though this sensation is mediated by myelinated mechanoreceptor units, it requires ongoing activity of primary afferent C-nociceptors (Torebjörk et al. 1996). The second type of mechanical hyperalgesia results in slightly painful pinprick stimulation being perceived as being more painful in the secondary zone around a focus of inflammation. This type has been called "punctate hyperalgesia" and does not require ongoing activity of primary nociceptors for its maintenance. It can persist for hours following a trauma, usually much longer than touch- or brush-evoked hyperalgesia (LaMotte et al. 1991).

In itch processing, similar phenomena have been described: touch- or brush-evoked pruritus around an itching site has been termed "itchy skin" (Bickford 1938; Simone et al. 1991). Like allodynia, it requires ongoing activity in primary afferents and is most probably elicited by low-threshold mechanoreceptors (Aδ-fibers) (Simone et al. 1991; Heyer et al. 1995). Also, more intense prick-induced itch sensations in the surroundings, "hyperknesis," have been reported following histamine iontophoresis in healthy volunteers (Atanassoff et al. 1999) (Fig. 3).

The existence of central sensitization for itch can greatly improve our understanding of clinical itch. Under the conditions of central sensitization leading to punctuate hyperknesis, normally painful stimuli are perceived as itching. This phenomenon has already been described in patients suffering from atopic dermatitis, who perceive normally painful electrical stimuli as itching when applied inside their lesional skin (Nilsson et al. 2004; Nilsson and Schouenborg 1999). Furthermore, acetylcholine provokes itch instead of pain in patients with atopic dermatitis

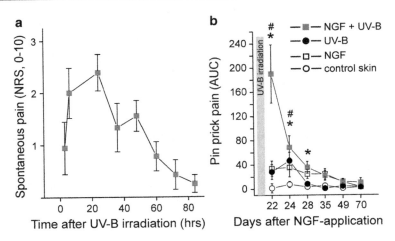

Fig. 3 (a) Pre-sensitization with nerve growth factor (NGF, 1 μg) injected 3 weeks before UV-B irradiation (threefold minimum erythema dose) provoked spontaneous pain ratings following the intensity of the UV-induced inflammation. (b) Hyperalgesia to pinprick stimuli develops following intradermal NGF injection and also for about 3 days after UV-B irradiation. Combined sensitization with NGF and UV-B irradiation causes a supra-additive increase of mechanical hyperalgesia. Modified from Rukwied et al. (2013b)

(Vogelsang et al. 1995), indicating that pain-induced inhibition of itch might be compromised in these patients.

The exact mechanisms and roles of central sensitization for itch in specific, clinical conditions have still to be explored, whereas a major role of central sensitization in patients with chronic pain is generally accepted. It should be noted that in addition to the parallels between experimentally induced secondary sensitization phenomena, there is also emerging evidence for corresponding phenomena in patients with chronic pain and chronic itch. In patients with neuropathic pain, it has been reported that histamine iontophoresis resulted in burning pain instead of pure itch which would be induced by this procedure in healthy volunteers (Birklein et al. 1997; Baron et al. 2001). This phenomenon is of special interest as it demonstrates spinal hypersensitivity to C-fiber input. Conversely, normally painful electrical, chemical, mechanical, and thermal stimulation is perceived as itching when applied in or close to lesional skin of atopic dermatitis patients (Heyer et al. 1995; Steinhoff et al. 2003).

Ongoing activity of pruriceptors, which might underlie the development of central sensitization for itch, has already been confirmed microneurographically in a patient with chronic pruritus (Schmelz et al. 2003a). Thus, there is emerging evidence, for a role of central sensitization for itch in chronic pruritus.

While there is obviously an antagonistic interaction between pain and itch under normal conditions, the patterns of spinal sensitization phenomena are surprisingly similar. It remains to be established whether this similarity will also include the underlying mechanism which would also implicate similar therapeutic approaches such as gabapentin (Dhand and Aminoff 2014) or clonidine (Elkersh et al. 2003) for the treatment of neuropathic itch.

2.2 Peripheral Sensitization

There is cumulative evidence for a prominent role of nerve growth factor (NGF)-induced sensitization of primary afferents in both chronic itch and pain: increased levels of NGF were found in chronic itch patients suffering from atopic dermatitis or psoriasis (Toyoda et al. 2002, 2003; Tominaga et al. 2009; Yamaguchi et al. 2009). Similarly, there is clear evidence for a major role of NGF in chronic inflammatory pain (Chevalier et al. 2013; Watanabe et al. 2011; Barcena de Arellano et al. 2011). Moreover, blocking NGF by specific antibodies proved to be analgesic in the chronic pain patients (Lane et al. 2010; Sanga et al. 2013). Anti-NGF strategies also were successful in animal models of chronic itch (Tominaga and Takamori 2014). It is therefore not surprising that intradermally injected NGF not only causes hyperalgesia to heat and mechanical stimuli in volunteers (Hirth et al. 2013; Rukwied et al. 2010) but also sensitizes for cowhage-induced itch (Rukwied et al. 2013c). Intracutaneous NGF injection does not induce visual inflammatory responses in human (Rukwied et al. 2010), but interestingly, when combined with an inflammatory pain model (UV-B sunburn), the subjects report of spontaneous pain (Fig. 3) and pronounced hyperalgesia (Rukwied et al. 2013b) that also includes axonal hyperexcitability (Rukwied et al. 2013a). These results nice match the analgesic effects of anti-NGF in chronic inflammatory pain that are not accompanied by reduced signs of inflammation (Lane et al. 2010). Therefore, it emerges that neurotrophic factors such as NGF can change expression patterns of primary afferent nociceptors such that their ability to signal pain or itch by local inflammatory mediators is increased. This increase might be based on higher discharge frequencies linked to sensitized transduction, but also to axonal hyperexcitability.

3 Perspectives: Mechanisms for Itch or Pain in Neuropathy and Chronic Inflammation

Finally, the current concepts differentiating itch and pain need to be evaluated in view of the obvious clinical questions concerning the development of itch or pain after neuropathy or in chronic inflammatory diseases. It is remarkable that some neuropathic conditions such as postherpetic neuralgia and diabetic neuropathy are primarily linked to pain symptoms whereas patients suffering from notalgia paresthetica or brachioradial pruritus mainly report chronic itch (Table 1).

It is important to note that more than 25 % of patients with neuropathic pain conditions such as postherpetic neuropathy also report itch (Oaklander et al. 2003). According to the specificity or selectivity theory, one would hypothesize that the mediators being released in diabetic neuropathy or postherpetic neuralgia

Table 1 Summary of neuropathic conditions and their dominant symptoms

	Pain	Itch
Postherpetic neuralgia	+++	++
Diabetic neuropathy	++(+)	+
Meralgia paresthetica	+++	(+)
Notalgia paresthetica	(+)	+++
Brachioradial pruritus	(+)	+++

determine to which extent itch-selective or itch-specific primary afferents are excited. Moreover, itching neuropathic conditions such as nostalgia paresthetica and brachioradial pruritus should be differentiated from painful meralgia paresthetica by primary activation of pruriceptors rather than nociceptors. However, it is completely unclear how such differentiation could be mediated for very similar peripheral neuropathic conditions. Possibly, specific pruriceptors only play a minor role under these conditions. In contrast, the spatial pattern of nociceptor activation might provide the crucial input: if only few scattered axons are spontaneously active, their input might mimic the one of scattered nociceptors being activated by cowhage spicules in the epidermis, whereas activation of numerous nociceptors of a peripheral nerve would result in pain. Thus, such itch sensation would be generated by the particular spatial code of activated nociceptors (Schmelz and Handwerker 2013; Namer et al. 2008). Accordingly, scattered activation of epidermal nociceptors might also occur in some chronic inflammatory diseases such as atopic dermatitis and explain the difference between itching and painful symptoms. If this hypothesis would be correct, the treatment of neuropathic itch and pain would have essentially identical therapeutic targets and mechanisms rather than itch or pain specific. Thus, the implications of theoretical concepts of itch are, unexpectedly, of major clinical relevance. It will therefore be of major interest for both clinicians and basic researchers to determine which fiber class generates the peripheral input for chronic itch conditions.

References

Ackerley R, Backlund Wasling H, Liljencrantz J, Olausson H, Johnson RD, Wessberg J (2014) Human C-tactile afferents are tuned to the temperature of a skin-stroking caress. J Neurosci 34 (8):2879–2883. doi:10.1523/JNEUROSCI. 2847-13.2014

Akiyama T, Carstens E (2013) Neural processing of itch. Neuroscience 250:697–714. doi:10.1016/j.neuroscience.2013.07.035

Andrew D, Craig AD (2001) Spinothalamic lamina 1 neurons selectively sensitive to histamine: a central neural pathway for itch. Nat Neurosci 4:72–77

Andrew D, Schmelz M, Ballantyne JC (2003) Itch—mechanisms and mediators. In: Dostrovsky JO, Carr DB, Koltzenburg M (eds) Progress in pain research and management. IASP Press, Seattle, pp 213–226

Atanassoff PG, Brull SJ, Zhang J, Greenquist K, Silverman DG, LaMotte RH (1999) Enhancement of experimental pruritus and mechanically evoked dysesthesiae with local anesthesia. Somatosens Mot Res 16(4):291–298

Barcena de Arellano ML, Arnold J, Vercellino GF, Chiantera V, Ebert AD, Schneider A, Mechsner S (2011) Influence of nerve growth factor in endometriosis-associated symptoms. Reprod Sci 18(12):1202–1210

Baron R, Schwarz K, Kleinert A, Schattschneider J, Wasner G (2001) Histamine-induced itch converts into pain in neuropathic hyperalgesia. Neuroreport 12(16):3475–3478

Bautista DM, Wilson SR, Hoon MA (2014) Why we scratch an itch: the molecules, cells and circuits of itch. Nat Neurosci 17(2):175–182. doi:10.1038/nn.3619

Bickford RGL (1938) Experiments relating to itch sensation, its peripheral mechanism and central pathways. Clin Sci 3:377–386

Birklein F, Claus D, Riedl B, Neundorfer B, Handwerker HO (1997) Effects of cutaneous histamine application in patients with sympathetic reflex dystrophy. Muscle Nerve 20(11):1389–1395

Blunk JA, Seifert F, Schmelz M, Reeh PW, Koppert W (2003) Injection pain of rocuronium and vecuronium is evoked by direct activation of nociceptive nerve endings. Eur J Anaesthesiol 20(3):245–253

Blunk JA, Schmelz M, Zeck S, Skov P, Likar R, Koppert W (2004) Opioid-induced mast cell activation and vascular responses is not mediated by mu-opioid receptors: an in vivo microdialysis study in human skin. Anesth Analg 98(2):364–370, Table

Braz JM, Juarez-Salinas D, Ross SE, Basbaum AI (2014a) Transplant restoration of spinal cord inhibitory controls ameliorates neuropathic itch. J Clin Invest 124(8):3612–3616. doi:10.1172/JCI75214

Braz J, Solorzano C, Wang X, Basbaum AI (2014b) Transmitting pain and itch messages: a contemporary view of the spinal cord circuits that generate gate control. Neuron 82(3):522–536. doi:10.1016/j.neuron.2014.01.018

Bromm B, Scharein E, Darsow U, Ring J (1995) Effects of menthol and cold on histamine-induced itch and skin reactions in man. Neurosci Lett 187(3):157–160

Brull SJ, Atanassoff PG, Silverman DG, Zhang J, LaMotte RH (1999) Attenuation of experimental pruritus and mechanically evoked dysesthesiae in an area of cutaneous allodynia. Somatosens Mot Res 16(4):299–303

Chevalier X, Eymard F, Richette P (2013) Biologic agents in osteoarthritis: hopes and disappointments. Nat Rev Rheumatol 9(7):400–410

Craig AD (2002) How do you feel? Interoception: the sense of the physiological condition of the body. Nat Rev Neurosci 3(8):655–666

Dhand A, Aminoff MJ (2014) The neurology of itch. Brain 137(Pt 2):313–322. doi:10.1093/brain/awt158

Elkersh MA, Simopoulos TT, Malik AB, Cho EH, Bajwa ZH (2003) Epidural clonidine relieves intractable neuropathic itch associated with herpes zoster-related pain. Reg Anesth Pain Med 28(4):344–346

Ferry X, Brehin S, Kamel R, Landry Y (2002) G protein-dependent activation of mast cell by peptides and basic secretagogues. Peptides 23:1507–1515

Geppetti P, Holzer P (1996) Neurogenic inflammation. CRC, Boca Raton

Gibson PG (2004) Cough is an airway itch? Am J Respir Crit Care Med 169(1):1–2. doi:10.1164/rccm.2310009

Hägermark O (1973) Influence of antihistamines, sedatives, and aspirin on experimental itch. Acta Derm Venereol 53(5):363–368

Han L, Ma C, Liu Q, Weng HJ, Cui Y, Tang Z, Kim Y et al (2012) A subpopulation of nociceptors specifically linked to itch. Nat Neurosci 16(2):174–182

Handwerker HO (2014) Itch hypotheses: from pattern to specificity and to population coding. In: Carstens E, Akiyama T (eds) Itch: mechanisms and treatment. Frontiers in neuroscience. CRC, Boca Raton

Heyer G, Ulmer FJ, Schmitz J, Handwerker HO (1995) Histamine-induced itch and alloknesis (itchy skin) in atopic eczema patients and controls. Acta Derm Venereol 75(5):348–352

Hirth M, Rukwied R, Gromann A, Turnquist B, Weinkauf B, Francke K, Albrecht P et al (2013) NGF induces sensitization of nociceptors without evidence for increased intraepidermal nerve fiber density. Pain 13:10

Ikoma A, Handwerker H, Miyachi Y, Schmelz M (2005) Electrically evoked itch in humans. Pain 113(1–2):148–154. doi:10.1016/j.pain.2004.10.003

Johanek LM, Meyer RA, Hartke T, Hobelmann JG, Maine DN, LaMotte RH, Ringkamp M (2007) Psychophysical and physiological evidence for parallel afferent pathways mediating the sensation of itch. J Neurosci 27(28):7490–7497

Johanek LM, Meyer RA, Friedman RM, Greenquist KW, Shim B, Borzan J, Hartke T, LaMotte RH, Ringkamp M (2008) A role for polymodal C-fiber afferents in nonhistaminergic itch. J Neurosci 28(30):7659–7669

Kamei J, Nagase H (2001) Norbinaltorphimine, a selective kappa-opioid receptor antagonist, induces an itch-associated response in mice. Eur J Pharmacol 418(1–2):141–145

Kardon AP, Polgar E, Hachisuka J, Snyder LM, Cameron D, Savage S, Cai X et al (2014) Dynorphin acts as a neuromodulator to inhibit itch in the dorsal horn of the spinal cord. Neuron 82(3):573–586. doi:10.1016/j.neuron.2014.02.046

Kjellberg F, Tramer MR (2001) Pharmacological control of opioid-induced pruritus: a quantitative systematic review of randomized trials. Eur J Anaesthesiol 18(6):346–357

Koltzenburg M, Handwerker HO, Torebjörk HE (1993) The ability of humans to localise noxious stimuli. Neurosci Lett 150(2):219–222. doi:10.1016/0304-3940(93)90540-2

Kumagai H, Ebata T, Takamori K, Muramatsu T, Nakamoto H, Suzuki H (2010) Effect of a novel kappa-receptor agonist, nalfurafine hydrochloride, on severe itch in 337 haemodialysis patients: a Phase III, randomized, double-blind, placebo-controlled study. Nephrol Dial Transplant 25(4):1251–1257

Lagerstrom MC, Rogoz K, Abrahamsen B, Persson E, Reinius B, Nordenankar K, Olund C et al (2010) VGLUT2-dependent sensory neurons in the TRPV1 population regulate pain and itch. Neuron 68(3):529–542

LaMotte RH, Shain CN, Simone DA, Tsai EFP (1991) Neurogenic hyperalgesia psychophysical studies of underlying mechanisms. J Neurophysiol 66:190–211

LaMotte RH, Shimada SG, Green BG, Zelterman D (2009) Pruritic and nociceptive sensations and dysesthesias from a spicule of cowhage. J Neurophysiol 101(3):1430–1443. doi:10.1152/jn.91268.2008

LaMotte RH, Dong X, Ringkamp M (2014) Sensory neurons and circuits mediating itch. Nat Rev Neurosci 15(1):19–31. doi:10.1038/nrn3641

Lane NE, Schnitzer TJ, Birbara CA, Mokhtarani M, Shelton DL, Smith MD, Brown MT (2010) Tanezumab for the treatment of pain from osteoarthritis of the knee. N Engl J Med 363 (16):1521–1531. doi:10.1056/NEJMoa0901510

Lewis T, Harris KE, Grant RT (1927) Observations relating to the influence of the cutaneous nerves on various reactions of the cutaneous vessels. Heart 14:1–17

Liu Y, Abdel Samad O, Zhang L, Duan B, Tong Q, Lopes C, Ji RR, Lowell BB, Ma Q (2010) VGLUT2-dependent glutamate release from nociceptors is required to sense pain and suppress itch. Neuron 68(3):543–556

Liu Q, Sikand P, Ma C, Tang Z, Han L, Li Z, Sun S, LaMotte RH, Dong X (2012) Mechanisms of itch evoked by beta-alanine. J Neurosci 32(42):14532–14537

McMahon SB, Koltzenburg M (1992) Itching for an explanation. Trends Neurosci 15(12):497–501

Mizumura K, Koda H (1999) Potentiation and suppression of the histamine response by raising and lowering the temperature in canine visceral polymodal receptors in vitro. Neurosci Lett 266(1):9–12

Namer B, Reeh P (2013) Scratching an itch. Nat Neurosci 16(2):117–118. doi:10.1038/nn.3316

Namer B, Carr R, Johanek LM, Schmelz M, Handwerker HO, Ringkamp M (2008) Separate peripheral pathways for pruritus in man. J Neurophysiol 100(4):2062–2069

Napadow V, Li A, Loggia ML, Kim J, Schalock PC, Lerner E, Tran TN et al (2014) The brain circuitry mediating antipruritic effects of acupuncture. Cereb Cortex 24(4):873–882. doi:10.1093/cercor/bhs363

Nilsson HJ, Schouenborg J (1999) Differential inhibitory effect on human nociceptive skin senses induced by local stimulation of thin cutaneous fibers. Pain 80(1–2):103–112

Nilsson HJ, Levinsson A, Schouenborg J (1997) Cutaneous field stimulation (CFS): a new powerful method to combat itch. Pain 71(1):49–55

Nilsson HJ, Psouni E, Carstam R, Schouenborg J (2004) Profound inhibition of chronic itch induced by stimulation of thin cutaneous nerve fibres. J Eur Acad Dermatol Venereol 18 (1):37–43

Nojima H, Cuellar JM, Simons CT, Carstens MI, Carstens E (2004) Spinal c-fos expression associated with spontaneous biting in a mouse model of dry skin pruritus. Neurosci Lett 361 (1–3):79–82

Oaklander AL, Bowsher D, Galer B, Haanpää M, Jensen MP (2003) Herpes zoster itch: preliminary epidemiologic data. J Pain 4(6):338–343

Pfab F, Valet M, Sprenger T, Toelle TR, Athanasiadis GI, Behrendt H, Ring J, Darsow U (2006) Short-term alternating temperature enhances histamine-induced itch: a biphasic stimulus model. J Invest Dermatol 126(12):2673–2678

Qu L, Fan N, Ma C, Wang T, Han L, Fu K, Wang Y, Shimada SG, Dong X, Lamotte RH (2014) Enhanced excitability of MRGPRA3- and MRGPRD-positive nociceptors in a model of inflammatory itch and pain. Brain 137(Pt 4):1039–1050. doi:10.1093/brain/awu007

Reddy VB, Iuga AO, Shimada SG, LaMotte RH, Lerner EA (2008) Cowhage-evoked itch is mediated by a novel cysteine protease: a ligand of protease-activated receptors. J Neurosci 28 (17):4331–4335. doi:10.1523/JNEUROSCI. 0716-08.2008

Ross SE (2011) Pain and itch: insights into the neural circuits of aversive somatosensation in health and disease. Curr Opin Neurobiol 21(6):880–887

Ross SE, Mardinly AR, McCord AE, Zurawski J, Cohen S, Jung C, Hu L et al (2010) Loss of inhibitory interneurons in the dorsal spinal cord and elevated itch in Bhlhb5 mutant mice. Neuron 65(6):886–898

Rukwied R, Mayer A, Kluschina O, Obreja O, Schley M, Schmelz M (2010) NGF induces non-inflammatory localized and lasting mechanical and thermal hypersensitivity in human skin. Pain 148(3):407–413

Rukwied R, Weinkauf B, Main M, Obreja O, Schmelz M (2013a) Axonal hyperexcitability after combined NGF sensitization and UV-B inflammation in humans. Eur J Pain 18(6):785–793

Rukwied R, Weinkauf B, Main M, Obreja O, Schmelz M (2013b) Inflammation meets sensitization—an explanation for spontaneous nociceptor activity? Pain 154(12):2707–2714. doi:10.1016/j.pain.2013.07.054

Rukwied RR, Main M, Weinkauf B, Schmelz M (2013c) NGF sensitizes nociceptors for cowhage-but not histamine-induced itch in human skin. J Invest Dermatol 133(1):268–270

Sanga P, Katz N, Polverejan E, Wang S, Kelly KM, Haeussler J, Thipphawong J (2013) Efficacy, safety, and tolerability of fulranumab, an anti-nerve growth factor antibody, in treatment of patients with moderate to severe osteoarthritis pain. Pain 13:10

Schmelz M (2002) Itch—mediators and mechanisms. J Dermatol Sci 28(2):91–96

Schmelz M, Handwerker HO (2013) Itch. Wall & Melzack's textbook of pain. Elsevier, Philadelphia

Schmelz M, Schmidt R, Bickel A, Handwerker HO, Torebjörk HE (1997) Specific C-receptors for itch in human skin. J Neurosci 17(20):8003–8008

Schmelz M, Michael K, Weidner C, Schmidt R, Torebjörk HE, Handwerker HO (2000) Which nerve fibers mediate the axon reflex flare in human skin? Neuroreport 11(3):645–648

Schmelz M, Hilliges M, Schmidt R, Orstavik K, Vahlquist C, Weidner C, Handwerker HO, Torebjörk HE (2003a) Active "itch fibers" in chronic pruritus. Neurology 61(4):564–566

Schmelz M, Schmidt R, Weidner C, Hilliges M, Torebjörk HE, Handwerker HO (2003b) Chemical response pattern of different classes of C-nociceptors to pruritogens and algogens. J Neurophysiol 89(5):2441–2448

Schmidt R, Schmelz M, Forster C, Ringkamp M, Torebjörk HE, Handwerker HO (1995) Novel classes of responsive and unresponsive C nociceptors in human skin. J Neurosci 15(1 Pt 1):333–341

Schmidt R, Schmelz M, Weidner C, Handwerker HO, Torebjörk HE (2002) Innervation territories of mechano-insensitive C nociceptors in human skin. J Neurophysiol 88(4):1859–1866

Seal RP, Wang X, Guan Y, Raja SN, Woodbury CJ, Basbaum AI, Edwards RH (2009) Injury-induced mechanical hypersensitivity requires C-low threshold mechanoreceptors. Nature 462 (7273):651–655

Shelley WB, Arthur RP (1957) The neurohistology and neurophysiology of the itch sensation in man. AMA Arch Derm 76:296–323

Sikand P, Shimada SG, Green BG, LaMotte RH (2009) Similar itch and nociceptive sensations evoked by punctate cutaneous application of capsaicin, histamine and cowhage. Pain 144 (1–2):66–75

Sikand P, Dong X, LaMotte RH (2011) BAM8-22 peptide produces itch and nociceptive sensations in humans independent of histamine release. J Neurosci 31(20):7563–7567. doi:10.1523/JNEUROSCI.1192-11.2011

Simone DA, Alreja M, LaMotte RH (1991) Psychophysical studies of the itch sensation and itchy skin ("alloknesis") produced by intracutaneous injection of histamine. Somatosens Mot Res 8 (3):271–279

Simone DA, Nolano M, Johnson T, Wendelschafer-Crabb G, Kennedy WR (1998) Intradermal injection of capsaicin in humans produces degeneration and subsequent reinnervation of epidermal nerve fibers: correlation with sensory function. J Neurosci 18(21):8947–8954

Steinhoff M, Neisius U, Ikoma A, Fartasch M, Heyer G, Skov PS, Luger TA, Schmelz M (2003) Proteinase-activated receptor-2 mediates itch: a novel pathway for pruritus in human skin. J Neurosci 23(15):6176–6180

Tominaga M, Takamori K (2014) Itch and nerve fibers with special reference to atopic dermatitis: therapeutic implications. J Dermatol 41(3):205–212. doi:10.1111/1346-8138.12317

Tominaga M, Tengara S, Kamo A, Ogawa H, Takamori K (2009) Psoralen-ultraviolet A therapy alters epidermal Sema3A and NGF levels and modulates epidermal innervation in atopic dermatitis. J Dermatol Sci 55(1):40–46

Torebjörk HE, Schmelz M, Handwerker HO (1996) Functional properties of human cutaneous nociceptors and their role in pain and hyperalgesia. In: Belmonte C, Cervero F (eds) Neurobiology of nociceptors. Oxford University Press, Oxford, pp 349–369

Toyoda M, Nakamura M, Makino T, Hino T, Kagoura M, Morohashi M (2002) Nerve growth factor and substance P are useful plasma markers of disease activity in atopic dermatitis. Br J Dermatol 147(1):71–79

Toyoda M, Nakamura M, Makino T, Morohashi M (2003) Localization and content of nerve growth factor in peripheral blood eosinophils of atopic dermatitis patients. Clin Exp Allergy 33 (7):950–955

Tuckett RP, Wei JY (1987) Response to an itch-producing substance in cat. II. Cutaneous receptor populations with unmyelinated axons. Brain Res 413(1):95–103

Vogelsang M, Heyer G, Hornstein OP (1995) Acetylcholine induces different cutaneous sensations in atopic and non-atopic subjects. Acta Derm Venereol 75(6):434–436

Vrontou S, Wong AM, Rau KK, Koerber HR, Anderson DJ (2013) Genetic identification of C fibres that detect massage-like stroking of hairy skin in vivo. Nature 493(7434):669–673

Wahlgren CF, Ekblom A (1996) Two-point discrimination of itch in patients with atopic dermatitis and healthy subjects. Acta Derm Venereol 76(1):48–51

Watanabe T, Inoue M, Sasaki K, Araki M, Uehara S, Monden K, Saika T, Nasu Y, Kumon H, Chancellor MB (2011) Nerve growth factor level in the prostatic fluid of patients with chronic prostatitis/chronic pelvic pain syndrome is correlated with symptom severity and response to treatment. BJU Int 108(2):248–251

Wooten M, Weng HJ, Hartke TV, Borzan J, Klein AH, Turnquist B, Dong X, Meyer RA, Ringkamp M (2014) Three functionally distinct classes of C-fibre nociceptors in primates. Nat Commun 5:4122. doi:10.1038/ncomms5122

Yamaguchi J, Aihara M, Kobayashi Y, Kambara T, Ikezawa Z (2009) Quantitative analysis of nerve growth factor (NGF) in the atopic dermatitis and psoriasis horny layer and effect of treatment on NGF in atopic dermatitis. J Dermatol Sci 53(1):48–54

Index

A
Acetylcholine (Ach), 201–202
Acute cutaneous inflammation
　pathways, 198
　pharmacology, 197–198
Acute pain, 148–149. *See also*
　　　Endocannabinoid (EC) system
Adenosine, 197
Adenosine monophosphate-activated protein
　　　kinase (AMPK), 22
α-amino-3-hydroxy-5-methyl-4-
　　　isoxazolepropionic acid (AMPA)
　　　receptor, 215
Amygdala, 7, 11
　circuitry, 262–264
　LA-BLA and CeA, 264
　pharmacology of
　　amino acid neurotransmitters, 266
　　CGRP, 274–275
　　CRF, 275–277
　　GABA, 272–273
　　ionotropic glutamate receptors,
　　　266–268
　　metabotropic glutamate receptors,
　　　268–272
　　NPS, 277–278
　plasticity
　　electrophysiological studies, 264
　　excitatory transmission, 265–266
　　neuronal activity changes, 265
　　output, 26
Analgesics
　new Nav1.7 channel, 52–53
　postsurgery pain, 30
　sex differences in, 227–228
Anandamide (AEA), 124, 125
Anterolateral tract (ALT), 173
2-arachidonoyl glycerol (2-AG), 124, 125

Astrocytes, 210–211
　classification, 148
　origin, 147
　physiological conditions, 148
ATP, 219–220
Atypical PKC (aPKC)
　intrathecal injection of, 25, 27
　LTP, 24, 25
　pep2m, 26
　PKMζ, 24
　role of, 25
　ZIP, 26

B
Blood–brain barrier (BBB)
　endothelial cells, 211
　permeable pharmacotherapies, 229
Brain derived neurotropic factor (BDNF),
　　24, 26–27

C
Calcitonin gene-related peptide (CGRP), 173
　antinociceptive effects of, 275
　electrophysiological data, 274
　functional receptors, 274
　during migraine attacks, 226
　stereotaxic administration of, 274–275
Cannabinoids
　CB_1 receptor, 123–124
　definition, 121
Cannabinoid type 2 receptor (CB_2 receptor)
　activation of, 124
　modulation of, 130
　role of, 128–129
Cardiovascular system, 243–244
Chemokine ligand 2 (CCL2), 217

Chemokine ligand 5 (CCL5), 223
Central immune cell
　non-stereoselective activation, 213
　signalling, 208
　synergy, 211
Central nervous system (CNS)
　astrocytes, 210–211
　CB_1 receptor, 123
　cytokines, 213–215
　microglia, 210 (see also Microglia)
　non-neuronal cell intracellular signalling in, 209
Central sensitization
　definition, 6
　in chronic pain
　　animal studies on, 81
　　conditions, 80–81
　　extraterritorial manifestations of, 82–83
　　local ipsilateral, 82
　　treatment of, 80
　　widespread manifestations of, 83–84
　itch (see also Itch)
　　exact mechanisms and roles of, 294
　　mechanical hyperalgesia, 293
　　pruriceptors ongoing activity, 294
　　punctate hyperalgesia, 293
　pain, 293–295
　Quantitative sensory testing (QST) for
　　assessment methods, 84
　　descending pain modulation, 90–91
　　different protocols, 84
　　localized vs. general hyperalgesia problem, 85–86
　　offset analgesia, 91–92
　　provoked facilitation, 86, 87
　　referred pain, 92–93
　　reflex receptive fields, 89–90
　　spatial summation, 89
　　temporal summation and aftersensation, 86–88
cGMP signaling, 108–110
Chemokines, 216–217
Cholecystokinin (CCK), 217–219
Chronic constriction injury (CCI), 62, 180
Chronic inflammation, 295–296
Chronic pain
　EC in, 128–129
　spinal cord, in neuropathic and inflammatory pain, 148–149
c-Jun N-terminal kinase (JNK)
　in astrocytes, 209
　phosphorylation of, 158
Collagen-induced arthritis (CIA), 159

Corticotropin-releasing factor (CRF)
　changes of, 276–277
　effects of, 277
　electrophysiological studies, 275
　endogenous receptor activation, 276, 277
　extrahypothalamic expression of, 275
　patch-clamp analysis, 276
　receptor antagonist, 275
　sources of, 275
CPEB. See Cytoplasmic polyadenylation element-binding protein (CPEB)
Crosstalk, receptor, 222–225
Cuff-algometry technology, 89
Cutaneous mechanical hyperalgesia, 60
CX3LC1 (Fraktalkine)/R1, 153–154
C-X-C chemokine receptor 1 (CXCR1), 223
C-X-C chemokine receptor 2 (CXCR2), 223
C-X-C chemokine receptor 4 (CXCR4), 224
CX3C (Fractalkine) receptor-1 (CX3CR1), 216, 224
Cytokine-mediated neuronal excitation, 215–216
Cytokines, 213–215
Cytoplasmic polyadenylation element-binding protein (CPEB), 23, 24

D
Delayed onset muscle soreness (DOMS), 63–65
Descending pain modulation, 90–91
Desensitisation, 222, 243
Diacylglycerol lipase-α (DAGLα), 125, 127
Dorsal root ganglia (DRG), 41
Dorsal root reflex (DRR)
　glial dependence of, 196–197
　spinal cord modulation of peripheral inflammation, 193–194

E
Endocannabinoid (EC) system
　acute pain processing, 127–128
　CB_2 receptors (see Cannabinoid type 2 receptor (CB_2 receptor))
　comparison of classical neurotransmitter systems, 121
　endogenous ligands, 124
　notional neuronal synapse, signalling, 122
　and pain, 125–126
　and peripheral pain processing, 126–127
　plasticity problems, 132–133
　supraspinal level, 130–132
　synthesis and degradation, 124–125

Endothelial NOS (eNOS or NOS-3), 104, 105
Epac signaling, 22
Extracellular signalregulated kinase (ERK), 21

F
Familial episodic pain (Na$_v$1.9), 48, 50–51
Fascia, 61–62
Fatty acid amide hydrolase (FAAH), 126–127, 133
Fractalkine, 216–217
Fragile X mental retardation protein (FMRP), 22–23
Functional magnetic resonance imaging (fMRI), 131, 132

G
GABAergic neurons, 183–184
Gamma(γ)-aminobutyric acid (GABA)
   amygdala, pharmacology of, 272–273
   immunostaining for, 175
   receptors, 215
   spinal cord modulation of peripheral inflammation, 193–194
Gastrointestinal system, 245, 252
Glia, 211
   activation, 214–215
   origin and function of, 146–148
   spinal changes (*see* Neuropathic pain)
Glycinergic circuits, 185–186
G protein-coupled receptor (GPCR), 121, 216, 222

H
Hereditary and sensory autonomic neuropathy type IID (HSANIID), 48
Heterologous desensitisation, 222
Heteromerisation, 223–224
Homologous desensitisation, 222
Human heritable sodium channelopathy
   familial episodic pain (Na$_v$1.9), 50–51
   inherited primary erythromelalgia (Na$_v$1.7), 47–49
   pain insensitivity (Na$_v$1.7 and Na$_v$1.9), 51–52
   pain-related, 47
   paroxysmal extreme pain disorder (Nav1.7), 49
   selective Nav1.7 analgesics, 52–53
   small fibre neuropathy (Na$_v$1.7 and Na$_v$1.8), 49–50
Hyperalgesic priming
   CNS regulation of
      atypical PKCs, 24–27
      BDNF, 24–27
      endogenous opioids, 27–29
      μ-opioid receptor constitutive activity, 27–29
      opioid effects, 29–30
      surgery as priming stimulus, 29–30
   in messenger signaling pathways, 19
   therapeutic opportunities, 31–32
   translational control pathways involved in, 21

I
IL-1 receptor antagonist (IL-1ra), 214
Immune signalling
   central, 214, 215, 229
   in central nervous system, 208
   homeostatic, 214
   neuron–glial central, 217
Immuno-active agents, 196–197
Immunocompetent cells, in opioid pharmacodynamics, 211
Inducible nitric oxide synthase (iNOS) isoform, 220
Inducible NOS (iNOS or NOS-2), 104
Inflammation
   acute cutaneous
      pathways, 198
      pharmacology, 197–198
   acute joint models, 196–197
   characteristics, 157
   chronic itch, 295–296
   chronic models of
      pathways, 200
      pharmacology, 199
   JNK phosphorylation, 158
   monoarthritis-kaolin/carrageenan knee, 195–196
   nervous system effects, 8
   and neuropathic pain, 149–150
   PAR1, 247
   PAR2, 248–250
   PAR3, 250
   PAR4, 250
   in pathophysiologic nociceptive pain, 3
   role of, 157
   spinal glia mechanisms, 158
   sympathetic terminals in, 200–201
Inherited primary erythromelalgia (IEM), 47–49
Interleukin 6 (IL-6), 22
Interleukin-1β (IL-1β), 156–157
Interleukin-1receptor (IL-1R), 156–157

Interneurons
  excitatory, 175–176
  inhibitory, 174–175
    loss, 182–183
    reduced excitation of, 184–185
Ionotropic glutamate receptors, 266–268
Itch
  central sensitization
    exact mechanisms and roles of, 294
    mechanical hyperalgesia, 293
    pruriceptors ongoing activity, 294
    punctate hyperalgesia, 293
  intensity and pattern theory
    activated nociceptors, 291–293
    non-histaminergic itch, 290–291
  mechanisms for, 295–296
  peripheral sensitization, 295
  specificity for
    antagonistic interaction, 288–290
    molecular markers, 287–288

L
Laterocapsular division of central nucleus of amygdala (CeLC)
  evoked responses, 268, 271
  excitatory synaptic transmission, 265
  glutamatergic inputs, 266–267
  hyperactivity of, 267
  stimulus-evoked activity, 264
  synaptic inhibition, 271–272
Long-term potentiation (LTP), 7, 23, 24
Low-threshold mechanoreceptors (LTMRs), 177

M
Manganese superoxide dismutase (MnSOD), 221
Mechanistic target of rapamycin complex 1 (mTORC1), 21
Medication-overuse headache, 226
Metabotropic glutamate receptors (mGluRs)
  activation of, 270, 271
  DCPG, 272
  electrophysiological analysis, 270, 271
  facilitatory effects of, 269–270
  pattern of, 269
  presynaptic receptors, 271
  role of, 268
  types, 268, 270, 271
  ZJ43, 271

Microglia, 210
  cell populations, 147–148
  monocytic/myeloid origins, 146
  phagocytes of, 147
  physiological conditions, 147
  roles, 147
Mitogen-activated protein (MAP), 202
Mitogen-activated protein kinase (MAPK) signalling pathway, 209
Monoacylglycerol lipase (MAGL), 127, 128
Monoarthritis-Kaolin/Carrageenan knee, 195–196
Monocyte chemoattractant protein-1 (MCP-1), 217
mTORC1. *See* Mechanistic target of rapamycin complex 1 (mTORC1)
μ-opioid receptor (MOR), 27–28, 209
Muscular mechanical hyperalgesia, 60, 61
Musculoskeletal system, 246–247
Myelinated nociceptors, 173

N
N-acylethanolamines (NAEs), 125
Nacyl-phosphatidylethanolamine-hydrolyzing phospholipase D (NAPE-PLD), 125
Naltrexone, 224
Natriuretic peptide receptor A (NPR-A), 108
$Na_v1.3$, 46–47
$Na_v1.7$
  inherited primary erythromelalgia, 47–49
  new selective analgesics, 52–53
  pain insensitivity, 51–52
  paroxysmal extreme pain disorder, 49
  rodent studies, insights from, 41, 45
  small fibre neuropathy, 49–50
$Na_v1.8$
  rodent studies, insights from, 45–46
  small fibre neuropathy, 49–50
$Na_v1.9$
  familial episodic pain, 50–51
  pain insensitivity, 51–52
  rodent studies, insights from, 46
Nerve growth factor (NGF), 22
  action mechanism
    acute sensitization, 68–69
    long-lasting sensitization, 69
    mechanical stimuli sensitization, 70–71
  cachexia pain, 67
  cancer pain, 67
  inflammatory pain, 62
  musculoskeletal pain
    cast immobilization, 65–66

DOMS, 63–65
  osteoarthritis models, 65–66
 neuropathic pain, 62–63
 nociceptive system
  development, 59
 nociceptor activities and axonal properties,
   67–68
 pain and mechanical/thermal hyperalgesia
   induced
  animals, 60
  humans, 60–62
 receptor, 58–59
 and receptor trkA, 9–10
 therapeutic perspective, 71
 visceral painful conditions, 66
Nervous system, 244–245
Neuroglia, 146
Neuronal NOS (nNOS or NOS-1), 104, 105
Neuropathic pain
 animal models of, 180–181
 astrocytic responses, to injury, 152
 cathepsin S, 153–154
 CX3CL1/R1, 153–154
 definition, 3
 IL-1β, 156–157
 IL-1R, 156–157
 inflammatory pain and, 149–150
 microglial responses to injury, 150–151
 neuronal mechanisms of, 5–8
 PAR1, 251
 PAR2, 251, 252
 PAR3 and PAR4, 252
 possible mechanisms, reduced inhibition of
   GABAergic neurons, 183–184
  glycinergic circuits role, 185–186
  inhibitory interneurons excitation,
     184–185
  inhibitory interneurons loss, 182–183
  inhibitory transmission effectiveness,
    185
 process of, 6
 reduced inhibitory synaptic transmission in,
   181
 TNF, 154–155
 TNFR, 154–155
Neuropathy, 295–296
Neuropeptide
 CGRP, 274–275
 CRF, 275–277
 NPS, 277–278
Neuropeptide S (NPS), 277–278
NGF. *See* Nerve growth factor (NGF)
NG-nitro-L-arginine (L-NOARG), 220

NG-nitro-L-arginine methyl ester (L-NAME),
   220
Nitric oxide (NO), 220
Nitric oxide (NO)-mediated pain processing
 in dorsal root ganglia, 104–105
 downstream mechanisms of
  cGMP signaling, 108–110
  NO-GC activation, 107–108
  peroxynitrite formation, 110–111
  S-nitrosylation, 110
 pro-and antinociceptive functions of,
   105–106
 in spinal cord, 104–105
Nitric oxide(NO)-sensitive guanylyl cyclase
   (NO-GC), 107–108
Nitric oxide synthase (NOS), 220
 inhibitors, 105–106
 isoforms, 104
N-methyl-D-aspartate (NMDA) receptors, 28
Nociceptive system
 acute sensitization
  direct phosphorylation by TrkA, 68–69
  indirect action of, 69
  membrane trafficking of TRPV1, 69
  sympathetic nerve involvement, 69
 central, 4
 descending system, 5
 development, 59
 long-lasting sensitization to heat, 69
 mechanical stimuli sensitization
  mechanical hyperalgesia, 70
  mechanical hypersensitivity, 70
  TrkA, 70–71
 molecular mechanisms of, 8–11
 peripheral, 4
 thalamocortical system, 4
Nociceptor priming
 animal models, 18
 chronic pain conditions, 18
 hyperalgesic priming (*see* Hyperalgesic
   priming)
 local translation, key mediator of, 20
  CPEB, 23
  epac signaling, 22
  experimental paradigm, 23
  FMRP, 23
  PKCε-induced priming, 24
  in sensory neurons, 22
  translation, 21
 naïve rodents, 18
 PKCε, crucial mechanism of, 19–20
 preclinical models, 16
 prostaglandins, 18

Non-neuronal cell intracellular signalling, 208, 209
Non-neuronal central immune cells, 209
  astrocytes, 210–211
  central immune synergy, 211
  microglia, 210

O
Offset analgesia, 91–92
Opioid-induced cytokine signalling, 225–226
Opioid-induced initiation
  analgesia opposition
    ATP, 219–220
    chemokines, 216–217
    cholecystokinin, 217–219
    cytokines, 213–215
    nitric oxide, 220
    potentiating/unmasking, 214
    proinflammatory cytokine-mediated neuronal excitation, 215–216
    sphingomyelins, 220–222
  gonadal hormone contribution, 228
Opioid-overuse headache, 225–226
Opioids, 208
  pharmacodynamics, 211
  tolerance, 208, 209, 218

P
Pain insensitivity (Nav1.7 and Nav1.9), 51–52
Paroxysmal extreme pain disorder (PEPD), 48, 49
Pathophysiologic nociceptive pain
  definition, 3
Periaqueductal grey matter (PAG), 130–131, 217
Peripheral nerve injury. *See* Neuropathic pain
Peripheral nervous system (PNS)
  CB_1 receptor, 123
  targeting, 31
Peripheral nociceptive system, 4
Peripheral pain processing, EC system, 126–127
Peripheral sensitization
  molecular mechanisms of, 5–6
  pain, 295
Peroxynitrite, 221
Phosphoinositide 3-kinases (PI3Ks), 249
Physiologic nociceptive pain, 3
Postherpetic neuralgia (PHN), 82
Primary afferent axons, 172–173
Primary afferent depolarization (PAD), 193–194
Proinflammatory cytokine-mediated neuronal excitation, 215–216
Proinflammatory cytokines, 214

Projection neurons
  in anterior lateral tract (ALT), 173, 176–177
  selective innervation of, 178
Proteases
  categories, 240
  proteolytic properties, 240
Proteinase-activated receptor (PAR), 10
  activating peptides and antagonists, 241
  activation, 240–242
  cardiovascular system, 243–244
  cleaving enzymes, 241
  definition, 240
  desensitisation mechanisms, 243
  drug target for pain, 253–254
  gastrointestinal system, 245
  and inflammatory pain, 247–250
  musculoskeletal system, 246–247
  nervous system, 244–245
  neuropathic pain, 250–252
  signalling
    PAR1, 242
    PAR2, 243
    PAR3 and PAR4, 243
Protein kinase A (PKA), 19
Protein kinase G (PKG), 108
Protein kinase M zeta (PKMζ), 24
Provoked central sensitization, 86, 87
Punctate hyperalgesia, 293
P2X4 receptors, 219

Q
Quantitative sensory testing (QST)
  assessment methods, 84
  descending pain modulation, 90–91
  different protocols, 85
  localized *vs.* general hyperalgesia problem, 85–86
  offset analgesia, 91–92
  provoked facilitation, 86, 87
  referred pain, 92–93
  reflex receptive fields, 89–90
  spatial summation, 89
  temporal summation and aftersensation, 86–88

R
Receptor activity-modifying protein 1 (RAMP1), 274
Receptor binding
  non-stereoselective, 212–213
Receptor crosstalk, molecular mechanisms of, 222–225

Index

Referred pain, 92–93
Reflex receptive fields, 89–90
Rheumatoid arthritis (RA) pain
　analgesic effect of, 160
　characteristics, 158
　clinical signs of, 158
　poly-arthritic rodent models, 159
　spinal microglia role, 160
　treatment of, 159
Rhizotomy, 192–193
Rostral ventromedial medulla (RVM)
　and cholecystokinin, 218
　excitatory projections to, 131
　microinjection, of cannabinoid agonists, 130

S

$SCN9A$ gene, 47, 48, 50
$SCN10A$ gene, 47
$SCN11A$ gene, 47, 50
Seven-transmembrane (7TM) receptors, 222
Small fibre neuropathy (SFN), 49–50
S-nitrosylation, 110
Sodium channels
　$Na_v1.3$, 46–47 (see also Nav1.3)
　$Na_v1.7$, 41, 45 (see also Nav1.7)
　$Na_v1.8$, 45–46 (see also Nav1.8)
　$Na_v1.9$, 46 (see also Nav1.9)
　Na_v transgenic mice studies, 41–44
Soluble guanylyl cyclase (sGC), 107
Spared nerve injury (SNI), 181
Spatial summation, 89
Sphingomyelins, 220–222
Sphingosine, 220
Sphingosine kinases (SphK) 1 and 2, 220–221
Spinal cord
　excitatory synaptic transmission in, 10
　mechanisms, 10–11
　nociceptive neurons in, 4
Spinal cord inhibitory mechanisms
　descending pathways, 176
　neurons and circuits
　　interneurons, 174–176
　　normal function of inhibitory mechanisms, 179–180
　　presynaptic inhibitory, 179
　　primary afferents, 172–173
　　projection neurons, 173–174
　　selective innervation of, 178
　　synaptic connections, 176, 177
　neuropathic pain
　　animal models, 180–181
　　possible mechanisms, 181–186
　　reduced inhibitory synaptic transmission in, 181
Spinal cord modulation of peripheral inflammation
　acute cutaneous inflammation, 197–198
　acute inflammatory models, 195–196
　chronic models of, 199–200
　dorsal root reflex, 193–194
　joint inflammation, 196–197
　rhizotomy, 192–193
　spinovagal circuitry, 201–202
　sympathetic effects on, 200–201
Spinal endocannabinoid system. See Endocannabinoid (EC) system
Spinal glia. See also Neuropathic pain
　during inflammatory pain, 157–158
　during rheumatoid arthritis pain, 158–160
Spinal immune cell function, 130
Spinal nerve ligation (SNL), 181
Spinal sensitization, 7
Spinovagal circuitry, 201–202
Stereoselective receptor binding, 212–213
Superoxide dismutase (SOD), 221
Sympathectomy, 200, 201

T

Temporal summation and aftersensation, 86–88
Terminal deoxynucleotidyl transferasemediatedbiotinylated UTP nick end labelling (TUNEL), 182
Thalamocortical system, nociceptive neurons in, 4–5
Toll-like receptor-4 (TLR4)
　LPS activation, 222
　medication-overuse headache, 226
　in non-stereoselective binding, 213
　opioids and, 228
Toll-like receptors (TLRs), 226
Tropomyosin-related kinase A (TrkA)
　direct phosphorylation by, 68–69
　mechanical stimuli sensitization, 70–71
　membrane trafficking of TRPV1 by, 69
Tumor necrosis factor (TNF)
　spinal cord, in neuropathic and inflammatory pain, 154–155
　spinal pretreatment with, 196

V

Voltage-gated sodium channels (VGSCs)
　alpha subunit, primary structure of, 40, 41
　mammalian, 40

CPSIA information can be obtained
at www.ICGtesting.com
Printed in the USA
LVOW02*1756010516
486186LV00001B/34/P